THE POLISH SCHOOL OF PHILOSOPHY OF MEDICINE

PHILOSOPHY AND MEDICINE

Editors:

H. TRISTRAM ENGELHARDT, JR.

The Center for Ethics, Medicine and Public Issues
Baylor College of Medicine, Houston, Texas, U.S.A.

STUART F. SPICKER

School of Medicine, University of Connecticut Health Center,
Farmington, Connecticut, U.S.A.

VOLUME 37

The Polish School of Philosophy of Medicine

From Tytus Chalubinski (1820–1889) to Ludwik Fleck (1896–1961)

Compiled, translated and Introductions

by

ILANA LÖWY

INSERM, Hôpital Necker-Enfants Malades, Paris

KLUWER ACADEMIC PUBLISHERS

DORDRECHT / BOSTON / LONDON

ISBN-13: 978-94-010-7461-2 e-ISBN-13: 978-94-009-2135-1
DOI: 10.1007/978-94-009-2135-1

Published by Kluwer Academic Publishers,
P.O. Box 17, 3300 AA Dordrecht, The Netherlands.

Kluwer Academic Publishers incorporates
the publishing programmes of
D. Reidel, Martinus Nijhoff, Dr W. Junk and MTP Press.

Sold and distributed in the U.S.A. and Canada
by Kluwer Academic Publishers,
101 Philip Drive, Norwell, MA 02061, U.S.A.

In all other countries, sold and distributed
by Kluwer Academic Publishers Group,
P.O. Box 322, 3300 AH Dordrecht, The Netherlands.

Printed on acid-free paper

TABLE OF CONTENTS

PREFACE vii

CHAPTER I. Introduction: Philosophy of Medicine in Poland 1

CHAPTER II. Tytus Chałubiński. The Last Follower of 'Medical Systems' or a Pioneer of a New Approach to Therapy? 13

CHAPTER II.a. Text of Chałubiński: Excerpts from *The Method of Finding Therapeutic Indications* (1874) 25

CHAPTER III. Edmund Biernacki on the Science of Diseases and the Art of Healing 37

CHAPTER III.a. Text of Biernacki: Excerpts from *The Essence and the Limits of Medical Knowledge* (1898) 49

CHAPTER IV. Władysław Biegański Between the Logic of Science and the Logic of Medicine 69

CHAPTER IV.a. Texts of Biegański: Excerpts from
General Problems of the Theory of Medical Sciences (1897) 91
Thoughts and Aphorisms on Medical Ethics (1899) 100
The Logic of Medicine or the Critique of Medical Knowledge (1908) 112

CHAPTER V. Zygmunt Kramsztyk and the Critical Evaluation of Medical Practice 121

CHAPTER V.a. Texts of Kramsztyk:
'Rational Treatment' (1897) 137
'Is Medicine an Art or a Science?' (1895) 143
'A Clinical Fact' (1898) 155
'On Being-up to Date' (1907) 160

CHAPTER VI. From *Medical Critique* to the *Archives of the History and Philosophy of Medicine*: The Institutionalization of Polish School of Philosophy of Medicine 163

CHAPTER VI.a. Texts of Wrzosek and Trzebiński:
Wrzosek: 'Trends in contemporary medicine' (1900) 185
Trzebiński: 'Rationality and 'Rationalism' in Medicine' (1925) 191
'Absurdity in Medicine' (1927) 200

CHAPTER VII. Ludwik Fleck. From Philosophy of Medicine to a Constructivist and Relativist Epistemology 215

CHAPTER VII.a. Texts of Fleck and Bilikiewicz
Fleck: 'Some Specific Features of the Medical Way of Thinking' (1927) 229
Fleck: 'On the Crisis of 'Reality'' (1929) 237
Fleck: 'Science and Social Context' (1939) 249
Bilikiewicz: 'Comments on Ludwik Fleck's 'Science and Social Context'' (1939) 257
Fleck: 'Rejoinder to the Comments of Tadeusz Bilikiewicz' (1939) 267
Bilikiewicz: 'Reply to the Rejoinder by Ludwik Fleck' (1939) 274

CHAPTER VIII. Conclusions: Philosophizing at the Bedside 277

NAME INDEX 285

PREFACE

My 'discovery' of the Polish School of philosophy of medicine stemmed from my studies in the genesis of Ludwik Fleck's epistemology. These studies, and my interest in the scientific roots of Fleck's epistemology were a nearly 'natural' result of my own biography: like Fleck I had been trained, an had worked as an immunologist, and had later switched to studies in the social history of medicine and biology. Moreover, it so happened that Fleck's book, *Genesis and Development of a Scientific Fact* - the description of a science as it is, not as it should be - was the first epistemological study in which I found echos of my experience in the laboratory. My interest in Fleck was also highlightened by the fact that in his works, and, as I discovered later, in the works of his predecessors of the Polish School of philosophy of medicine, was formulated the problem that had stimulated my interest in the history of medicine and biology, and is still central to my present investigations: the relationships between biological knowledge and clinical practice.

The writing of the book was made possible through to the help of many colleagues and friends. The unfailing support for my research, whatever its subject might be, from my colleagues from Unit 158 of INSERM and in particular from its head Patrice Pinell, has made my study of the Polish School possible. I am indebted to those who assisted me through discussion, suggestions and encouragement: Gad Freudenthal, Barbara Markiewicz, Anne-Marie Moulin, Elżbieta Pakszysz, Barbara Tuchańska, Thomas Schnelle and Włodzimierz Wypych. Maria Cubrzyńska, head of the ancient books department at the Library of Warsaw University, and Krzysztof Turowski at the Polish Library in Paris, were of great aid in finding the original works. Alexander Rodzewicz helped with the translations from Polish. Robert Sayre not only painstakingly corrected my imperfect English - a thankless task! - but in many cases contributed to the clarification of my thought as well. Stuart Spicker patiently read several imperfect versions of the manuscript and made numerous, highly valuable editorial comments and Suzan Engelhardt carefully proofread the final draft. Marie-Jo Bézard with her usual competence and kindness dealt with all the technical aspects of the production of this work. I am indebted to Martin Scrivener and to the Kluwer Academic Editions for the efficient and sympathetic processing of the manuscript. Finally, during the writing of this book my children - Tamara, Daniel, Naomi and Rachel - were a permanent, and for me an indispensable source of joy.

CHAPTER I

INTRODUCTION: PHILOSOPHY OF MEDICINE IN POLAND

"One can feel the philosophy of medicine in the air" ["dass eine Philosophie der Medizin in der Luft läge"].... This rather over-optimistic declaration - the philosophy of medicine became institutionalized as a distinct branch of philosophy only in the mid-1970's [6] - was made in 1930 by the German theoretician of medicine Richard Koch, after hearing a paper on the state of the philosophy of medicine in Poland, presented by the historian and philosopher of medicine Władysław Szumowski at the International Congress of the History of Medicine (Rome, 1930) [22].

At that time the situation of the philosophy of medicine in Poland was indeed quite exceptional. In other countries the 'philosophy of medicine' (or medical philosophy) was seen, at best, as a non-formalized activity of philosophically-minded physicians, and, at worst, as a synonym for personal opinions, polemical excursions and laments over lost glory [18].[1] In contrast, in Poland chairs of History and Philosophy of Medicine existed from 1920 on in five major medical schools; the journal *Archives of the History and Philosophy of Medicine, [Archiwum Historii i Filozofji Medycyny]* (founded in 1924) published articles on philosophical aspects of medicine; and medico-philosophical subjects were debated in the meetings of the Polish Society of the History and Philosophy of Medicine [4, 22, 30].

This unique development and institutionalization of the philosophy of medicine in Poland after the First World War may at first seem quite surprising. The young Polish Republic, struggling to reconstruct its system of higher education after a long period of foreign domination, had in all probability more urgent tasks to accomplish than the development of the rather esoteric domain of the philosophy of medicine. This evolution is less surprising, however, if one takes into consideration the fact that the institutionalization of the teaching of philosophy of medicine was the ultimate step in a long history of interest in the philosophy of medicine in Poland.

This discipline was first introduced into Poland in the second half of the 19th-century by Tytus Chałubiński (1820-1889). Chałubiński's efforts to find a rational basis of therapy were the starting point for the reflections of the next generation of philosophers of medicine: The most important among these physician-philosophers were Edmund Biernacki (1866-1908), Władysław Bieganski (1857-1917) and Zygmunt Kramsztyk (1848-1920). Kramsztyk was also the founder and editor-in-chief of *Medical Critique [Krytyka Lekarska]*, a journal which for eleven years (1897-1908) largely contributed to the growth of the Polish School of Philosophy of Medicine [17].

Some leading Polish physicians at the beginning of the 20th-century were influenced by the essays in *Medical Critique*, and developed an interest in the philosophy and history of medicine. After the proclamation of the independence of Poland in 1918, several such historically- and philosophically-minded physicians found themselves holders of high governmental offices. They were able to utilize their political influence to promote the teaching of the history and philosophy of medicine. Thus the historian and philosopher of medicine, Władysław Szumowski, who became the head of the department of scientific publications in Poland's Ministry of Health, strongly advocated the introduction of history of medicine but also of philosophy of medicine in the medical school curriculum [23, 24]. Szumowski's point of view was adopted by Adam Wrzosek, a historian of medicine, who in 1920 was the head of the division of higher education in the Ministry of Education, and who played an important role in the creation of chairs of history and philosophy of medicine in five medical schools in Poland [4, 22]. In 1924, Wrzosek, who in the meantime left the Ministry of Education and became a professor of history and philosophy of medicine in Poznan, founded the *Archives of History and Philosophy of Medicine*. In the editorial of the first issue of this journal, Wrzosek affirmed that philosophy of medicine should play an important role in the new journal, in order to promote the future development of the Polish philosophico-medical school [30].

The institutionalization of studies of philosophy of medicine in Poland between the two World Wars can thus be explained by the existence of a local tradition of studies in this discipline, and by the specific historical circumstances which brought some of the followers of this tradition to decision-making positions during the reconstruction of Polish universities in the post-war period in Poland (1919-1920). But what were the origins of the philosophical traditions of Polish medicine? Why were Polish physicians attracted by philosophical, historical and methodological investigations? Why did they form an important (and self-concious) school of thought which maintained its activity through three consecutive generations of physicians?

The study of the origins of the philosophico-medical tradition in Poland is complicated by the fact that the social history of Polish medicine (and, in general, the social history of other sectors of Polish society in the 19th and 20th-centuries) has yet not been written. Its absence makes the study of the socio-cultural background of the Polish school of philosophy of medicine somewhat difficult. Nevertheless, I will attempt to propose at least a partial explanation of the rather puzzling fact that an important school of philosophy of medicine developed in a peripheral country, not even known for its particular achievements in medicine. My hypothesis is that the interest of Polish physicians in theoretical questions developed as a consequence of their relative marginality as well as the great distance separating their professional aspirations from the practical possibilities of realizing them. After the division of Poland between Russia, Prussia, and Austria in 1795, and the resulting deterioration of the system of higher education in the occupied country, Polish students often chose to study medicine abroad. Consequently, many Polish physicians were trained in the best European (mostly German and Austrian) universities, and they became well-versed in the most advanced medical science. However, a decision to return to their country meant very limited opportunities to use their knowledge and skills. The often unescapable result was a "brain drain". The most brilliant among Polish-born physicians chose to remain abroad and establish their entire scientific career there. Thus, some of the well-known German and Austrian physicians (e.g., Robert Remak, Joseph Dietl) were born and educated in Poland. Those who chose to return to Poland often remained frustrated in their professional aspirations. The difficulties of some of the central figures of the second generation of the Polish School (Biernacki, Nusbaum, Hoyer) to find an adequate framework to conduct research, and Bieganski's struggle to found a bacteriology laboratory in a provincial town hospital can illustrate this point. Although migratory movements of medical students to important scientific centers existed in many peripheral countries, such movements were particularly important in occupied Poland, which was deprived of the possibility of developing an adequate local university network.

The foreign occupation also strongly limited the possibilities for Polish physicians to take part in the organization of medical education and health services in their country. Polish physicians were unable to play an important role in the planning of hygienic and prophylactic activities, in sanitary education, and even in the management of medical schools. They remained entirely dependent on the good (and often not so good) will of the occupant [14, 20]. In addition, even the most advanced medical science was of little use when the majority of health problems observed by physicians had their roots in the social conditions of the patient's life. In 19th-century Poland, the majority of the peasant population and important sectors of the urban population lived in extreme poverty and backwardness. There was a sharp need for actions aimed at changing the living conditions of the poorest sectors of the society. This need was strongly felt by the famous German pathologist Rudolf Virchow, who as a young man was sent by the Prussian government to study the impact of a typhus epidemic among the Polish peasants in Silegia (1848). In his report to the Government, Virchow explained that the only way to prevent future epidemics was to improve the living conditions of the peasants, to provide them an adequate education, and to secure their political rights [26]. The study of typhus in Silezia led the young Virchow to the conviction that medicine was a social science, and that medical problems could be solved by political means. He became involved in the political struggle in Germany, participated in the 1848 revolution and the post-'48 radical movements, and finally served as a liberal deputy to the Reichstag (1880). His Polish colleagues did not have similar opportunities, however, to continue their medical action by political means. Although many Polish physicians in the 19th-century were aware of the importance of the social aspects of health problems, the political circumstances in which they were obliged to work sharply reduced the scope of the possibilities of activity in this domain [21].

The political and economic conditions in occupied and divided Poland thus severely limited the possibilities for a gifted and often well-educated physician to find avenues of expression for his creativity. Philosophical reflection on the art and the science of medicine may have been such an expression. Polish philosophy of medicine should not, however, be viewed solely as a pleasant hobby of physicians who lacked more stimulating ways to express their talents. Polish philosophers of medicine were physicians who were well-acquainted with the latest developments in medicine and biology, but who found themselves outside the mainstream of medical research. This specific situation gave them a unique observation point from which to view medicine during a crucial period of its history. This outsider status, their provincialism, their practical preoccupations, and the fact that they did not belong to any of the important national medical schools, enabled them to perceive clearly the great achievements, but also the limitations of the scientific revolution in 19th-century medicine (and, in particular, the difficulties in applying new knowledge to medical practice), and to develop a highly original and rich reflection on the relationships between theory and practice in medicine.

Some of the Polish philosophers of medicine are known today in Poland as "great doctors" of the past: Chałubiński - who helped to develop the well-known health resort, Zakopane, in the Tatra mountains; Biernacki - the discoverer (or co-discoverer) of blood sedimentation; Biegański - a pioneer in social medicine. However, the philosophical activity of the Polish School is not very well known even within Poland, and it is practically unknown abroad. There is, however, one exception: the epistemological studies of Ludwik Fleck.

In the 1920's and 1930's Fleck published several highly original epistemological works, the best known among which is his book of 1935, *Genesis and Development of a Scientific Fact* [7]. In recent years, there has been a growing appreciation of Fleck's works. He is viewed today as a pioneer of the constructivist-relativist trend in the philosophy of science and of the sociological approach to the study of the evolution of scientific and medical knowledge. His works are largely discussed by philosophers and sociologists of science. But Fleck was neither a professional philosopher nor a sociologist. He was a physician, and in the period in which he carried out his epistemological reflections he worked as clinical bacteriologist in a routine analysis laboratory.[2] Although Fleck never mentioned his ties with the Polish School of Philosophy of Medicine, he was an active member of the Lwow Circle of Amateurs of the History of Medicine (affiliated to the Polish Society of History and Philosophy of Medicine), and published his first epistemological study in the *Archives of History and Philosophy of Medicine* [8, 10]. Moreover, a careful analysis of Fleck's early philosophical studies reveals that subjects discussed by the Polish philosophers of medicine - e.g., the conventional aspects of medical taxonomy, the ever-changing nature of medical knowledge and its dependence on larger socio-cultural factors - occupied a central position in the evolution of Fleck's epistemological thought. A study of the relationships between Fleck's ideas and the reflections of other philosophically-minded Polish physicians can thus help us to set Fleck's highly original epistemology into a broader historical and intellectual context.

Fleck's philosophical thought stemmed from his practical experience as a physician and clinical laboratory worker. This is even more true for other Polish philosophers of medicine. All of them - and this is perhaps their greatest originality - were physicians, not scientists or philosophers. They kept private practices and seldom held university positions. The fact that in Poland the philosophy of medicine was born out of clinical practice [25] had a deep impact on their philosophical investigations. For the thinkers of the Polish School, medicine was not only a science but also, or rather above all, the art of healing. Their principal goal was not to learn what diseases are, but what can be done about them. They sometimes discussed (as did their colleagues who wrote about the philosophy of medicine in other countries), the general principles of medical science, but they were mainly interested in therapy - the most neglected and most backward aspect of the new scientific medicine.[3] The fact that even their most theoretical writings (including those of Fleck) are illustrated with numerous detailed examples from their daily work clearly attests to the central place of medical practice in their reflections.

Their practice-centered orientation stimulated their interest in subjects such as reductionism versus holism in medicine, the methods of evaluation of new therapies, the psychological and sociological aspects of illness, and medical ethics. More specifically, this practice-centered orientation led them to situate at the heart of their theoretical reflections the debate on the specificity of medical knowledge: 'Is medicine a science or an art?', and to bring some original answers to this traditional question.

Several thinkers of the Polish School - Biernacki, Biegański, Kramsztyk, Fleck - attempted to affirm the legitimacy of their clinical practice (and in the case of Fleck, clinical laboratory practice) and to demonstrate that their therapy-oriented activity was not inferior to fundamental scientific research. But why was such a demonstration necessary in the first place? Professionals, even those active in fields directly dependent on scientific knowledge, usually do not feel the need to prove that their activity is as valid as fundamental scientific research, or that it contributes to the development of scientific knowledge. As Kramsztyk put it, although an engineer's work is based on physical knowledge, he is judged on the quality of his concrete achievements, not on his contributions to theoretical physics [13]. This was not the case, however, for physicians. From the mid-19th-century on, many leading physicians developed a scientist's self-image, and considered that the physician's contribution to scientific knowledge should be the principal critierion for his appraisal. Why were physicians different from other professionals?

Biegański, Biernacki, and Kramsztyk explained that the tendency of physicians to develop a scientist's self-image was a result of the "scientific revolution" in medicine in the second half of the 19th-century and the resultant modification of both medical knowledge and medical practice.[4] One of the most important consequences of the importation of scientific methods into medicine was the growing acceptance of the view that the bulk of therapy used by doctors in previous periods was, if not harmful, at least useless. The recognition of this fact probably led to a redefinition of the occupational role of the physician.

In the majority of the professions, professionals seek gratification from their clients. This gratification usually parallels the gratification obtained from their usual "reference group" - their professional peers. Thus a successful lawyer who is able to win many trials, or a successful builder who is able to attract many powerful clients is usually esteemed by the members of his own profession. The rationale behind the behavior of the peer group is the shared assumption that success with clients should, and often does, reflect a high level of professional commitment, and that pressure exerted by clients who demand better and more competent service is usually a powerful stimulus for professional progress. This assumption is also adopted in disciplines relying on scientific knowledge, such as agronomy, architecture, engineering, mechanics, etc. In these professions, a steady (albeit not always linear) progress preceded the scientific revolution and the incorporation of its achievements into professional knowledge. Scientific research undoubtedly allowed for significant progress in many practical disciplines (e.g., botanical research led to improvements in agriculture and chemical research gave new perspectives to the textile industry). It did not, however, invalidate the previously accumulated empirical professional knowledge.

The situation was quite different in medicine. In the late 19th-century, thanks to the progress of the fundamental sciences, leading physicians gradually realized that many of the traditional and supposedly time-proven treatments which for centuries had fully satisfied patients, were in fact worthless. The introduction of science into medicine led them to the conclusion, strongly expressed for example in Biernacki's writing, that there was little if any cumulative progress in the art of healing.[5] The internal mechanisms that ensured progress in other practical disciplines were thus clearly insufficient in medicine. This analysis probably led some of the leaders of the medical profession to address themselves to an external reference group, the scientists.[6] The latter were perceived as the only group able to ensure the future progress of medicine. However, professionals cannot be expected to rely forever on an external professional group for evaluation and gratification. A new professional ideal and a new occupational role were therefore developed: those of the physician-scientist.[7]

The evolution of this new professional ideal was far from being shared by all physicians [28]. The thinkers active in the Polish School were among those who refuted the uncritical acceptance of the ideas of scientific medicine. Thanks to their up-to-date medical education thinkers such as Chałubiński, Biegański, Biernacki, and Kramsztyk were aware of the crisis of traditional therapy, and recognized the importance of the "scientific revolution" in medicine. However, they were reluctant to admit that medicine should be reduced to the science of pathology. Moreover, their personal position as physicians in a peripheral and occupied country often gave them no choice but to adopt the discredited occupational role of "just a practician". The need for legitimation of this supposedly secondary role may have been an important stimulus for the development of their original reflections on the specificity of medical activity, a topic which remains central to the present debates in the philosophy of medicine.

NOTES

1 The historian of medicine, Henry Sigerist, the head of the Institute of History of Medicine of Leipzig University in the years 1925-1932, defined the program of this institute as the alliance of medicine, history, and philosophy. But in all probability Sigerist's approach was quite exceptional in the 1920's and 1930's. See [27].

2 I have argued elsewhere that Fleck's original epistemology has its roots in his professional practice. In this work I argue that, in addition, Fleck's transition from the bacteriology laboratory to epistemological reflection was less abrupt than it would appear at first sight. My thesis is that there was an intermediate stage in this process: a critical evaluation cf his own professional experience as a physician and as a bacteriologist in light of the topics discussed by the Polish School of Philosophy of Medicine.

3 On the contradictions between the science of physiology and the art of healing in the late 19th-century, see [11, 28].

4 It is true that the expression "scientific revolution in medicine" is a simplification. The modification of the everyday practice of many physicians was slow and the diffusion of innovation was very unequal. In many places, particularly in those far away from the leading teaching centers, the physician's practice changed very little during the 19th-century. The slow penetration of new ideas should not, however, obscure the fact that the main characteristics of 20th-century scientific medicine had their roots in the 19th-century evolution of this discipline ([19], pp. 142-165).

5 This was not the case however for surgery and obstetrics. Those medical disciplines, which always maintained a separate tradition, were not as affected by the 'scientific revolution' of the 19th-century as was internal medicine. Innovations such as anesthesia or asepsis undoubtedly increased the ability of surgeons to cure disease; they did not invalidate, however, their previous practical knowledge. The evolution of surgery was therefore similar to the evolution of other professions which rely on scientific knowledge, and surgeons did not feel the need to develop a new occupational role.

6 This process was particularly important in the German-speaking countries. See [5, 9]. The evolution of the new occupational role of physician-scientist was much slower in France. Although the "scientific revolution in medicine" had begun with the studies of the French Clinical School, fundamental scientific research in this country was confined to science faculties, 'Grandes Ecoles', and later independent institutions such as the Pasteur Institute, and only very slowly did it penetrate the medical schools and the hospitals. For a long time (in some medical disciplines, until the Second World War),

medical scientists remained marginal to mainstream French medicine [1, 12].

7 The differences between the two professional roles in medicine, the medical scientist who obtains his rewards from the scientific community and the medical practitioner who gets his rewards from his patients, has been studied by Joseph Ben David. See [2, 3].

BIBLIOGRAPHY

[1] Ackerknecht, E.: 1967, *Medicine at Paris Hospital 1794-1848,* The Johns Hopkins Press, Baltimore.

[2] Ben David, J.: 1958, 'The Professional Role of the Physician. A Study in Role Conflict', *Human Relations* **11**, 255-274.

[3] Ben David, J.: 1960, 'Roles and Innovations in Medicine', *American Journal of Sociology* **65**, 557-568.

[4] Bilikiewicz, T.: 1961, 'Introduction', in W. Szumowski, *History of Medicine [Historia Medycyny],* Panstwowy Zakład Wydawnictw Lekarskich, Warsaw, pp. VII-VIII.

[5] Billroth, T.: 1924, *The Medical Sciences in German Universities,* New York.

[6] Engelhardt, H.T. Jr.: 1986, 'From Philosophy *and* Medicine to Philosophy *of* Medicine', *The Journal of Medicine and Philosophy* **11**, 3-8.

[7] Fleck, L.: 1979 (1935), *Genesis and Development of a Scientific Fact*, translated by F. Bradley and T.J. Trenn, The University of Chicago Press, Chicago.

[8] Fleck, L.: 1927, 'About Some Specificities of the Medical Way of Thinking' ['O niektorych swoistych cechach myslenia lekarskiego'], *Archiwum Historji i Filozofji Medycyny* **6**, 55-64, transl. in R.S. Cohen and T. Schnelle (eds.), 1986, *Cognition and Fact - Materials on Ludwik Fleck,* Reidel, Dordrecht, pp. 39-46.

[9] Flexner, A.: 1912, *Medical Education in Europe,* New York.

[10] Fritz, J.: 1925, 'Report on the Foundation Meeting of the Society of Amateurs of Medicine in Lwow' ['Sprawozdanie z posiedzenia organizacyjnego Towarzystwa Miłośnikow Historji Medycyny w Lwowie'], *Archiwum Historji i Filozofji Medycyny* **4**, 154-155.

[11] Geison, G.L.: 1979, 'Divided we Stand', in M.J. Vogel and C.E. Rosenberg (eds.), *The Therapeutic Revolution,* University of Pennsylvania Press, Philadelphia, pp. 67-90.

[12] Jamous, H.: 1973, 'Professions ou systèmes autoperpétués? Changements dans le système hospitalo-universitaire français', in *Rationalisation, mobilisation et pouvoir,* Centre de

Sociologie de l'Innovation, Paris, pp. 5-56.

[13] Kramsztyk, Z.: 1899, 'Medicine - a Science or an Art ?', [Medycyna nauka jest czy sztuką?], in Z. Kramsztyk, *Critical Notes on Medical Subjects, [Szkice krytyczne z zakresu medycyny]*, E. Wende, Warsaw, pp. 110-119.

[14] Kus, M.: 1966, 'Medical Sciences in Poland, 1864-1914' ['Nauki medyczne w Polsce w okresie 1864-1914'], *Studia i Materiały z Dziejow Nauki Polskiej* seria B, **10**, 121-135.

[15] Löwy, I.: 1986, 'The Epistemology of Science of an Epistemologist of Science: Ludwik Fleck's Professional Outlook and its Relationship to his Philosophical Works', in R.S. Cohen and T. Schnelle (eds.), *Cognition and Fact - Materials on Ludwik Fleck*, Reidel, Dordrecht, pp. 421-442.

[16] Löwy, I.: 1988, 'The Scientific Roots of Constructivist Epistemologies: Hélène Metzger and Ludwik Fleck', in G. Freundenthal (ed.), *Etudes sur Hélène Metzger*, Corpus (Paris) 8-9, 219-235.

[17] Ostrowska, T.: 1970, 'The Dynamics of the Development of Polish Medical Publications in the 19th-Century' ['Dynamika procesu rozwojowege polskiego czasopiśmiennictwa lekarskiego w XIX stuleciu'], *Studia i Materialy z Dziejow Nauki Polskiej* seria B, **18**, 33-59.

[18] Pellegrino, E.D.: 1986, 'Philosophy *of* Medicine: Towards a Definition', *The Journal of Medicine and Philosophy* **11**, 9-16.

[19] Rosenberg, C.E.: 1987, *The Care of Strangers*, Basic Books, New York.

[20] Skulimowski, M.: 1966, 'Medical Sciences in Poland, 1795-1864' ['Nauki medyczne w Polsce w latach 1795-1864'], *Studia i Materialy z Dziejow Nauki Polskiej* seria B, **10**, 105-120.

[21] Słowiecki, M.: 1981, *History of Polish Science [Dzieje Nauki Polskiej]*, Interpress, Warsaw, pp. 192-196.

[22] Szumowski, W.: 1949, 'La philosophie de la médecine - son histoire, son essence, sa dénomination et sa définition', *Archives internationales d'histoire des sciences* **2**, 1097-1139.

[23] Szumowski, W.: 1919, 'The History and Philosophy of Medicine as a University Discipline' ['Historia i filozofia medycyny jako przedmiot uniwersytecki'], *Polska Gazeta Lekarska* **11**.

[24] Szumowski, W.: 1920, 'The Philosophy of Medicine as a University Discipline' ['Filozofia medycyny jako przedmiot uniwersytecki'], *Przeglad Filozoficzny* **23**.

[25] Szumowski, W.: 1926, 'The Most Urgent Tasks of the Philosophy of Medicine' ['Najbliższe zadania filozofii medycyny'], *Archiwum Historji i Filozofji Medycyny* **4**,

70-73.

[26] Taylor, R., and Rieger, A.: 1984, 'Rudolf Virchow on the Typhus Epidemic in Upper Silesia', *Sociology of Health and Ilness* **6**, 203-215.

[27] Temkin, O.: 1977, 'The Double Face of Janus', in O. Temkin, *The Double Face of Janus and Other Essays*, The Johns Hopkins University Press, Baltimore, pp. 3-37.

[28] Warner, J.H.: 1986, *The Therapeutic Perspective: Medical Practice, Knowledge and Identity in America*, Harvard University Press, Cambridge, Mass., pp. 235-257.

[29] Weisz, G.: 1987, 'The Medical Elite in France in the Early Nineteenth Century', *Minerva* **25**, 150-170.

[30] Wrzosek, A.: 1924, 'Tasks and Goals of the Polish Archives of History and Philosophy of Medicine' ['Zadania i zamierzenia polskiego Archiwum Historji Medycyny i Filozofiji Medycyny'], *Archiwum Historji i Filozofji Medycyny* **1**, 1-13.

CHAPTER II

TYTUS CHALUBINSKI THE LAST FOLLOWER OF 'MEDICAL SYSTEMS' OR PIONEER OF A NEW APPROACH TO THERAPY?

Tytus Chałubiński (1820-1889), a physician, scientist and philosopher, is usually viewed as the founder of the Polish school of philosophy of medicine, and the date of publication of his book, *The Method of Finding Therapeutic Indications,* 1874 [10], as the establishment of this school.

Born in Bialograj, son of a judge in a local court, Chałubiński began medical studies in Vilnaus, in 1838. At that time, shortly after the death of Jedrzej Śniadecki (1768-1838), the Vilanus Medical School was still under the influence of its most famous teacher. Śniadecki, a professor of chemistry at Vilnaus University, was known for his *Theory of Organized Beings* (1804), one of the pioneering works in physiological chemistry, and an attempt at the construction of a chemical theory of life based on mechanisms of assimilation/dissimilation.[1] In all probability Chałubiński became acquainted with Śniadecki's writings at Vilnaus. When the Russian government closed the Vilnaus university in 1840, Chałubiński decided to continue his studies in Germany. He soon moved to Dorpat with the intention of completing his medical studies. In Dorpat he became interested, however, in the philosophy and history of science, and instead of studying medicine he wrote a philosophical thesis on the discovery of sexual reproduction in plants [11]. Later (1844-1846) he went to Würzburg to finish his medical studies there.

Returning to Warsaw, Chałubiński continued his studies in the history of botany,[2] and at the same time started a clinical practice at the Evangelic Hospital under the supervision of the hospital's director, Fendynand Dworzaczek. Dworzaczek (1804-1877) was the founder of the Warsaw school of internal medicine. He worked in Paris (1831-1835) and brought back to Poland the methods of the French Clinical School [3, 22]. According to Chałubiński's collaborator and friend, Baranowski, the clinical methods taught by

Dworzaczek had a decisive influence on the evolution of Chałubiński's ideas on medicine.[3] Dworzaczek was also interested in the philosophy and history of medicine. These subjects became his principal occupation when, in 1847, he lost his sight and was obliged to leave his clinical work.[4] Chałubiński, who rapidly acquired a reputation as a gifted physician, took over Dworzaczek's clinical functions in 1847. In 1857 he was appointed to the chair of pathology and of specific therapy at the newly-opened Medico-Surgical Academy of Warsaw, which in 1862 became a part of the reconstructed University of Warsaw [Szkoła Główna]. At the same time, he continued his hospital duties and kept an important private practice. He was a very successful physician, and had among his patients some of the leading personalities of Warsaw. He also gave free consultations to the poor, and was one of the rare Polish physicians who treated poor Jewish patients free of charge. Chałubiński's clinical and pedagogical activity was disturbed by the Polish insurrection of 1863, and came to a definitive end in 1873, when the Russian government fired him in the process of russification of Warsaw University. He then turned to botanical and mineralogical studies. Later Chałubiński became deeply interested in the Tatra mountains and contributed to the development of the health and tourist resort Zakopane, and to the growing interest of Polish intellectuals in the specific folklore of the peasants of this region.[5]

In 1874, one year after Chałubiński lost his professorship in Warsaw, he published his book, *The Method of Finding Therapeutic Indications* [10]. In this book, intended mainly for medical students, he summarized his ideas on medical science and the art of healing. Chałubiński described his book as an attempt at systematization of the practical methods used by (good) physicians when they decided what should be the appropriate treatment of a given clinical case. He underscored the point that medical textbooks usually gave only insufficient and abstract practical information concerning therapeutic indications. They only seemed to reproduce the conventional classification of therapeutic indications: vital indications, causal indications, symptomatic indications, fundamental indications, etc. Such classsifications were, however, of little use for the practician. First, valid classifications of therapeutic indications made sense only when the physician was absolutely sure of his diagnosis and dealt with a disease whose nature was well understood. But such a situation was (and one might add, still is) far from being the most frequent one. Second, textbook classifications (e.g., the distinction between symptomatic and vital indications) were sometimes scholastic exercises with little or no relationship to real-life phenomena. For example, in some diseases the treatment of a symptom (e.g., fever in typhoid fever) influenced one of the essential features of the disease itself,[6] and in other diseases (e.g., heart disease) the so-called "symptomatic" treatment could reinstate normal function ([10], pp. 7-8; 10). Third and most important, the classificiation of therapeutic indications was a *post-factum* event, an adjective which described the previously-found therapy. Such post-factum classification was of little practical value for the physician who needed to know in the first place how to find an adequate treatment for every specific case.[7]

Chałubiński's method of finding therapeutic indications aimed precisely at satisfying the concrete needs of the physicians. This method had its roots in his holistic view of medicine and his vision of disease as perturbed function. Chałubiński defined himself as a materialist and firmly believed that physico-chemical laws are sufficient to explain all the phenomena of organic life. But on the other hand, he rejected reductionism and adopted a holistic view of biological phenomena. He tried to reconcile the two by stressing the complexity of living organisms. The combination of all possible modifications in the conditions of the forces and the bodies participating in the physico-chemical reactions of living matter created, he explained, a practically infinite number of possible variations. For this reason (although all living organisms were shaped by similar physico-chemical forces), each organism should be viewed as a unique entity ([10], p. 23). In addition, Chałubiński viewed disease as a disturbance of normal physiological function. He considered the physician's task to be the restoration of the organism's lost equilibrium.[8] Chałubiński's definition of disease as disturbed function allowed for a highly individualized perception of pathological phenomena, because each perturbation strongly depends on the underlying individual characteristics of the sick organism ([10], pp. 32-35).

The holistic approach, that stressed the individuality of each living organism and each pathological phenomenon, led Chałubiński to adopt a constructivist vision of diseases as an abstract classification created by physicians and thus to reject a model of disease as some natural phenomena existing "out there". For him, "the existence of specific organs and their functions should allow, at least to a certain degree, for the uniformity of representations of pathological disturbances. We can thus understand that their great number and infinite variability nothwithstanding, disease processes can be studied within a more general framework, and some classifications, necessary for the understanding of pathology, are possible. These general pictures of disease are not sufficient, however, to contain all the richness of pathological manifestations.... There are no two identical epidemics, nor two identical cases of disease.... The disease pictures are indispensable and it is not necessary to insist on their utility. One should not forget, however, that while representations in natural history - of animals, plants, minerals - refer to real entities, pictures of disease are abstract representations of certain phenomena separated from their natural context of individual life" ([10], p. 39).[9]

But how is the physician to adapt these abstract and schematic representations to the study of pathological phenomena? Chałubiński proposed an original solution: to replace the rigid notion of 'disease entities' by the much more flexible concept of 'disease moments'. Each disease moment should represent a single pathological manifestation of a given patient at a given moment of his disease ([10], pp. 42-43). The physician's aim should be to deal one by one with all the disease moments (i.e., clinical symptoms) present at a given time, and in this way restore the lost equilibrium of the organism. The physician should therefore ignore all the textbook recipes and view each patient as a completely new, unique case. He needs to enumerate carefully all the pathological symptoms, then to use his knowledge of physiology and of general pathology in order to determine what is the most distant link in the casual chain of each of the disease moments still reachable by the available therapeutic means.

Theoretically, the physician should deal with each one of the disease moments separately. Chałubiński stressed, however, the importance of taking into consideration the impact of a treatment destined to eliminate or to modify a given disease moment (e.g., the administration of drugs to diminish fever) on other disease moments present in the same patient (e.g., irregular heart beat) ([10], pp. 92; 60; 51-53). In practice, the physician can seldom treat simultaneously all the disease moments present in a given patient. He should therefore first attend to the most important disease moment, i.e., the one which is the most dangerous to the patient's health or is the most decisive in the evolution of a given morbid state. The determination of this most important disease moment, Chałubiński explained, depends *in fine* on the individual characteristics of a given case: "*we should never forget that we are not treating diseases but patients*" ([10], pp. 84; 58-67).[10]

Although Chałubiński's theory was based on the perception of illness as a unique phenomenon, he did not negate the value of general statements about pathology and therapy. He simply strongly warned against the abusive use of such general statements. He explained that in each treatment one should distinguish between 1) *what* should be done and 2) *how* it should be done. The first point depends nearly exclusively on the medical knowledge of a given period. The second point is, in addition, conditioned by the personal experience of a given physician and by the individual characteristics of the case he is treating. It is thus perfectly possible to achieve a consensus on the nature of therapeutic indications in a given pathological state. However, the precise way of conducting a therapy can never be entirely codified and made uniform ([10], p. 12).

Chałubiński's method of finding therapeutic indications was a reaction against two phenomena, judged by him to be extremely harmful to medicine: 1) the routine approach to therapy based on "textbook-recipes" and 2) "therapeutic nihilism" based on an over-estimate of medical science and the neglect of medical practice. Although Chałubiński believed that every treatment that is not helpful is harmful, and that simplicity is always a great advantage in therapy, he strongly attacked therapeutic nihilism, describing it an absurd, and accusing it of preventing the development of medical science ([10], pp. 86; 92; 81-83; 93). This critique reveals that Chałubiński viewed medicine solely as a practical science, by identifying the progress of medical science with the progress of therapy. In his book on malaria (1875), he affirmed that medical theories should always be closely related to clinical practice. Medical research, "should have, at least to a certain degree, a specific character, the essence of which is utilization of every disease-related detail for practical goals where the emphasis would always be placed on the therapeutic aspect of every investigation. *A physician should accept, investigate, and verify every new fact, new discovery, and new theory from the point of view of its importance for the cure of diseases*" ([13], p. 13).[11]

For the followers of therapeutic nihilism the very opposite was true. They justified their therapeutic attitude by the need to transform medicine into an exact science. In fact, therapeutic nihilism was born as a result of the rapid evolution of biological and medical knowledge. It was also a reaction against the "medical systems" which reduced all disease symptoms to a single cause (or at best a few) and advocated the excessive use of "heroic" medications (bleedings, emetics, mercury) without adequate experimental proof of their efficiency. The largely justified criticism of the excessive methods of "heroic medicine" did not, however, always lead to the same conclusions. Chałubiński felt close to the physicians of the French Clinical School, who based their refusal of heroic therapies on meticulous clinical observations. They were therapeutic sceptics, but not therapeutic nihilists: they advocated prudence, not the total absence of therapy. In contrast, the physicians of the Viennese school, led by Karl Rokitansky (1804-1878) and Joseph Skoda (1805-1878), based their refutation of previous methods of medicine on the growing knowledge of anatomy and histology. They observed the lack of effect of the majority of usual treatments on pathological modifications of tissues, and concluded that nearly all those therapies were useless and that withholding treatment was often better than inefficient, if not harmful, treatment. Their idea was reinforced by the observation that, in many cases, leaving the patient with no treatment at all produced better results than applying the traditional treatments (blood-letting, emetics, cataplasms) [2, 9].

The physicians of the Viennese school gave high priority to laboratory investigations and justified this approach by the claim that an adequate understanding of the scientific nature of disease is a precondition of efficient therapy. One of the spokesmen for this school, the Polish-born Joseph Dietl (1804-1878), explained in 1845 that "physicians should be judged according to their knowledge, not according to the results of their therapy. One should appreciate the physician as nature's investigator and not as an individual dedicated to the art of healing.... Our strength lies in our knowledge, not in our actions. Our actions stem from our knowledge, like a fruit growing from a flower. If the natural sciences blossom, practical medicine, their fruit, will also be established" [16].

Chalubinski's approach was diametrically opposed to the one professed by Dietl. Although in *The Method of Finding Therapeutic Indications*

Chałubiński briefly discussed his ideas on the broad principles of biology ([10], pp. 14-49),[12] he made no attempt whatsoever to link clinical knowledge to the latest progress in physiology, histology, cellular pathology, etc., in all probablility because he did not view this progress as relevant to the practical art of healing.

The radical nature of Chałubiński's views, and the fact that he paid no attention to the new directions in medical science, was considered by his critics as the principal reason for the oblivion into which his teachings fell after his death. In a critical evaluation of Chałubiński's ideas written in 1900, the philosopher of medicine Edmund Biernacki (1866-1911) asserted that Chałubiński was the last of the followers of "systems" in medicine ([6], pp. 42-46; 50-58; 307).[13] Biernacki based his claim on an analysis of the practical examples of therapy proposed by Chałubiński, particularly on his frequent recommendation of antipyretic drugs and on his insistence on the importance of gastric disorders. This therapeutic approach betrayed, according to Biernacki, the influence of Broussais who viewed gastro-intestinal inflammation as the source of nearly every pathological state.[14]

Nevertheless, Biernacki, in agreement with the spirit of Chałubiński's therapy-centered theories, affirmed that Chałubiński's writings remained pertinent, their (false) theoretical principles notwithstanding, because they retained their pratical value. Chałubiński had adopted an outdated system, but he "'unlearned' this system in practice and became an excellent physician". The old-fashioned theoretical ideas found in Chałubiński's writings, Biernacki explained, were largely counterbalanced by the importance of his clinical indications, still of great practical value to physicians ([6], pp. 37-38; 46-47; 76; 92-94). The respect for Chałubiński's teaching was shared by other philosophers of the second generation of the Polish School - Biegański, Biernacki and Kramsztyk. They attempted, however - each in his own way - to "modernize" Chałubiński's teachings and to reconcile them with the scientific knowledge of their time.

NOTES

1 Among the Polish philosophers of medicine, only Adam Wrzosek viewed Jedrzej Śniadecki (1768-1838), and not Chałubiński, as the true founder of the Polish school of philosophy of medicine. Other thinkers of the Polish School viewed Śniadecki as a biologist and philosopher of biology, not as a philosopher of medicine. See [4, 32, 38].

2 In 1846, Chałubiński published his second book on history of botanics [12]. He also translated into Polish the botany textbook of Adrien de Jussieu.

3 The French Clinical School, which stressed the importance of accurate clinical observations, included among others, Gabriel Andral (1791-1876), Pierre Louis (1787-1872) and François Magendie (1783-1855). It was active in the first half of the 19th-century [1].

4 According to Szumowski, the Polish expression "philosophy of medicine" [filozofia medycyny] was first introduced by Dworzaczek in his lecture 'Introduction to the philosophy of medicine', which was presented at the meeting of The Medical Society of Warsaw (Towarzyswo Lekarskie w Warszawie), January 20th, 1856 [33].

5 For biographical notes on Chałubiński, see [3, 19, 20, 35, 36, 37].

6 On the problem of fever in typhoid fever and its treatment with antipyretic drugs in the 19th-century, see [25].

7 The philosopher of medicine E. Biernacki (1866-1911) claimed that the refutation of the classifications of therapeutic indications by Chałubiński was a reaction against the textbook of pathology and therapy of Niemeyer (published in 1859), which was translated into Polish shortly before the publication of Chałubiński's book. Biernacki thought that Chałubiński over-reacted to Niemeyer's book. Niemeyer, Biernacki affirmed, did not pretend to introduce absolute classifications of therapeutic indications but rather to stress the limits of the physician's intervention by showing that many of the therapeutic means used by physicians did not influence the evolution of a given disease but only alleviated its symptoms ([6], pp. 14-15).

8 Chałubiński's idea that the physician's task was the restoration of a lost equilibrium was perhaps influenced by the ideas of Śniadecki, who developed a dualistic vision of disease as either over-stimulation or inability to respond to stimuli ([36; 20], pp. 7-8). Śniadecki's dualist perception of disease reflected the influence of "brownism", one of the medical systems of the late 18th century. Śniadecki spent two years in Edinburgh studying John Brown's theory of disease. This theory, which viewed every pathological condition as a result of either sthenia (over-stimulation) or asthenia (inability to

respond to stimulation), was probably one of the principal sources of inspiration of Sniadecki's biological theories ([31; 22], pp. 466-467).

9 In all probability Chałubiński limited his constructivist vision solely to the description of disease entities. Unlike the philosophers of the second and the third generation of the Polish School, he never pointed to the possibility that other aspects of medical knowledge could have been constructed by physicians, and he provided no historical dimension to his reflections on medicine.

10 Chałubiński wrote his book shortly before the "bacteriological revolution" in medicine (in the 1880's and 90's) and the consequent shift to a strictly causal model of infectious diseases - a model that also strongly influenced other domains of medicine (e.g. [14; 15; 8], pp. 55-77). From this new point of view, stressing specific causes of pathological states, Chałubiński's ideas on disease and therapy may seem to be a simple reflection of an outdated approach to pathological phenomena. This is, however, not the case. Discussions on the nature of classification of diseases and on causality in medicine have never lost their pertinence for physicians. See, e.g. ([21], pp. 126-127; 204-223). Recently, some physicians have advocated the abandonment of the notion of causally defined 'diseases' and a return to a symptomatic definition of "clinically-expressed tissue reactions" which should be recognized and treated on several separate levels [17, 29].

11 For the present debate on medicine as a practical science, see, e.g. [27, 28, 34].

12 Chałubiński's ideas on biology were rather outdated. For example, he considered the formation of cells as a mechanical process, quite similiar to the formation of crystals ([10], p. 17). This view, in all probability inspired by Schleiden's cellular theory (Chałubiński was Schleiden's student in Würzburg in the 1840's), was strongly outdated in 1874, nearly 20 years after Virchow's famous affirmation, "omni cellula cellula". On cellular theory in biology and medicine, see, e.g. ([4, 24, 26; 30], pp. 67-69).

13 "Medical systems" were created by physicians who, like Brown (1735-1788) or Broussais (1772-1838), attempted to find a single system able to explain the totality of pathological phenomena. See, e.g. ([7], pp. 18-31 ; [1], pp. 84-108).

14 Biernacki's affirmation that Chałubiński was influenced by Broussais was strongly refuted by other Polish physicians and philosophers of medicine. Chałubiński's collaborator, Ignacy Baranowski, remembered that Chałubiński always recommended to his students the pathology textbook written by Andral, one of the principal opponents of Broussais's theories. Chałubiński's student Zygmunt Kramsztyk affirmed that Chałubiński had no affinity whatsoever with Broussais's theories [23], while the

philosopher of medicine, Wladyslaw Biegański, explained that Chałubiński's method was probably directly inspired by the "rational therapy" advocated by the Viennese clinician Wunderlich (1815-1877). This approach, a reaction to clinical nihilism, was much debated when Chałubiński studied in Würzburg ([5], p. 245).

BIBLIOGRAPHY

[1] Ackerknecht, E.: 1967, *Medicine at the Paris Hospital 1794-1848,* The Johns Hopkins Press, Baltimore.

[2] Ackerknecht, E.: 1982, *A Short History of Medicine,* The Johns Hopkins Press, Baltimore.

[3] Baranowski, I.: 1907, *Tytus Chalubiński, 1820-1889,* Drukarnia L. Staszica, Warsaw.

[4] Bechtel, W.: 1984, 'The Evolution of our Understanding of the Cell: A Study in the Dynamics of Scientific Progress', *Studies in History and Philosophy of Science* **15**, 309-356.

[5] Biegański, W.: 1908, *The Logic of Medicine [Logika medycyny],* E. Wende, Warsaw.

[6] Biernacki, E.: 1900, *Chałubiński and the Present Goals of Medicine [Chałubiński i obecne zadania lekarskie],* Wydawnictwo Przegladu Filozoficznego, Lodz.

[7] Canguilhem, G.: 1981, 'Une idéologie médicale exemplaire, le système de Brown', in G. Canguilhem, G., *Idéologie et rationalité dans l'histoire des sciences de la vie,* Vrin, Paris.

[8] Canguilhem, G.: 1981, 'L'effet de la bactériologie dans la fin des "Théories Médicales" au XIXe siècle', in G. Canguilhem, *Idéologie et rationalité dans l'histoire des sciences de la vie,* Paris, Vrin.

[9] Canguilhem, G.: 1966, *Le normal et le pathologique,* Presses Universitaires de France, Paris.

[10] Chałubiński, T.: 1874, *The Method of Finding Therapeutic Indications [Metoda wynajdywania wskazań lekarskich],* Gebethner & Wolf, Warsaw.

[11] Chałubiński, T.: 1843, *Historical Review of the Ideas Concerning Sexuality and Fertilization in Plants [Historyczny przeglad mniemań o płciowosci i sposobie zapładniania roślin],* Biblioteka Warszawska, Warsaw.

[12] Chałubiński, T.: 1846, *A View of the History of Botanics and its Relationships with Other*

Natural Sciences [Rzut oka na dzieje botaniki i na stosunek jej do innych umiejetności przyrodzonych], Biblioteka Warszawska, Warsaw.

[13] Chalubinski, T.: 1875, *Malaria - A Study From a Practical Point of View [Zimnica - Studyum ze stanowiska praktycznego],* Gebethner & Wolf, Warsaw.

[14] Codell-Carter, C.: 1985, 'Koch's Postulates in Relation to the Work of Jacob Henle and Edwin Klebs', *Medical History* **29**, 353-374.

[15] Codell-Carter, C.: 1977, 'The Germ Theory, Beriberi, and the Deficiency Theory of Disease', *Medical History* **21**, 119-136.

[16] Dietl, J.: 1960 (1845), 'Praktische Wahrnehmungen', in J. Dietl, *Collected Works,* Krakow, quoted by W. Szumowski (1928), *Historia medycyny,* Warsaw.

[17] Fessel, W.J.: 1983, 'The Nature of Illness and Diagnosis', *The American Journal of Medicine* **75**, 555-560.

[18] Galland, M.R.L.D.: 1986, 'Common Sense Models of Health and Disease', *New England Journal of Medicine* **314**, 652.

[19] Gyöffry, I. : 1935, 'Dr. Tytus Chałubiński', *Archiwum Historji i Filozofji Medycyny* **15**, 45-56.

[20] Jarosinski, T.: 1978, *Tytus Chałubiński: His Life and His Scientific and Social Activities, [Tytus Chałubiński: życie, dzialałnosc naukowa i społeczna],* Nauka Dla Wszystkich, Krakow.

[21] King, L.: 1982, *Medical Thinking: A Historical Preface,* Princeton University Press, Princeton.

[22] Kośminski, S. (ed.): 1883, 'Ferdynand Gotard Karol Dworzaczek', in *Dictionary of Polish Physicians [Słownik lekarzy polskich],* Warsaw, pp. 101-102.

[23] Kramsztyk, Z.: 1900, 'Review of Biernacki's book, *Chałubiński and the present tasks of medicine'* ['Recenzja ksiażki Biernackiego, *Chałubiński i obecne zadania lekarskie'], Krytyka Lekarska* **2**, 59-65.

[24] McMenemey, W.H.: 1968, 'Cellular Pathology', in F.N.L. Poynter (ed.), *Medicine and Sciences in the 1860's,* Wellcome Institute of the History of Medicine, London, pp. 13-44.

[25] McTavish, J.R.: 1987, 'Antipyretic Treatment and Typhoid Fever', *Journal of the History of Medicine* **42**, 486-506.

[26] Maulitz, R.: 1971, 'Schwann's Way: Cells and Crystals', *Journal of History of Medicine* **26**, 431-436.

[27] Maull, N.: 1981, 'The Practical Science of Medicine', *The Journal of Medicine and Philosophy* **6**, 165-182.

[28] Munson, R.: 1981, 'Why Medicine cannot be a Science', *The Journal of Medicine and*

Philosophy **6**, 183-208.

[29] Scadding, J.G.: 1988, 'Health and Disease: What can Medicine do for Philosophy', *Journal of Medical Ethics* **14**, 118-124.

[30] Schiller, J.: 1980, *Physiologie et classification,* Maloine, Paris.

[31] Skulimowski, M.: 1966, 'The Medical Sciences in Poland in the Years 1795-1864' ['Nauki medyczne w Polsce w latach 1795-1864'], *Studia i Materialy z Dziejow Nauki Polskiej* seria B, **10**, 105-120.

[32] Szumowski, W.: 1933, 'L'école polonaise médico-philosophique', in *La Pologne au VIIe Congrès International des Sciences Historiques,* Société Polonaise d'Histoire, Warsaw, pp. 3-9.

[33] Szumowski, W.: 1947, 'The Sources of *The Method of Finding Therapeutic Indications* of Chałubiński' ['Źródła *Metody wynajdywania wskazan lekarskich Chałubińskiego'],* *Archiwum Historii Medycyny* **18**, 273-278.

[34] Widdershover-Heerding, I.: 1987, 'Medicine as a Form of Practical Understanding', *Theoretical Medicine* **8**, 179-185.

[35] Wrzosek, A.: 'Tytus Chałubiński', *Polish Biographical Dictionary [Polski slownik biograficzny],* vol 3, pp. 253-259.

[36] Wrzosek, A.: 1970, *Tytus Chałubiński,* Panstwowy Zaklad Wydawnictw Lekarskich, Warsaw.

[37] Wrzosek, A.: 1937, 'La jeunesse de Tytus Chałubiński' ['Mlodość Tytusa Chałubińskiego'], *Archiwum Historji i Filozofji Medycyny* **16**, 1-12.

[38] Wrzosek, A.: 1924, 'Tasks and Goals of the Polish Archives of History and Philosophy of Medicine', [Zadania i zamierzenia polskiego Archiwum Historji i Filozofji Medycyny], *Archiwum Historji i Filozofji Medycyny* **1**, 1 -13.

[39] Wrzosek, A.: 1959, 'Ignacy Baranowski on Tytus Chałubiński' ['Ignacy Baranowski o Tytusie Chałubińskim'], *Archiwum Historii Medycyny* **22**, 577-584.

CHAPTER II.A

TEXT OF TYTUS CHALUBINSKI
THE METHODS OF FINDING THERAPEUTIC INDICATIONS (1874)*

Recovery. The physician's attitude to illness. The finding of indications

Having previously described the proper point of view to take when considering the state of illness, and having made clear how important the tracing of the interrelations of pathological moments is, let us now discuss how we should interpret the ability of organisms to compensate for pathological disturbance, i.e., to recover.

We have found that a man can sometimes pass through all the ages of his life, and through many modifications of the environment and of conditions of life, in perfect health. We have also learned that an organism can be hurt by various pathological disturbances and yet its life not necessarily be destroyed; on the contrary, its functions may return to some extent or even completely to a state of equilibrium.

As we look at how some of the man-designed and man-constructed machines operate, we can see that even in these there is the possibility of preventing certain incidental failures or irregularities, and of compensating for others and removing their effects. All these are practicable either by means of auxiliary mechanisms, i.e., regulators, or by suitably applied devices or forces which are provided for the normal and purposeful operation of these mechanisms, such as brakes, engine-slowing devices, safety valves, etc. We have previously identified the human body as a true biological machine. Hence, in order to keep a state of health in varying environmental conditions, and to recover from the most severe sickness, health should be understood as the effect of automatic regulation of this biological machine and the compensation for disorders through the same devices and functions which maintain the normal state of life.

---

* pp. 50-55; 67-73; 84-87; 92-94.

Let us see now what the task of medicine is when confronted with irregular states of life, that is, with illness. From our point of view, life should be considered as the state of health and the state of illness. We may readily formulate a general idea about the limits of the useful action of a physician. Medicine is based on natural knowledge, and thus is dependent on the development of natural science; it investigates and highlights the following problems:

1• Under what conditions can individuals or communities attain optimal normal life and avoid health disturbances, i.e, hygiene and diet.

2• What conditions are needed to compensate for health disturbances and to recover, or to maintain the optimally possible state of health, i.e., therapy.

Here we shall engage the latter problem, and in particular the way of finding correct therapeutic indications.

Having closely examined the patient and estimated all the relations between the existing pathological moments, having identified the disease, we should then review the pathological moments and rank the one which, out of all those existing or possible, is the most endangering for the state of health -- or life. Then, we should estimate to what extent this moment is therapeutically accessible to us, that is, what are our possibilities to have it removed, or reduced, or modified. Let us call it the major moment. Now, since we want to remove or to modify it, we should find out if the major moment is dependent on any other moment, and whether this moment is accessible to us as well. This other moment could be called the higher moment. If we follow this path step by step, we shall reach the highest therapeutically susceptible moment possible to reach. The most important is to remove -- or to modify -- the last or the highest therapeutically accessible moment possible. For we can be sure that the whole group of moments depending on this highest moment will be removed -- or modified -- as much as possible.

We should estimate generally the weight and direction of the therapeutic procedure concerned with this primary indication; thus let us forecast the expected effect of our action on the whole condition of illness. Should we find in this forecast that a certain group of pathological moments would not be affected by our general procedure but still contains some important health- or life-endangering moments, we have to apply the same method of conduct to this group, in other words again remove or modify the last or the highest therapeutically accessible moment which exerts its influence on this group.

Next, we must analyze whether the realization of this secondary indication would not be contradictory with the primary one, and whether it will be acceptable or desirable to apply it simultaneously with or following the primary one.

Then, if in our analysis of the prospective effects of our therapeutical action we find that there will still remain a group of important moments unaffected, we should then repeat the above-described procedure until all pathological moments have been reached.

Keeping to this method, we can be sure that none of the indications is overlooked. However, when using these indications to construct the general plan of therapeutic action -- which we shall discuss later -- they will not always be fulfilled in the same order as they were developed. Many times we must begin with the removal of the moment which is specifically less important. This will happen in circumstances in which this moment, even though less important, prevents effective action for modification or removal of a moment of higher importance. For an explanation one may look at some examples below.

The above method of estimation of pathological moments, and the way of formulating of the respective indications, must be repeated by the doctor every time he sees the patient, the more so as the importance of pathological moments, or their effect on the state of health -- or life -- can change in the process of illness, not to speak of the course of therapy.[1]

It goes without saying that every time we can see, from both the individual condition of the patient and our knowledge of the pathological process, the prospect of full compensation for the disorder (which may even be crucial for life) by natural powers, there is no need to undertake any positive action; but even then, one can usually discover a means of proper maintenance of the patient which allows for even smoother completion of the compensation.

If we contemplate the ways of reaching various pathological moments through therapeutic means, we arrive at the conclusion that only a few of them can be removed completely, in most cases through surgery, while in the overhelming majority of the cases pathological moments are only susceptible to modifications. The simplest phenomena occur naturally only in certain particular conditions. As long as even one of these particular conditions has been removed or changed, the phenomenon will not occur or will be modified.

Each phenomenon in the human body -- or moment in the state of illness -- depends on many conditions. We call these "particular conditions of pathological moments". Actually, the term "particular moments" would be more correct; however, we prefer to refer to "conditions" in order to avoid the repetition of the word "moments".

If we have it in our power to remove or modify any one or more of these particular conditions or moments, the phenomenon -- or the more general moment -- will either cease, or vary, or be modified. Now, when proceeding to combat a pathological moment we have to contemplate what are the particular conditions under which this moment came into being, and which ones we would be able to reach by therapeutic methods, and calculate how much the major moment would be changed or modified in a desirable direction as a result of the action considered by us.

In order to indicate closely this desirable direction, the case must be compared with the processes known to us from similar cases, where we know the conditions under which the case was directed to recovery either by natural powers or through active therapeutic conduct. Besides, one should take into consideration the condition and the physiological relations of the respective functions. Sometimes these functions constitute the only target that we can head for in the midst of the disease's deviations and vicissitudes.

Let us take a few examples for a better explanation of the above-described procedure.

Let us imagine a developing pleuritis case, clearly due to a cold, in a middle-aged patient with no other abnormalities in his history. By comparison with the known pathological process we can be temporarily assured that there is no danger to the patient's life, because in similar cases such sickness was not a direct cause of death. However, the intensity of the process and its consequences are not trivial to the patient.

The major moment in this process is exudation : the type of exudent, its abundancy, its effect on the adjoining lung and its more or less complete reabsorption. Hence, this moment should be viewed as major, since its effect may develop in a most unwholesome way. Thus we are setting the stage for our therapeutic action.

Remembering the general rule of medical conduct, which requires that the patient should be placed in the best possible conditions for compensation of the pathological moments, we have to review the conditions of which the exudation is the consequence. We know that exudation is one of the indispensable elements of the inflammation process. Tracing down the direct moments in which the process has originated, we arrive in this case at the moment the patient caught cold. However, this etiological moment, which (at the point when we are carrying on the examination) has already ceased, is out of reach of our medical action. We must thus limit our conduct to subsidiary moments.

If the weight of particular pathological moments and their interrelations is estimated, and the physician's attitude to illness, formulated above, is assumed, we may be sure that any doctor, even one who has just started his practice -- but of course possessing theoretical knowledge of medical science -- will immediately find the right indication or the answer to the question: "What should be done?"

Surely, no experienced practitioners seeing a consumptive in a state of heavy pneumorrhagia would ever start working on the patient with an application of cod-liver oil, or koumiss, or by sending him to a mountain spa for air, or with whey or mineral water treatment. Neither would he apply any drugs to soothe the cough of the bronchitis patient in danger of suffocation due to heavy clogging of bronchi with mucus, nor would he use them at the very beginning of this illness in its acute form if he had found that antiphlogistic means should be applied. Neither would he trust the pacemaking properties of foxglove, which is also a diuretic means, in the case of highly developed effects of heart disease involving heavy dropsy and stagnation within the main organs. But I should hardly imagine a beginning doctor finding his way to the starting point of a therapy with therapeutical suggestions such as those usually found in medical textbooks. With our method, however, anyone will readily aim at the correct indication.

Having presented this method for use by students, I must express my opinion that none of the correct, individually-invented indications have been made by any other method than this one. For this is the one and only method possible. The point is that so far the judgments have been created in doctors' minds, as the logicians say, synthetically. The reasons for these judgments, however true in themselves, have never been clearly described, and have never been clearly and correctly expressed. This is because the notions of illness and pathological moments have not been sufficiently defined. There is no clearly-stated criterion according to which pathological moments can be evaluated. Finally, the physician's attitude to illness has been incorrectly formulated.

One can be a great orator or lawyer without studying rhetoric or logic. One can provide correct therapeutical indications without knowing the method explained above. But when utilized, this method will prevent everyone from giving wrong indications and will provide effective means for producing the indications in a rational and comprehensive way, making the main lines of reasoning on the problem more distinct. Hence, I neither can nor want to claim credit for anything else than formulating in general terms the process which in each individual case must be carried out by each independent mind.

Plan of treatment. Accomplishment of the plan

When watching how an efficient practitioner can examine in a relatively short time a considerable number of patients and prescribe for each one an appropriate, rational, and individually-devised treatment, it might seem that this is quite an easy procedure, at least for the more common cases.

But it is not so. First, because the subject itself is not easy to handle, and then because of the responsibility for human life and health which exerts a heavy burden on the consciousness of the doctor. Such responsibility makes our judgment more prudent, and constrains us to even more reasonable conduct in solving problems. Who can forget the first patients in his practice, and in particular the first days when he had the exclusive responsibility for the entire hospital service? Such things are never easy. They become somehow easier later on, thanks to the experience attained. But it would become much easier yet if various notions and functions involved in analysis of this problem were shown in a clear light. Only when these notions and functions are clearly determined in the mind of the physician will he be able to handle the problems that arise in the twinkling of an eye, meeting them with success if the patient recovers, or with sorrow, but at least with a clear conscience if there is nothing that can help.

Hence we must take some time to explain these notions and functions. We shall sometimes even reach out to apparently distant problems, but the reader will find in retrospect that even in such cases we have still aimed at our target.

In the preceding section we tried to describe the general line of reasoning leading to an answer to the question of what should be done to put the patient in the best possible conditions for compensation of pathological disturbances. Here we shall take a wider look at how in general this should be done, viz: how the indications obtained must be used in order to prepare the plan of treatment, and how this plan is to be accomplished. We have seen that the indications are reached either through estimation of the importance of pathological moments and review of their interrelations; or on the exclusive basis of the data of the details of the pathological process; or on the data derived from the object of examination itself, i.e., from the patient. But knowledge of the pathological process is only a groundwork for finding the direction of our reasoning and prospective action. To go further, the physician must bring to mind entirely different areas, that is, facts and notions related to the effects of therapeutic means on the human body.

The concept of therapeutic means necessarily involves the notion of effect or action on the human body. Anything viewed as a therapeutic means must result in a series of modifications of vital phenomena, with a great proportion of these resulting in substantial disturbances.

Now, as we proceed with treatment of a patient, we have to remember that in every case, in addition to the existing pathological disturbances, we are about to develop other phenomena -- or disturbances -- resulting from the application of therapeutic means, and in every case we have to evaluate the effects of the latter in order to arrive at compensating - as much as possible - for pathological disorders.[2]

When we see a child with diphtheric croup, who is in danger of suffocation because of exudation to the larynx, we shall not fear to provoke, beside the dreadful existing disturbances, the artificial disorder of coerced vomiting or the detrimental effects of an emetic -- e.g., cupric sulphate -- on the stomach walls. For we reckon that although the sum of disturbances in the patient's body will temporarily increase, not only will the danger be restrained, but our action will probably substantially soothe and speed up the disappearance of the pathological process.

However, because therapeutic means affect the organism in a physical and chemical way, our knowledge of their action is limited (with only a few exceptions) to the final changes that are caused by these means either in individual organs or in certain systems of the body.

Let us see then how we can pass from the indications to real therapeutic action.

An indication, being the reasoned basis of medical conduct, points to both the starting point and the line of action of a treatment. We may see the starting point because through review of the relative weight and interrelations of pathological moments we find the specific moment which, when removed or modified, will certainly be the cause of favorable change in the pathological process of a given patient. We also perceive a line of action by the analysis of the disease moments, in particular when these moments are broken down into particular conditions -- or particular moments -- leading us to a clear picture of the path by which the pathological moments can be reached and how they can be changed, or modified, or completely removed.

1• Usually moments that can be completely removed are the causal moments, and the major proportion of moments that can be treated through surgery. One can remove noxious vapors, or unwholesome food or drink, join separated parts, or separate parts that have been pathologically joined, amputate members of the body, cut out cancers, let out pus, etc. All these result directly from the accurate wording of the indication, and with adequate knowledge of medical means there should be no confusion in the selection or application of these.

2• It is possible to change or to modify pathological moments by augmenting (or stimulating), or reducting (or slowing down) the functions of the organs, or groups of organs, or systems of the human body.

The practical pharmaceutical textbooks classify drugs according to their effects on particular organs, or groups of organs, or systems -- e.g., expectorantia, emetica, eccoprotica, drastica, narcotica, excitantia, etc.

Thus, every time the function of one or another organ, or group of organs, or system is indicated to be augmented or reduced -- stimulated, or inhibited, or slowed down -- our mind naturally turns towards the area in which we can find the appropriate medicine. The more detailed is the indication and the deeper is our knowledge of pharmacodynamic phenomena, the better will selection of the drug and the method of its administration be.

We receive the data on which our plan of treatment can be built from the indication we arrive at, and from the finding of the appropriate drug; and, providing we have some experience, this step alone should not use up much of our time. Where we have one indication only, the therapeutic plan is very simple, and no problems are expected. Where the indications are more numerous, but impracticable for simultaneous application, it is obvious that those which are more important will have to take place before the others. The exceptions consist of the cases where the moment which is less important by itself constitutes a relative counter-indication for the fulfillment of the more important one, as has been mentioned when we have dealt with the methods of finding the indications. When planning the means of application of the next indication, we must try to harmonize our action with the application of the previous ones.

After the determination of an indication and the finding of adequate means for its application, it remains to suitably moderate its therapeutic effect, that is, to adjust the intensity of the action needed for a given case. To do this, we must remember that we are not treating diseases, but patients. We should be aware that a disease is the abstract concept that describes similar disturbances. The type and nature of these disturbances may constitute a simplified way of finding the therapeutic means by turning our mind toward analogous methods of action. But as it has been said before, if the selection of therapeutic means should depend on the most detailed analysis of the particular conditions of each pathological moment we intend to affect, then the intensity of the action can also be moderated by the individual status of the disturbed organism.

Any severe inflammation will require antiphlogistic means, but their application will vary from case to case, depending on whether it is applied to a well-nourished or a lean organism, a young or an elderly man, or a child, a patient with a lymphatic or a sanguine constitution.

We should take into consideration the fact that different individuals who suffer from a disease classified by us within the same notion, should be treated by the same method, but the therapeutic means will be different and they will be applied in different degrees, or different doses, if drugs are involved.

The science of dosages, and the general rules of application of various therapeutic methods and means, provide the student with an approximate starting point on this question. Frequent visits to hospitals and clinics, personal experience -- cautious estimate of the action of a given drug and of the effects obtained by its use -- will smooth out the path for a physician, subsequently.

In every indication, irrespective of the drugs that are suggested, we should always consider what conduct may create the best possible conditions for accomplishing this indication. This can be done rapidly in a blink of an eye. It is, however, critical for treatment to use all the possible circumstances for advantage of the patient. Rest or activity, the position of the body, ambient temperature, kinds and quantities of food and drink, all have been observed to influence the state of the disorder. And thus the purely scientific evaluation of these conditions is apt to be very helpful for the doctor, first because more means of action become available to him, and second because he can thus avoid thousands of detailed questions from the patient and his family. Everything that does not help is harmful. This is a general rule to which the overall behavior of the patient must be adjusted. And the more serious are the disturbances we have to counteract, the more consistently this rule should be followed. But how often do we witness the ignorance of the effect of diet by doctors! How often does a doctor employ drugs to fight tachycardia, but at the same time allow the patient to drink coffee or tea since he is accustomed to it! But there are no facts or experiments proving that the effect of these drinks, which strongly stimulate the action of the heart, will stop. It has become a common practice in recent times to provide substantial nourishment in addition to mercury treatment in syphilitic diseases, out of fear of anemia -- a bugbear we shall speak about later -- which, when overestimated, is very often the cause of irrational conduct in present-day therapies. How can the method based on intensive disintegration (in which an activation of reabsorption is required) be logically reconciled with the substantial nourishment of the patient? This cannot be explained or justified by any physiological or pathological rule. It is not surprising that the results are not in favor of these methods of treatment.

Again, every time we see the patient we should draw the most accurate picture of the existing pathological moments and closely observe all the changes as they occur, and at the same time rethink the reasons for the indication, and adjust our action accordingly. Such review of the progress or development of the illness, of the changes involved in it, and of the effects of the therapeutic means applied, will not only enrich our mind and add to our own experience, but often will become a source of useful indications related to the observed disease process.

Finally, we should take into consideration that the use of drugs is justified insofar as it is advised on the grounds of the strict need for active therapeutic performance. Simple therapy will always constitute the highest merit of a doctor.

Leading the reader along a purely clinical route, and having once described the task of a physician as consisting in placing the patient in the most advantageous conditions for the compensation of a pathological disorder, we have analysed what should be done to attain the goal, and how it should be done. But I still think that it is useful to remind the reader that not long ago otherwise very respectable persons tried to convince us that therapeutic intervention was of no avail and that the best way was to leave the disease to run its own course. By denying the rules of therapy, skepticism had to lead to nihilism. A review of the circumstances that resulted in this doctrine reveals that this error may be considered somewhat justifiable given the way it was born. The first advocates of therapeutic skepticism were men of extremely high merit in the field of diagnostics and anatomical pathology. It is easy to understand that, as they were devoting their lives to experiments that no individual could complete, and at the same time were fighting against a tendency toward superficiality which, along with polypharmacy, was the characteristic feature of contemporary medicine in central Europe, it was not possible for them to equally thoroughly investigate therapeutic tasks. The skepticism of these pioneers of modern pathology in Germany is, as I said before, justifiable to some extent. But its transformation into a school had very negative consequences. An entire generation of physicians found itself in a twofold false position: first, toward the patients who, presumably, should be grateful not for prospective help but for scientific description of their physical sufferings, and second, toward medical science, to which they became involuntarily disloyal by not trying to draw aside the curtain which had been laid over their eyes by this school.

There is almost no need to prove that nihilism has no basis. If one takes as a starting point only a single clear case of healing a given disease, e.g., healing some form of malaria, even if this is achieved through the most empirical conduct, then the task of tracing to what degree any other form of disorder is curable or modifiable should become the strictly scientific task of medicine.

Considering the progress of science and the rapidly growing stock of experimental means, each new generation has not only the right but also the obligation to criticize the scientific achievements of the earlier generations. Each school should instill to the highest possible degree critical tendencies in young doctors, but none has the right to produce scientific dogmas in any direction.

It is natural that pure nihilism was impracticable. On the one hand, diseases and sufferings have their laws and logic. On the other hand, there exist undeniable facts of the effects of therapeutic means on the human body. The nihilists were thus constrained to cure, but due to the lack of a reasonable method of testing the effects of therapy, or due to the extreme distortion of the notions on which their science was founded, they were unable to go beyond a few distinct pharmaceutical facts, which were used without giving them enough thought. This was the background of that simple-minded therapy: whoever had a fever -- no matter what kind -- he received quinine; whoever grew pale -- no matter for what reason -- was given iron; whoever suffered pains -- no matter of what origin -- had to have morphine.

NOTES

1 There is no need to prove that the same method of estimating pathological moments has to be applied every time as the basis for the prognosis.

2 One may remind the reader that many poisons make extremely effective medicine when used in proper doses and in appropriate circumstances. But even neutral bodies, or very common physical life-stimuli, such as food or drink, may cause substantial disorder in an organism when used in certain circumstances as a therapeutic means.

CHAPTER III

EDMUND BIERNACKI ON THE SCIENCE OF DISEASES AND THE ART OF HEALING

Edmund Biernacki (1866-1911) was born in Opoczno, in a family of impoverished Polish nobility. He studied medicine at the Warsaw University. During his medical studies Biernacki demonstrated a strong inclination for laboratory research. He worked first in a physiology laboratory directed by Prof. Tumas, studying there fermentation and putrification phenomena and the physiological effects of subcutaneus injection of saline. The results of this last study were published in 1888 in *Medical Review [Przegląd lekarski]*, and for this work, Biernacki - who was still a medical student - was awarded the Gold Medal of Warsaw University. After graduation in 1889, Biernacki worked for a year at the University Hospital, and obtained a scholarship to complete his training in Germany and France. He attended the lectures of Erbe in Heidelberg and of Charcot in Paris, and specialized in physiology and in hematology in the laboratories of Kühne in Heidelberg, of Reigel in Geissen and of Hayem in Paris. Returning to Warsaw, he was unable to obtain a university position. He accepted work as a clinician in the Wolski Hospital, and started a private practice at the same time. His main interest remained, however, hematological research. The Poles attribute to him priority in the discovery of blood sedimentation (1894) [13]. Biernacki was not satisfied with his clinical work and with his private practice in Warsaw. In addition, his first philosophico-medical studies were strongly criticized by Warsaw physicians, making his presence in that city difficult. He attempted to improve his professional status by moving to Lwow (then under Austrian occupation) in 190I. There he was more successful in obtaining a university position. He was appointed docent in pathology in 1902. At the same time, he started a private practice in the health resort of Karlsbad, where he worked during the summer season. In 1908, Biernacki was appointed to the chair of general pathology at Lwow University. He died three years later, when only in his 45th year.[1]

In the years 1899 to 1902, Biernacki published three books dealing with the philosophy of medicine: *The Essence and Limits of Medical Knowledge; Chalubinski and the Present Goals of Medicine, and The Principles of Medical Understanding.*[2] In the introduction to his third book, Biernacki affirmed that this work was the last of a series of three. The three books were, he explained, an attempt to construct a philosophy of medicine without metaphysics, or rather an "outline of synthetic medicine" ([3], pp. 1-4). The starting point of Biernacki's philosophical reflections was the problem of relationships between medical science and the art of healing. This problem was central in the writings of the second generation of Polish philosophers of medicine. While for Chalubinski the essence of medicine was therapy, and the progress of medical science properly subordinated to the needs of medical practice, such a simplified vision could not satisfy the second generation of thinkers of the Polish School, active during the years 1890-1910, the period of the triumph of "scientific medicine".

In his first book, *The Essence and Limits of Medical Knowledge,* Biernacki tried to solve the dilemma: "Medicine - an art or an science?" by a radical separation between the theoretical and the practical aspects of medicine. He affirmed that the theoretical science of pathology and the practical knowledge of healing were completely distinct disciplines. The first aim of the science of medicine was the exact recognition of a disease, i.e., the diagnosis. But, Biernacki claimed, diagnosis is not a necessary condition for successful treatment. All physicians' affirmation to the contrary notwithstanding, very little is known in fact about the causal relationships between the pathological phenomena and the effects of the drugs used to treat them. Moreover, the great majority of treatments proposed by physicians are purely symptomatic, and an understanding of the nature of the disease is not really necessary for a successful symptomatic therapy.

In the second part of *Chałubiński and the Present Goals of Medicine, and in The Principles of Medical Understanding,* Biernacki modified his radical point of view on the relations between diagnosis and healing. His later ideas were closer to the usual vision of interactions between theory and practice in medicine. In *The Principles of Medical Understanding,* Biernacki reaffirmed that medical science is quite distinct from the art of healing, but he expressed the hope that the gap between the two would narrow in the future. This hope was based on a faith in the ability of medical science to modify medical practice and make it more scientific. Although therapy was still viewed by him essentially as practical knowledge, Biernacki affirmed that, thanks to the progress of science, the medicine of his time was becoming more efficient ([3], pp. 42-46; 57-61). Biernacki summarized his point of view on this question in his last book, *What is Disease?* (1905). He explained that in the 19th-century, for the first time in history, the practice of medicine had been influenced by medical science. But, he added, until now it has been mainly a negative influence: scientific medicine has deeply modified therapy by showing the inefficiency or even the danger of many of the older treatments. At present, the practice of healing is often still "at a medieval level", and with the majority of diseases there is no direct relationship between diagnosis and treatment. However, Biernacki believed that this was bound to change. In a growing (albeit still very limited) number of diseases, diagnosis was becoming an important factor in therapeutic decisions, and he hoped that this trend would continue in the future([4], pp. 8-9).

The shift in Biernacki's outlook, which took place around 1900, was noticed by another thinker of the Polish School, Zygmunt Kramsztyk. In his review of *Chałubiński and the Goals of Medicine,* Kramsztyk explained that in the first part of Biernacki's book criticized Chalubinski from a sceptical and pessimistic point of view, while in the second part he presented a more optimistic faith in the future of medicine [10]. Biernacki himself did not explain the reasons for this shift in his views, nor did he explicitly acknowledge its existence. One can only guess that the first, more sceptical works may have reflected the disillusionment of a gifted, research-oriented physician who was obliged to turn to private medical practice and only then realized that all his up-to-date knowledge and scientific experience did not make him a better or more successful physician than his less educated colleagues.[3] The later works, which point to the possibility of concilliation of the future medical science and the practice of healing, were perhaps influenced by Biernacki's hope of returning to a university setting and there to continue his research work. This possiblility was already present in 1900 - Biernacki was then a candidate for the chair of pharmacology in Lwow - and was concretized two years later with his appointment as lecturer of pathology at the Lwow University [11]. The shift in Biernacki's views may also have been stimulated by his growing interest in serotherapy, viewed by him as the first therapy entirely based on the progress of medical science.[4]

The central thesis developed by Biernacki in his writings, and in particular in his first book, *The Essence and Limits of Medical Knowledge*,[5] is that medicine is in fact composed of two different disciplines: the science of diseases and the science (or rather the art) of healing. For him the history of medicine clearly revealed that these two disciplines evolved separately, and the analysis of the practice of medicine in his time pointed to the fact that they remained distinct ([1], pp. 104-105).[6] Moreover, Biernacki affirmed that the permanent association of the science of diseases and the art of healing under the common heading of "medicine", was more harmful than beneficial to both. The art of healing, he explained, was a much older discipline, an empirical knowledge which evolved through trial and error, and which contained an important element of "know-how". A good physician, like every good craftsman, had first of all to learn recipes by rote, to acquire technical knowledge and manual dexterity. Experience rather than logic is the best guide in therapy. When physicians tried to find "logical therapies", forgetting how incomplete their knowledge about diseases and their therapy was, they became the victims of various medical systems. For this reason, the recent growth of knowledge about diseases had in some cases negative impact on the art of healing. Impatient physicians attempted to base general therapeutic conclusions on small and incomplete fragments of knowledge, sometimes with disastrous effects for their patients ([1], pp. 107-108; 133-135).[7] For example, the attempts to cure cholera by injection of saline (Biernacki was, in 1888, involved in the unsuccessful clinical testing of this method) were based on a belief that the principal pathological manifestation of this disease was the "thickening of the blood". But, Biernacki explained, bacteriological investigations later proved that in fact cholera was induced by bacteria and by their toxins ([1], p. 134).[8]

Unlike the art of healing, the science of diseases was, for Biernacki, a true science. It was based on logical induction and used scientific methods of verification through experimentation. Not only anatomy and physiology, but also general pathology (i.e., the science of diseases) should therefore be classified among the biological sciences, and taught at the university alongside botany and zoology. One could even imagine the inclusion of studies of general pathology in the high school curriculum ([4], pp. 4; 116-117). It was, however, a relatively young science. It became codified and formalized only in the 19th-century, as a result of the rapid development of anatomy, physiology, and bacteriology. The relative youth of the science of pathology could explain its limitations and its shortcomings. But, Biernacki added, there was another reason for the relative underdevelopment of medical science, namely, its close association with medical practice. This association often hampered the development of general pathology. Ideally, the physician should be able to approach a pathological case in the same manner in which a botanist studies a plant. In practice, however, he faces patients who, for obvious reasons, have little interest in the development of the science of diseases and are interested solely in the alleviation of their sufferings. This "nightmare of utilitarianism" in Biernacki's words - i.e., the pressure of the sick on physicians - often led to the abandonment of promising paths of investigation in favor of approaches which were able to give immediate satisfaction to the patient's demands. Many negative aspects and inconsistencies found in medical science, and absent from other natural sciences were, according to Biernacki, the result of the pressure of medical practice on the science of pathology ([1], pp. 132-133). The physican who tried simultaneously to understand the nature of a disease and to evaluate the effect of a treatment, endeavored to solve a single equation with two unknowns, clearly an impossible task. Only when, thanks to the boldness of the Viennese medical school, physicians started to observe disease in a "natural", unmodified state, was it possible to state that in fact diseases usually are healed without the physician's intervention, and that the majority of the traditional treatments are at best useless, and often harmful ([1], p. 137).

The separation of medical science and of observation of pathological states from the practical art of healing should be, Biernacki claimed, beneficial to both. It should diminish the present confusion in medicine by allowing for an unbiased study of disease and a clear-cut perception of the possibilities of therapeutical intervention. For example, both physicians and patients gave high priority to adequate diagnosis. But, Biernacki explained, while accurate diagnosis is very important for the progress of medical science, it is much less important for therapy ([1], p. 11; 26). The layman, however, believes that if only the physician can find the right name for his patient's complaint, he will be able to treat it effectively. This idea, Biernacki explained, is based on two false assumptions. The first false assumption is the belief that there are such things as concrete, fixed and well-defined natural entities called "disease units". But the study of the history of medicine points to the fact that "the notion of a classificatory unit is a fluid one, and is dependent on the state of evolution of science.... Various older disease units - usually abstract creations - are no longer separated 'diseases' for us. But in the meantime, we have been looking for and sometimes we continue to look for specific medications for these ideal entities, which do not exist in reality" ([3], p. 64).[9]

The second false assumption is the idea that there was an obligatory relation between diagnosis and therapy, and that there exists an adequate medication for every disease. Patients believe that determining the correct diagnosis, i.e., the right name of their disease, is the most important step toward administering the right treatment. They forget, however, that the progress of the science of pathology is not matched by parallel progress of the art of healing: "the great progress of medicine in the last century notwithstanding, the real efficiency of medicine is much lower than was previously assumed, and than is believed by lay persons today.... What we call "treatment of disease" is, above all, the alleviation of suffering, and it usually has little effect on the disease process itself" ([1], pp. 96-97).[10]

The physician, Biernacki explained, usually has only a very limited ability to act directly on the pathological process. There are a few exceptions to this rule, e.g., quinine for malaria, or derivatives of salicylic acid in rheumatism. However, in the great majority of pathological states, only symptomatic treatments exist, and in all these cases diagnosis is not an indispensable precondition for effective medical intervention.

The confusion between medical science and medical practice, Biernacki explained, is also at the origin of the confusion concerning the problems of causality in medicine. Physicians, faithful to what they believe to be a scientific approach, attempt to reproduce in a clinical setting the behavior of a scientist in a research laboratory. They try first to find the causes of a given pathological state, then to act on these causes. But in doing so they forget the complexity of the notion of causality in medicine. Even in relatively simple cases such as infectious disease, it is not easy to prove that a given microorganism is the sole *cause* of a disease. The presence of a microbe is not necessarily followed by the presence of a disease, and, on the other hand, human beings, unlike artificially infected laboratory animals, often suffer from complex infections.

In pathology, Biernacki affirms "it is very difficult to prove causal relationships. Usually one can, at best, obtain proof that in a concrete case a given factor participated in the development of a specific pathological condition" ([3], pp. 172-175; 244).[11]

Moreover, the notion of causality in medicine is complicated by the fact that, unlike other scientists, the practicing physician always deals with unique individual cases. Biology and general pathology (which for Biernacki are biological disciplines) are complicated enough, and cannot be reduced to purely physico-chemical considerations. But at least they can be studied with experimental methods, and one can formulate general laws concerning these disciplines ([3], pp. 27-29).[12] This is not the case, however, for the practice of medicine. In this discipline the establishment of general laws is not possible, because "each patient has his own illness". Following Chałubiński, who considered that the individual evaluation of each case was the only important indication for the physician, Biernacki explained that the evolution of a given disease was always influenced by the specific biological constitution of the patient, by "this very obscure thing that we do not know how to investigate, nor how to measure, i.e., individuality" ([3], p. 306). The reaction of each organism to these chemical modifications depends on the patient's hereditary constitution, on past diseases, and on the immediate environment. Therefore, if one drew a graphic representation of the biological evolution of a given disease in various patients, each one would have a distinct, individual curve for each case. Biernacki, making explicit the logic of Chałubiński's reasoning, affirmed that in order to find the adequate therapy for each pathological case the physician usually did not need to attempt to make a diagnosis, i.e., to find the adequate classification of the pathological disorder observed by him. He should only look for the specific therapeutic indications for each individual pathological symptom, applying to this task the principles of general physiology, rather than those of the specific science of diseases ([3], p. 277; [4], pp. 79-88).

Biernacki maintained a chemical vision of biological individuality and a functional vision of pathology. For him, what was called "disease" was first and above all an "intoxication", i.e., a harmful modification of the metabolism of the organism. This intoxication was fought by the organism by chemical means. Reversible or irreversible anatomical lesions induced by the disease (and which constituted the essence of pathological states for followers of an anatomo-pathological approach to disease) were for Biernacki solely the secondary effects of prolonged modifications of the normal metabolism of the body by the disease ([4], pp. 104-105). One should note, however, that Biernacki's functional, or rather, in his case, chemical vision of disease and of resistance of the organism to it, was not in all probablity related to the older concept of disease as "irritation" or "over-excitation". These concepts did not appear at all in his writings. Biernacki's chemical or "metabolic" vison of disease, which he opposed to the earlier anatomic-pathological vision, in all likelihood evolved under the influence of the new discoveries in bacteriology and immunology and, more specifically, the discovery of bacterial toxins and antitoxins (i.e., the antibodies able to neutralize these toxins). In Biernacki's words: "the new science realizes that one should oppose a chemical concept to another chemical concept. Therefore we hear more and more about 'alexins', 'antitoxins', etc., as the chemical substrate of individual susceptibility and of immunity" ([13], p. 114).[13]

The variability of individuals as well as of pathological manifestations and their consequences - e.g., the impossibility of formulating general laws in medicine - were not, however, the only obstacles to the transformation of therapy into an exact science. For Biernacki, another important factor made the transformation of medicine into an exact science impossible: the crucial importance of psychological factors in every human illness. The role of the physician was also - for Biernacki often above all - to act in response to his patient's mind: "The patient receives usually only one kind of real help from his physician - the psychological help." Biernacki affirmed that the physician's personality was a decisive factor in healing ([1], p. 114; 117). Many drugs were effective mainly due to the force of suggestion that was attached to them. "Suggestion can bring better mental equilibrium and better feeling, and can influence in a positive way even the most material perturbations of vegetative functions of the organism". For this reason Biernacki also thought that a scientific verification of the effectiveness of drugs was not possible. Ideally, he explained, one should give the drug to two series of patients: to one the tested drug should be administered "with suggestion", and to the other "without suggestion". But "how can drugs be administred without suggestion?" ([3], p. 297).[14]

Biernacki made a clear-cut distinction between the role of the medical scientist (for him, similar to the role of a natural scientist) and the role of a physician. According to him, the physician should never forget that the profession of medicine exists in order to heal the sick. He should remember that the worst harm done by disease is human suffering, and that "what exists in this world are sick persons, not diseases" ([4], pp. 109-111). The physician is dealing not with abstract entities, but with suffering fellow men. Thus, his actions cannot be guided solely by the search for maximal effectiveness. Biernacki's conclusion was that therapeutic activity has an important *normative* dimension. The physician should not only ask if he *can* act in a given way but also if he *should* act in this way ([2], p. 31).[15]

The physician, Biernacki explained, should also always keep in mind the limits of his therapeutic activity. The most obvious limits were the socio-economic conditions of the patient. Physicians know that many diseases can be avoided or cured by adequate hygiene, the right kind of food, and sufficient amount of rest, but "unfortunately the practical application of these simple therapeutic means is impossible, because of economic obstacles.... I once heard that the best treatment for tuberculosis is money" ([1], p. 92).[16] But even if the socio-economic problems of health care were miraculously solved, medicine would still have to face its natural limits: the impossibility of eliminating disease and death. The physician will always be obliged to face cases in which he is unable to restore health or to prevent death. For this reason Biernacki did not believe that the progress of medicine would lead to the elimination or even to an important reduction in the activity of "medical sects" (such as magnetism, homeopathy, magic practices). Some patients facing a medical verdict of incurable illness will always try to find help, or at least its illusion, elsewhere ([1], pp. 161-162).

NOTES

1 For the biography of Biernacki, see [1, 17, 20, 21].

2 E. Biernacki's book, *The Essence and Limits of Medical Knowledge*, was translated into German (Biernacki, E.: 1901, *Die moderne Heilwissenschaft*, transl. by S. Ebel,

B.G. Teubner, Leipzig) and into Russian: it appeared as an annex of *Wiestnika Znaja,* St. Petersburg, 1903.

3 Szumowski affirmed that Biernacki was too research-oriented and failed to develop a substantial private practice in Warsaw [17].

4 Biernacki's interest in immunology and in serotherapy probably stemmed from his previous studies of blood disorders. He taught hematology at Lwow University, translated into Polish the book *On Human Nature,* written by Elie Metchnikoff, one of the founders of the science of immunology, and published in 1905 an essay on serotherapy ([4], pp. 147-158). On the impact of serotherapy on the modification of physicians' outlooks on scientific investigations, see [14].

5 Biernacki's book, *The Essence and Limits of Medical Knowledge,* was first read by the author at three meetings of the Warsaw Medical Society [Warszawskie Towarzystwo Lekarskie] (January-February 1899), where Biernacki's unorthodox ideas provoked angry reactions from the audience.

6 For a recent formulation of the same idea, see [16].

7 For a more recent formulation of the same idea, see ([7], pp. 134-135).

8 Today, however, Biernacki's example can be used to illustrate the opposite view: the importance of empirical deductions in therapy. Present medicine has fully rehabilitated the principle, if not the technical details of the method - first proposed by Cantani - that Biernacki tried to investigate in his youth. Loss of water (or "thickening of blood") is viewed today as the principal cause of mortality in cholera, and rehydration - often by intravenous injection of saline - together with antibiotics, is the principal treatment for this disease [12].

9 Biernacki is in all probability indebted to Chałubiński for developing this idea. But Chałubiński, unlike Biernacki, did not found his notion of the abstract nature of disease units on an historical approach.

10 In 1905, Biernacki modified his point of view on this subject. He explained that although the number of diseases in which diagnosis was indispensable for adequate treatment was still very limited, their number would in all probability increase in the future ([4], p. 8). On the low therapeutic effectiveness of medicine as late as the 1920's and 1930's, see, e.g. ([18], pp.19-35).

11 Biernacki did not distinguish between "causes" and "necessary causes". The concept of "necessary cause" (i.e., a cause that, if eliminated, would prevent a given event: if not A, not B) was crucial to the evolution of bacteriology and, more generally, to the development of the search for specific disease-inducing factors. See, e.g. [5, 19].

12 Biernacki, unlike Chalubinski, was not a mechanist, but believed in the existence of a particular form of energy, specific for living beings.

13 For the role of immunology in the evolution of a unified chemical vision of disease, see, e.g. ([8], pp. I-VII), and for discussion of this evolution [9].

14 Randomized 'double blind' therapeutic essays, in which the patients participating in the test, and the physician who administers the tested treatment do not know if a given patient is receiving the medication or a placebo, were introduced only in the 1940's, and were fully developed in the 1950's and 1960's [6].

15 Biernacki attributes the priority for this affirmation to Chałubiński, but his own formulation is much more explicit.

16 G.B. Shaw also explained in 1911 that "what most of the patients really need is not medicine but money" ([15], p. 72).

BIBLIOGRAPHY

[1] Biernacki, E.: 1899, *The Essence and the Limits of Medical Knowledge* [*Istota i granice wiedzy lekarskiej*], Biblioteka Dzieł Wyborowych, Warsaw.

[2] Biernacki, E.: 1900, *Chalubinski and the Present Goals of Medicine, [Chalubinski i obecne zadania lekarskie]*, Wydawnictwo Przegladu Filozoficznego, Lodz.

[3] Biernacki, E.: 1902, *The Principles of Medical Understanding [Zasady Poznania Lekarskiego]*, Wende i Ska, Warsaw.

[4] Biernacki, E.: 1905, *What Is a Disease ?* [*Co to jest choroba ?*], H. Altenberg, Lwow.

[5] Codell-Carter, C.: 1985, 'Koch's Postulates in Relation to the Work of Jacob Henle and Edwin Klebs, *Medical History* **29**, 353-374.

[6] Doll, R.: 1982, 'Clinical Trials, Retrospect and Prospect', *Statistics in Medicine* **1**, 337-344.

[7] Dubos, R. and Dubos, J.: 1987 (1952), *The White Plague*, Rutgers University Press, New Brunswick and London.

[8] Duclaux, E.: 1898, 'Introduction générale', in *Traité de microbiologie*, Masson, Paris.

[9] Kohler, R.: 1973, 'The Enzyme Theory and the Origins of Biochemistry', *Isis* **64**, 181-196.

[11] Ochorowicz, J.: 1899, 'Introduction', in E. Biernacki, *The Essence and the Limits of Medical Knowledge [Istota i granice wiedzy lekarskiej]*, Biblioteka Dzieł Wyborowych, Warsaw, pp. 1-20

[12] Pollitzer, R., Swaroop, J., and Burrows, W.: 1960, *Cholera*, WHO Monograph 43, Geneva.

[13] Rytal, A.: 1957, 'The Sixtieth Anniversary of the Development of the Sedimentation Test by Edmund Biernacki', *Bulletin of Polish Medical History and Medical Science* **1**, 3-7.

[14] Salomon-Bayet, C. (ed.): 1986, *Pasteur et la révolution pastorienne*, Payot, Paris.

[15] Shaw, G.B.: 1982 (1911), 'Preface on Doctors', in *The Doctor's Dilemma*, Penguin Books, Harmonsworth.

[16] de Solla Price, D. : 1965, 'Is Technology Historically Independent of Science? A Study in Statistical Historiography', *Technology and Culture* **6**, 553-568.

[17] Szumowski, W.: 1936, 'Edmund Faustyn Biernacki', *Polish Biographical Dictionary*, vol. 2, pp. 78-79.

[18] Thomas, L.: 1984, *The Youngest Science: Notes of a Medicine Watcher*, Oxford University Press, Oxford.

[19] Whitbeck, C.: 1977, 'Causation in Medicine: the Disease Entity Model', *Philosophy of Science* **44**, 619-637.

[20] Wrzosek, A.: 1912, *Edmund Biernacki, 1866-1911*, Drukarnia Uniwersytetu Jagiellonskiego, Krakow.

[21] Zwoźniak, W.: 1964, 'The History of the Medical Departement of Lwow University' ['Historia wydziału lekarskiego uniwersytetu lwowskego'], *Archiwum Historii Medycyny* **27**, 193-226.

CHAPTER III.A

TEXT OF EDMUND BIERNACKI
THE ESSENCE AND THE LIMITS OF MEDICAL KNOWLEDGE
(1898)*

If we analyze how the people generally view the significance of medicine and its representatives, it should become apparent that many questions in this realm are explained in a specific and simple way, which, it must be added, is not altogether unreasonable. Thus, an average patient finds out the following: there are diseases in the world and there exist respective treatments which cure the former. In order to make a right "guess" what medication is needed to treat the disease correctly, one should first of all "know" the disease. The doctor who is able to "know" diseases will treat patients successfully. They say: oh, this or that case has been attended by so many doctors without success, but eventually another doctor was called, who seemed to know this disease, because his treatment has immediately helped. The belief in relation between successful treatment and correct "knowledge" of the disease occurs even in the minds of the least educated. It is not uncommon that doctors have to listen to the opinions of patients in hospitals, of navvies or of poor housewives: "I have seen so many doctors but nobody could help me. Perhaps you will at last know what my disease is and cure me."

At the same time, in the minds of lay persons, there exists a very high opinion which I should add is perfectly logical, about the efficiacy of drugs and treatment methods. Such opinions as: "He did harm to the patient" or even: "He killed the patient" by the administration of that but not of other medicine are frequent. Even in the most highbrow circles, even in the coldest minds, the death of a patient is often attributed to wrong treatment or to the lack of skill of the doctor. "If I had listened to the X or Z advice, I would have been dead long ago." "If he had been given that medicine -- even if it was castor oil -- but not

* pp. 27-30; 40-45: 77-79; 105-119.

this one, [the patient] would be living another hundred years." Saying that such and such doctor "prolonged" illness, or that he was intentionally -- for personal profit -- refraining from curing the patient rapidly, is perhaps the most common allegation, while even the most popular and famous doctors are not absolutely free from such epithets as "killed" or "ruined".

Such talks reveal highly reasonable claims of people, who expect cure to be the task and the goal of medicine. But... let us state the problem clearly right from the beginning. Such claims are a far cry from the actual capacity of medical knowledge, which is unable to satisfy the substantially legitimate requirements of patients. Medicine, with only a few exceptions, has no adequate means to cure particular diseases, the healing methods are in general not related to the "knowledge" of the disease, the treatments by doctors are not sufficiently effective to save or to kill a patient, and doctors have no power to prolong illness.

In the reasoning of the non-medical public there is first of all constant ignorance about one basic notion. The public is familiar with only one term "knowing the disease" and this term is used to describe two different notions, *viz* : "knowing the disease" and "recognizing the disease". The term "recognition" is almost totally unknown to general public and its use is substantially limited to doctors. But "recognition" of the disease is something quite different from "knowing" it. We are able to recognize many diseases, and diagnostics is a highly developed craft, much more developed than the ability of "knowing", i.e., to understand the diseases. A great majority of diseases totally lack detailed explanation, many of them are only partially explained (or rather start to be explained), and just a few of them are sufficiently explained.

What does "knowing" a disease mean? It certainly means knowing the reasons and the nature of the disease, that is, knowing the agent without which the state of illness could not exist, or the conditions under which such an agent becomes active. It also means the finding of specific changes by which the state of illness could be differentiated from the state of health and from other disorders. To know a disease is to explain in detail which one among the observed changes is the primary lesion or, in other words, which change is of primary importance, and which changes are the effects of the former and have a secondary significance only.

To recognize a disease is, however, something different. To recognize is to show usually by reference to the pathological signs or symptoms that the case we deal with is only a given nosologic entity and is not one of the other nosologic units. A pathological disorder provokes various ailments or sensations -- subjective symptoms -- or readily noticeable external signs -- objective signs. Such symptoms are for example: pain, weakness, cough, accelerated pulse, temperature, hyperthesia, hypokinesia, etc. Now, it happens very often, that similar symptoms occur in the diseases whose cause and nature might be different. A headache may result from painful head muscles or from various pains in meninges or brain, or from general nevralgia, or from kidney trouble, etc. A cough may indicate various mild or severe disturbances in the respiratory tract but also in some other organs which do not belong to the respiratory system. However, the number of signs and symptoms and their mutual relations and sequence are different in various diseases; hence, by timing the signs and symptoms and considering their relations it becomes possible to state that one patient is suffering from a certain disease while a second patient is suffering from another disease. It should be, perhaps, added that the recognition of and the understanding of disease may meet at a certain point, usually when the cause of the disease remains within the patient's body and is used to recognize the disease.

Here we have arrived at a point which is quite difficult to explain, and to understand correctly throughout the science of medicine. Therapy does exist, notwithstanding all the deficiences in the knowledge about diseases. It existed even when these deficiencies were still deeper; when the knowledge about diseases could not be compared with the science of the present day; when even limited knowledge about the human body did not exist, that is, when anatomy was yet in an embryonal state. In other words, there has been [clinical] practice and therapy when almost nothing else was present. Such an extraordinary phenomenon cannot be found in any natural, pure, or applied science. In these cases, the practice, or at least fundamental applications, usually follow discoveries and the adequate understanding of natural phenomena. Aniline inks and the dyestuffs industry have followed the studies on properties of certain carbon compounds; steam machines have followed the studies of steam properties; telegraph and telephone systems have followed the studies on electric current, etc.

Why is medicine developing in a different manner? Why therapy has existed in spite of all deficiencies in the knowledge about diseases? Of what consequence was therapy in the past and today? The best way to address these questions will be to review the origins of medical knowledge and its history. We are going to deviate from the line of our reasoning for a while, because it is necessary for an adequate understanding of the keynote of our work.

Medicine is a strange branch of science indeed, which, we could say, has developed backwards, but... it could not be the other way round. Medicine was born because there always have been diseases and because sick men have always asked for their fellow-man. For this reason therapeutic methods had to be found first. Primary medical knowledge has been limited to therapy. Therapy has been and is the basis of medical knowledge, even among nations at the most primitive levels of civilization.

It can be readily explained what such basic medicine or primitive therapy might have been. Why does a patient call a doctor nowadays? Because he has got headache, bellyache, cough, pain in his side, diarrhea, or the like. The troubles for which doctors have been called by the natural men were the same. And for both the present and the prior patients bellyache or cough has always been the total disease. Primitive men that were asked by their contemporary brothers for help, considered the pains and coughs as "maladies" and measures for these maladies were sought. How these measures were found, what methods were used to find them, is very often lost in the obscurity of history. Nevertheless, various aids had eventually been developed for suffering humanity. There are means to soothe pains, to reduce cough, to arrest diarrhea, etc., -- the means which have removed or alleviated the maladies of primitive men.

But what are all these aches, coughs, and diarrheas to us? We do not consider them as diseases, but only as symptoms of diseases. A headache or bellyache, as we said before, could make symptoms of a dozen kinds of disorders that vary in cause and essence, which may be serious or mild. If primitive man has found and possessed suitable means to soothe his pains or coughs, these were not means to cure diseases but rather to alleviate symptoms. Now, this primitive treatment is the so-called symptomatic treatment, and is not the cure of disease but the alleviation of the specific symptoms which are declared by the patient and for which the doctor has been called to help. There was no need to begin with the detection of signs and symptoms, or to examine the patient, but means had to be sought to remove the subjective signs of disease. The civilized patient of the present day, too, is seeking first of all medical help for the pains from which he is suffering, and he asks that these pains or symptoms be removed as fast as possible. Historically, symptomatic treatment was the first phase of therapy and also the first phase of medical knowledge, and could never vanish because of the requirements of patient. Curing the symptom is the basis of therapy in the present days as well. Symptomatic healing can, however, be practiced without any further knowledge, that is, without knowledge about the nature of the disease.

There is no doubt that even in the most ancient history people had come to the conclusion that symptomatic treatment is not satisfactory. Say a patient sought remedy for cough: in one case application of various medicaments resulted in reduction of cough, but the same treatment did not help another patient or only helped him for some time, after which the cough relapsed and the patient eventually died. This was obvious proof that healing the symptom could not be considered to be the same as curing disease. Attention has been gradually turned to accessory symptom until some limited idea about the state of illness has been created and, at the same time, the question about the nature of the illness itself has raised. As soon as the notions of causes and nature of disease have been basically recognized, efforts were made to remove these causes and to affect the disease itself.

This was how, besides symptomatic treatment, attempts have been made to practice radical or rational treatment, that is, treatment proper, aimed at removal of the cause of disease or at modification of its basic harmful effects. But the rational treatment in the distant past was quite specific. Primitive man, who acknowledged the presence of good or evil gods in the surrounding nature, could not think that these gods would not be present in such a natural phenomenon as illness. Consequently, many diseases, in particular those of epidemic nature the causes of which have been a mystery, were explained by the action of evil spirits, or by the presence of evil spirits in the man, or else as a punishment imposed by good spirits for some offence. One of the "divine" diseases was epilespy. It was called " *morbus sacer, divinus*" -- a divine, sacred malady. Hence, if the nature of illness consisted in settlement of evil spirits, or was a punishment sent by good spirits, then the only remedy for such illness would be an invocation or a prayer to good spirits so that they might defeate and drive off the evil ones, or suspend the punishment. Consequently, prayers for the patient were considered the most important treatment and were used more abundantly, the worse, the longer, and the more mysterious was the illness. Sometimes, even the simplest and mildest disturbances would not go without prayers. Nations had their "gods of good health" for which temples and altars were built up, and where the patient could recollect and wait for recovery. No wonder that, as a consequence of such concepts of illness, treatments were performed by those who were the go-betweens the gods and ordinary men, that is, the priests. They healed by prayers and sacrifices offered to the gods in search for material help. They also had utilized the gods themselves by making use of either the words spoken by Pythia sitting on the tripod, or of the patient's dreams, or the like.

Distinct signs of such rational therapy can be seen at present days in practice of quack-doctors, active among uneducated society classes, in which spells and sorcery make indispensable curing means. Such therapy had survived untill the Middle Ages or even until the recent times. Various hysterics were recognized as persons possessed by the evil sprit or the devil. Hence, the only treatment possible could be applied in the church where, by means of exorcisms, the devil had been driven out, that is, the cause of illness has been removed. Such treatment has been rational indeed.

Besides the symptomatic therapeutical aids, besides this specific kind of rational therapy, some real therapeutic methods have come to be used over the ages. These involved certain external, local disturbances, but limited within the range that we tend today to classify as surgical cases. It was easy enough to detect the correct method of curing in cases of broken or dislocated bones -- to apply spints, or to locate a joint. When a kid has loaded peas in his nose or ear, it was certain that these should be removed to prevent pains in the future. When someone had swallowed poison, there was no doubt that the stomach should be purged from this poison -- or the cause -- as soon as possible.

This is the framework in which medicine had developed, and it would be difficult to believe that this process could have ever taken place in a different manner.

Now, coming back to the question that was put forward before: "If medicine knows nothing about the essence of diseases, and can hardly provide means to recognize diseases, how can we treat diseases?" Here is the answer: "Curing diseases is possible, even if the knowledge about diseases is incomplete, because diseases are simply self-cured." The autotherapy compensates all deficiencies in diagnosis and science, and is the first and the most important therapy. This statement can sound paradoxical or fallacious for many readers, or even as trivial, since sayings such as: "It is not medicine but nature that cures diseases" are commonly heard. But its meaning will become clearer if we try to answer the following questions.

If it is certain that diseases are cured by autotherapy, but still people die of these diseases, even of diseases for which the autotherapeutic effect has been best explained -- *viz:* infectious diseases, then, perhaps autotherapy is not a very reliable therapeutic means? What should be done if it fails? What should be done to make it reliable?

Only science can give us the answer to these questions. Not only we do not know the essence of diseases (at most we know it inadequately), but in addition we do not understand the action of therapeutic means and we do not know which therapeutic means could directly and doubtlessly affect the cause and the essence of disease either.

There are, however, a few exceptions to this rule. These include the action of mercury and iodine preparations on syphilitic infections, and quinine on malarial infection. We have no doubt that both these diseases can sometimes be cured or at least improved without any treatment due to autotherapy, since in so many cases both mercury and quinine are so rapidly and strikingly effective that we have to recognize these medicaments as the "disease-healing therapeutic means". We accept this as a fact because we are convinced by hundreds of cases, although we do not know what is the healing mechanism of mercury, or of iodine, or of quinine. We also do not know what is the nature of the affliction, e.g., malarial fever, and what could be the target of action of these medicaments. The salicylic preparations which have been introduced to treat severe rheumatism cases a few years ago also have specific effects. Here, the rheumatic modifications of the joints improve so quickly that we can add the salicylic preparations to the list of "disease-healing medication", although here too we do not understand the principles on which their effects are based.

It should be noted that the latter medicaments are not viewed as highly reliable therapeutic means as are mercury, iodine, and quinine.

There is also no doubt that the iron preparations are effective in curing the chlorosis of women; although chlorosis can be self-curing and we do not hear about chlorosis-caused deaths, we know that with iron preparations in many cases cure has been attained faster than through the self-curing process. Recently we have made substantial progress in our knowledge about chlorosis, but the mechanism controlling the therapeutic activity of iron still may be only guessed at by us. It should be remembered that iron has been recently brought into disrepute, because it sometimes was and still is applied in cases similar to chlorosis, or in secondary anemia resulting from affliction of various organs, or in so-called "pseudoanemia"-- or apparent anemia -- while these cases seem to be totally unsusceptible, or susceptible in very low degree only, to the action of iron. Finally, the positive effect of arsenic treatment has been observed in certain disorders and other pathological states, such as in goiter, or in lesions of vessels that are affected by iodine. We have thus only a limited number ot therapeutic means in the strict sense of the word, that is, the therapeutic means that has an effect on the essence of disease. We may add to this list a number of drugs used to deliver bowels from ascaria or tapeworms which disturb the alimentary duct, or certain nervous symptoms, or sometimes anemia conditions, and also various drugs are very effective in skin diseases. But we have no specific drugs to cure typhoid fever, or cholera, or plague; we have no such medication that, if administered, would stop the disease or remove it within a limited period. We have no drugs which would reverse lesions in kidneys, liver or heart, or TB lesions in lungs. To do this some people do not have therapeutic means but... illusions. The widespread curiosity about various new medication which are supposed to actually cure these diseases is a sufficient proof of the absence of such therapeutic means.

All that we have said so far leads us to one basic conclusion indispensable for the understanding of our further arguments: that the history of medical knowledge clearly shows that medical knowledge is not a uniform structure but, on the contrary, consists of two sections which were generated irrespective of one another. One section is the knowledge about the means to cure diseases. This consists of the essence of medical science and this section only might be called a science; the latter is not entirely entitled to be called "science".

Present-day knowledge of diseases, which includes also the practical means of diagnosis, has been constructed stepwise, is based on general principles within the natural/biological science, and is closely related to other branches of science. However, contemporary therapeutics, the major part of which had existed before any fundamental medical sciences, such as anatomy or physiology, originated mainly by chance or through experience. It owes little to logical reasoning, and usually cannot be depicted as a scientific system. The contemporary therapeutics can claim to be a science only because scientific criticism has been applied to this domain as well (resulting in a serious limitation of its scope, as compared with the past ages) and because, in order to relate its own empirical achievements to the findings of natural sciences, we are trying to find rational reasons to these achievements through scientific research.[1] Therefore, no genetic relationship between these two sections of medicine exists (with perhaps a few negligible exceptions) and it can be concluded that knowledge about diseases and therapeutics are independent of one another, and in fact, they exist in the doctor's mind as two different kinds of knowledge. These two kinds of knowledge are acquired in different ways and under different conditions.

First of all, it should be noted that as soon as knowledge about disease and medical knowledge became effective, and was applied at the patient's bed, there appears an indispensable element of art. This element of art has much greater weight in therapeutics than in knowledge about diseases. It has, however, gained some importance in the latter, following the development of modern examination methods: percussion and auscultation. The rapid development of modern therapeutics also increased the art element in therapy. It was not without reason that during the past ages, when medicine consisted almost exclusively of therapy, that it was called an "art", and Plato compared Hippocrates with Polyclitus and Phidias.

What is the meaning of the "element of art"? This means that a skill in performing various, necessary operations is required in order to recognize a disease and to cure the patient. Skill in performing percussion and auscultation is needed so that the information obtained through these procedures would not be erroneous. Trying to help a patient one should learn to use manual/mechanical aids. The number of such techniques has increased in recent years. In addition, various manual manipulations, e.g., massage, which had not existed before, are performed now, while a great number of lengthy prescriptions have vanished altogether. This, as I just have mentioned, enhances the element of art in therapy.

In order to learn these various methods of examination and to provide help for the patients, the most important thing is of course not training of the mind, or quality of logic reasoning, but the training of the senses and, first of all, the training of the hands. A surgery is also called an art. As a matter of fact, this is a little unfair to what we nowadays call an art, which requires special talents or the spark of divinity. The medical art has rather more in common with crafts. To carry out percussion and auscultation, or even -- contrary to what the general public thinks -- to perform a surgery, no particular gifts or special talents are needed. Surely enough, one man may be more gifted than another, but eventually, within the range of therapeutic skills, not much more talent is required than to plane smooth a plank of wood or to knit well a stocking. After all, many doctors can equally well apply the stomach drain, equally well deliver a baby.

Beside the element of art, there is not much left in therapy for intellectual practice. Being chiefly a collection of empiric (knowledge) data, this science requires mainly a good memory. One should remember descriptions of many therapeutic means, the doses that are usually applied, what are the effects, and when one drug is preferred to another. These drugs often do not require much intellectual effort, since they are symptomatic treatments, while the basic symptoms can be noted even by a layman; patients indicate them and ask for help by them. Premises for logical reasoning can of course be formulated during the process of deciding which treatments should be used to help patients, but, curiously enough, this happens most frequently when the doctor begins to follow the apparently rational but actually false and sometimes even pernicious healing "systems" of our ancestors.

The knowledge about disease is, however, something entirely different. Here we already have an edifice of science; an edifice which still requires many efforts in many domains to reach the roof, but it is very ample and high. Here not only memory is needed, but many other talents too, such as highly developed critical judgement, and good ability to draw conclusions. Here, the ability to cover more or less accurately the whole problem, to guess what is the right plan among the mass of fermenting material, among contradictory views and subjective or mystic opinions, runs in parallel with the [physician's] personal talents. And, when facing the problem of the application of our knowledge about disease to diagnose, the doctor finds himself before puzzles which he has to solve at every step. The science of diagnosis of the present day is without doubt too complex and must become simpler in future: this will happen as we learn the true nature of diseases and, as we are able to recognize diseases directly from the natural data on its nature. But now, because we have to identify disease from a kaleidoscope of signs and symptoms -- with not so many signs at hand -- the pressure on a doctor in many cases impels his mind not toward a routine, automatic functioning, but toward an activity which is in certain relation to an art. The science of diagnosis is exhausting, while, on the other hand, it is important for the development of the physician's individuality. It seems to me that there is no better way to train the human mind in logical speculation and reasoning than in medical studies and, in particular, the science of diagnosis, because it very frequently requires an ability to solve puzzles. If doctors enjoy the reputation of being intelligent people, this should be undoubtedly attributed to the fact that they continuously exercise their brain, while in many other trades the performance is more automatic. If all the problems of diagnosis were (as is often thought) limited to the question of memory, doctors would probably not enjoy such a reputation.

Therefore, as far as knowledge about diseases, skill in the identification of disease and estimation of individual signs and symptoms of disease are concerned, doctors must differ from one another as much as they differ in their personal talents. But, when we pass to practical actions, that is to therapy, the differences rapidly disappear, and the doctors, provided they have good practical training and experience in their trade, and are self-dependent, would not differ much from one another. When a more clever doctor meets a less clever one at the bed of a patient, they might be very different from each other; one of them will be aware how far his diagnosis may be extended; he may even see something new, or so far unknown, which would be totally out of reach of the other one, who might either abandon the idea of further investigation or... might be as sure of himself as a god on Olympus. But when it comes to treatment, they become equal, they both will not hesitate to determine what therapeutic means usually -- purely symptomatic -- should be suggested to the patient. Gone are the days when doctors earnestly threw the prescriptions of their colleagues down the drain. Today, such a deed would almost exclusively be produced by ill or indecent competition.

It follows that the doctors, who need to make their living from their trade, and from whom the society requires first of all that they possess good knowledge of therapy, enter the contest armored materially with almost equal weapons. And, what is most disgusting, they are competing also with medical laymen, people entirely foreign to medical science but acquainted only with therapeutic means. This therapeutic "knowledge" can be obtained without much intellectual effort by, e.g., surgeon attendants who watch closely what kind of medicine is administered by doctors in cases of pain, or cough, or learning how a number of physical or surgical means (which, I repeat, have been in use since the time when nobody even heard about the science of medicine) are applied. And if the country folk, or the so-called simple people, go to surgeon assistants -- whether they should is another question -- it is not only because of ignorance: who can expect anyone to go to surgeon assistants if no help was obtainable from them? However, in many cases relief can be obtained from an unqualified person, exactly as from the learned doctors: patients can readily find remedies for their various pains or coughs or the like. In addition, patients will get well in the presence of a surgeon assistant, for they usually get well even without any medical help. The relatively few cases in which true science is required, that is, when real help is available only from a intelligent doctor who is very versed in his subject, are muffled by the great number of common, everyday cases, while the cases in which the ignorance of the person who applied the treatment had really harmed the patient, can readily be explained by "God's will". The low classes are therefore unable to recognize the supremacy of qualified physicians over those who are unqualified.

Moreover, in an educated community too, a doctor will have similar experience. Here, the doctors seldom can demonstrate their supremacy over each other. For each of them has, in the opinion of laymen, "cured" a number of patients, and each of them had fatal cases: none of them has any material means of treatment to reduce the number of fatal cases much below the average.

There are, no doubt, doctors who have almost no fatal cases among their patients: but those are usually physicians dealing with diseases that seldom are the cause of death, e.g., eye diseases. On the other hand, there are doctors who work mostly or exclusively in such pathologic realms, that have always yielded -- according to the statistics -- higher fatality rates than the others, and sometimes deaths occur very quickly, after short illness, for example, with childhood diseases. It should seem that this latter kind of doctor is providing the proof of his personal inability in public, because he is unable to cure one or another patient; one might think that suffering persons will avoid such a doctor, and that the opinion of the community concerning this doctor should be rather negative in comparison with another one who specializes with diseases which do not lead to death, or at least present less danger of death to the patient.

But this is not the case. The fact is that suffering people willingly seek help from the doctors who, by their own choice, act on a certain range of diseases -- e.g., heart or lung diseases -- and statistically have, to have a higher fatality rate among their patients than the others, who work with less dangerous objects, e.g., nervous diseases. The former can be and are the benefactors of their patients as are the latter. In other words, it turns out that, when looking from a certain distance, the fatality rate has not much influence on the opinion of the community concerning the supremacy of one doctor over another, in spite of the fact that one of them may be called more readily, because people are convinced that the former is saving patients from death more frequently and efficiently than the other. Perhaps, as an afterthought, one could find enough proof against the above allegation. A young doctor, coming to a place for the first time, might have to leave it after some time because of the public's opinion that he was not able to save people from death, that he was killing his patients, etc. We may actually come to find out that there were numerous fatal cases at the beginning of his practice due to unfortunate coincidences, or due to epidemics of diphtheria or scarlet fever, etc., but not due to his incompetence. On the other hand, a number of personal examples can be cited to show that out of two or three competing doctors, the one who did not experience the high fatality rate at the beginning of his practice had enjoyed the least popularity. There are other examples, too. A patient, after visiting two doctors, becomes convinced that the first one did not help him at all, or even did him some harm, while the second one made him recover or even saved his life, although both doctors had administered the same medicine (probably having no healing effects whatsoever), under two different pharmaceutical labels.

What is it, then, that creates the supremacy of one doctor over another? First of all, it is the psychological factor which, as I have said before, is always present in medical practice.

This can be clearly seen from the last example. If a patient had received the same pharmaceutical from two doctors, but arrived at the conclusion that the results were different, then the action of the doctors must have been different and this difference must have been something immaterial, probably the effect of the doctors' personalities. This immaterial factor becomes a real therapeutic means, for in many cases out of the entire therapeutic system the only available help a doctor can give to a patient is his personal influence. It is so because in many diseases recovery or improvement occurs thanks to autotherapy, while the therapeutic means which are applied have no effect at all on the illness. In such cases the only real help provided by the doctor to the patient is the moral help, that is, the assurance of improvement, or the assurance that the patient will recover. The practice shows that this moral help is the *primum movens* of the attitude of the doctor to the patient, that a sick person would require first of all this kind of help and that he appreciates it greatly. Let us remember why a doctor is usually called by the patient or his family? Because the patient wants to be "reassured", that is, because his mind is anxious about the state of his health and he wants to find peace of mind. It often happens that a patient starts his treatment only when he has been "reassured", when he becomes convinced, when he begins to trust that he will improve. Very often, when the patient has not been positively reassured, he would not even initiate the suggested treatment, but will look for another doctor instead. However, the exclusively verbal statements about the positive effects of the suggested material treatment, as made by the doctor, is by no means sufficient. It often happens that the patient or his family would not become convinced by the doctor's explanations about the positive effects of the treatment but would have confidence in the doctor who has prescribed a certain treatment without any comment.

The most important proof of the importance of the psychological factor in medical practice is the reason provided by patients when they choose a doctor. The patients always go to a doctor whom they trust. It is due to this trust that one doctor, but not any other, is being recommended by a patient to his friends; even when the patient can see positive results of the activity of another doctor, he would not go to him, for he has no trust in him. Strangely enough, when looking for an architect, or an engineer, or a chemist, etc., such "trust" is not the first factor taken into consideration; the customer goes to a given person because he has been told or he just knows, that this person is a clever, or diligent, or enterprising person. Among all trades, beside the doctors, perhaps only clergymen are selected generally on the basis of trust. The greater the confidence of a patient in the doctor, the higher the dose of the psychological factor available for the treatment; the less, the lower. Only the doctor who can provide a higher dose of moral help, or a better assurance of positive treatment results to the patient, will be considered (with equal material therapeutic means) as "better", and the other one as "worse". Thus, in medical practice we meet a phenomenon that does not exist, or is very limited, in the practice of other trades.

To work as a chemist, one needs to possess a certain sum of knowledge; the higher that sum, and the higher the talent and personal ability in making use of the information, the better would be the professional activity of the man. But in the case of a doctor, the sum of information concerning therapy is not enough, and something else is required as well -- which in fact should be required first; something that has nothing to do with knowledge, but which makes possible the practical application of the non-material aspects of therapy. The supremacy of one chemist over another results from the difference in their intellectual capacities or personal talents, but supremacy of one doctor over another results not only from the difference in their knowledge of therapy and in their talent for making use of that knowledge -- as these differences very often do not exist -- but from the difference in the outer and inner personal features which create greater confidence among suffering mankind.

From where, then, does a doctor acquire this psychological therapeutic factor? Well, it is acquired from various sources which are found neither in the Academies nor the Universities since, being out of the range of science, they cannot be lectured on. Besides, these sources are always related to the personality of the therapist. First of all, to obtain the patient's confidence, the doctor must know how, as the laymen say, to approach a patient. What actually is the meaning of this "approaching" the patient, and what kind of approach induces a patient to trust the doctor, cannot be described in detail here. It is a known fact that "to know how to approach a patient" is sometimes something one must be born with. There are doctors who are trusted by many sick persons, who just attract suffering individuals, who create even a hundredfold as great faith as the others whose conduct is apparently identical. This congenital confidence consists of a number of various indistinguishable physical and mental features of the doctor and a certain subconscious or subtle style of conduct which very quickly wins him the great confidence and popularity among simple people, even if he does not originate from the lower classes. Such doctors exert their influence not only on the patients, but on their colleagues and students, too. They have appeared in various nations and ages. At the beginning of this century such a doctor was Professor Schönlein of Berlin, who has been described by his very critical student, Greisinger, in the following words: "Alles schien er mir zu wissen, alles am Krankenbette zu können." It seemed to me that this man knew everything, that he could do everything at the patient's beside. And, before, in the Middle Ages there was the Dutch doctor Boerhaave, and such a person was -- as many readily confirm -- the Polish doctor Chalubinski, although no one of them was better able to materially [therapeutically] help the patient any more than could any of his not-so-famous colleagues.

Besides such a congenital disposition, there are some personal features which very often are critical for the psychological effect of a doctor on a patient. No doubt that, e.g., self-confidence, self-possession, and steadiness add to confidence in a doctor, while uncertainty and lack of seriousness undermine it. But since the present-day well-learned and critical doctors must very often feel uncertain, and have little faith in the effects of their methods, it seems to me that the persons gifted with some mystical disposition, who expect their hopes to come true in spite of the contradictory scientific proof, can often gain the upper hand over realistic minds and can more readily than the latter exert a psychological effect on their patients. For the mystic mood of a doctor may generate spiritual ties between the doctor and the patient, and such spiritual ties distinctively add to the confidence of the patient in the doctor, and to the moral influence of the doctor on the patient. Thus, it is almost always so that the signs of sympathy with the patient will win preference for one doctor over his less compassionate or dry colleagues.

To specify all the personal features or qualities of character and ways of conduct that produce a moral influence of a doctor over patients is not simple. Besides, these features are not permanent, their effectiveness may be different in different social classes, and they may depend on the education level of these classes, on characters of a given race, and of suffering individuals. On the other hand, it is absolutely clear that many purely external conditions may add to the psychological component of medical practice. These conditions may be classified as those which in our social system make one individual subordinate to another. Are there not many people who yield to the effects of differences of social levels? Does not a poor man in worn-out shoes feel like someone worse than a man dressed in fashionable clothes? It naturally follows that for many patients a doctor who has a more distinguished social position than his colleagues will be considered a "better doctor", and will inspire patients with more confidence than another lacking in such esteem. A doctor living in a fine house will make a better impression on many patients than one living in conditions similar to those in which his patients live. A doctor "in a coach" will make a better impression than one walking on foot, etc. For this reason, the well-known doctor will be better able to provide "psychological therapy" than one whose name still remains unknown to the public. Empty waiting rooms will in most cases reduce a doctor's popularity, while a crowd of patients will increase his popularity in geometric progression.

And then, if we have a closer look at doctors from the social point of view, what, in addition to the above-mentioned requirements, would increase the confidence in the sons of Aesculapius in different social groups or among different people? A lusty size, a look of self-assurance, an ample beard or a Prince Athis' shrug. All these things, being the mediations of moral factor in therapy, often play the role equivalent to the one of deep knowledge and skill!

Once, when Trousseau, the well-known clinician, was asked what should be done to develop a practice, he said: "Beaucoup de bon sens, un peu de tact et du courage." -- "Et la science?" -- "La science?... Oui, elle peut quelquefois servir." ["A lot of common sense and a litte tact and courage." -- "But what about knowledge?" -- "Knowledge?... Oh, yes, it can be of use sometimes."] You can believe me, this answer has not been merely a witty expression, but the description of the actual state of the art.

There is no doubt that medical science has substantially progressed since the time of Trousseau. Nowadays the knowledge of disease and good diagnosis based on science and the doctor's talent, might be extremely important for the patient. But even in the present time, doctors can hardly demonstrate scientific superiority over one another -- even now the suffering public has to select the doctor by the degree of "trust", which means that superiority of one doctor over another consists of the sum of various features totally foreign to [scientific] knowledge.

Therefore, it is possible even now that popularity might be gained by doctors of poor mind or even of limited therapeutic knowledge, doctors who would not be able to help or to save the life of a patient in cases in which this could be done through talent and science, provided they have sufficient inner and outer features to establish trust. Sometimes self-assurance and solid conduct, so important in building up the confidence of patients, may result from poor intellect, while the lack of talent or of a critical spirit would allow the doctor to believe in and apply with true faith a therapy which has no impact on the disease but only improves the morals of the patients.

On the other hand, if a clever and well-educated doctor creates confidence among the suffering public, it is very often not because he could do anything more in connection with the disease than a doctor of limited talent, but simply because his being clever allows him to judge the value of different ways of winning trust and to learn and train himself in the correct behavior with patients, i.e., medical *savoir-vivre*.

NOTE

1. So far, this is the only rational division of medical knowledge, because it is based on the history of its development. The current division of medicine into internal and external medicine -- called by the general public "medicine" and "surgery" -- results substantially from a practical reason; the capacity of one person to grasp everything known about diseases, or the impossibility of training one person equally well in both diagnosis and the practical application of therapeutical means.

Indeed, so-called "internal" medicine -- with its subsections of nervous and mental diseases -- mainly aims at the recognition of diseases, while external medicine -- which includes the so-called "surgical" diseases, ophthalmic disorders, gynaecopathies -- usually is centered on manual treatment methods performed by doctors. If, however, from the point of view of applied medicine, doctors' specialization seems -- within the main branches and subbranches -- reasonable enough for both internal and external medicine, from a scientific as well as a purely practical point of view such undue specialization as can be observed nowadays, is unreasonable if not pernicious, because patients are trying to find an independent "specialist" for each of their health problems!

CHAPTER IV

WLADYSLAW BIEGANSKI BETWEEN THE LOGIC OF SCIENCE AND THE LOGIC OF MEDICINE

Władysław Biegański (1857-1917) is probably the best known among the Polish philosophers of medicine, and the only one (besides L. Fleck) who was known abroad. He was also the only thinker of the Polish School whose reflections on the philosophy of medicine and biology were his starting point for the development of a general philosophical system. His fame as a philosopher was eclipsed by the evolution of Polish philosophy between the two World Wars, in particular by the birth of the Polish school of logic (Łukasiewicz, Tarski, Chwistek). He is, however, still remembered in Poland, albeit mostly as an exemplary physician and as a pioneer of social medicine [11, 33].

Biegański was born in the small town of Grochów, to a working-class family. He studied medicine in Warsaw. At that time (the late 1870's) the Medical School of Warsaw University still had some prestigious teachers (the histologist Hoyer, the physiologist Nawrocki, the clinician Baranowski), but the overall level of medical studies was low. During his medical studies Bieganski, who was bored with the medical curriculum, became interested in philosophy. He attended the lectures of the philosopher Struve, and participated in a competition organized by the philosophy department on the comparison between Locke's and Leibniz's ideas. After his graduation in 1880, he worked for two years in central Russia. There he became acquainted with the difficulties of being a country doctor in a backward area. He had spent the year 1882 abroad, studying medicine in Berlin and in Prague. In 1883 he settled in the provincial town of Częstochowa, which at that time was becoming an important industrial center, and he remained there until the end of his life. Biegański worked in the local hospital, and was physician for two of the most important factories of the town. According to the testimony of his wife Mieczysława, Biegański had first chosen to work in these factories mainly in order to be able to keep regular working hours, and to have free time for his investigations. His contact with industry stimulated, howevern his interest in occupational medicine, and later he published several works on diseases of industrial workers and on the differential pathology of various social and ethnic groups. At the same time, he published numerous clinical observations, mostly on neurological disorders, and struggled to found a small biochemical and bacteriological laboratory in the hospital in Czestochowa - the first of its kind in a provincial hospital in Poland. In 1891, Biegański published his *Differential Diagnosis of Internal Diseases,* which became a standard textbook on this subject for several generations of Polish students [3, 16, 21, 24, 25, 26, 32].

From the 1890's on, Biegański developed a growing interest in the philosophy of medicine. In 1894, he published his first philosophical-medical work, *The Logic of Medicine, or The Principles of General Methodology of Medical Sciences* [4]. Biegański's book explained the principles of general logic (as expressed in the writings of J.S. Mill), and tried to show how those principles should be applied to medicine. The book was criticized for being too abstract, and not sufficiently grounded in medical practice. The philosopher of medicine Kramsztyk defined it as "a textbook of general logic illustrated with medical examples". Biegański, Kramsztyk affirmed, tried to adapt the reality to his pre-conceived ideas. However, all of its shortcomings notwithstanding, Bieganski's book undoubtedly influenced the studies of other Polish philosophers of medicine. Many of the works published later referred to this work and discussed its main theses [27].[1]

Four years later, Biegański published his second medico-philosophical book, *General Problems of the Theory of the Medical Sciences* [5]. This book attempted to explain pathological phenomena from the point of view of a functional understanding of diseases. Bieganski opposed his dynamic vision of pathology to the static vision developed by the anatomic-pathological school. In 1908, he published a second, enlarged and modified edition of his *Logic of Medicine* under a new title, *The Logic of Medicine, or a Critique of Medical Knowledge* [9]. The new edition was very different from the original. When writing it, Biegański took into consideration not only the criticism of the first edition of this book, but also the results of 15 years of debates on the philosophy of medicine in Poland. While the first edition of the *Logic of Medicine* was in all probability one of the events that started a period of intensive activity of the

Polish school of philosophy of medicine, the second edition summed up the discussion which had taken place during the peak period of this activity. The second edition of *The Logic of Medicine* was later translated into German [10]. In general, the book was favorably received by critics in Poland and abroad, and had a short period of glory, but was practically forgotten after the First World War [16, 17, 18, 35].

From 1897 on, Biegański gradually left his medical occupations (with the exception of his private practice, which he maintained to the end of his life) and devoted his time to philosophical studies. He published, in 1903, his *Principles of General Logic;* in 1907, a*Textbook of Logic* for high schools; in 1910, *Tractatus on Knowledge and Truth;* in 1912, *Theory of Logic;* in 1915, *The Theory of Knowledge from a Teleological Point of View;* and in 1918, *General Ethics* [3, 19, 21, 32].

Biegański's philosophical ideas had their roots in his reflections on biology and on medicine. His general philosophical approach, which he called "previsionism" ['prewidyzm'], and which claimed that the goal of knowledge is not to understand reality but to predict the future, originated in his study of the importance of teleology in biology on the one hand, and his efforts to prove that the medical practice of his time can be transformed into a scientific discipline, on the other. His vision of all knowledge as goal-oriented, and his ideal of the fusion of practice and theory in the formation of knowledge, can be seen as an attempt to extend his ideas on the formation of practical medical knowledge to the process of the formation of knowledge in general. In addition, Biegański's interest in logic and ethics - both, according to him, normative disciplines - was, in all probability, anchored in his reflections on the normative aspects of medical practice.[2]

In the first, 1894 edition of *Medical Logic,* Biegański tried to make explicit the scientific basis of medicine. He affirmed that the starting point of his own reflections was Chałubiński's book, *The Method of Finding Therapeutic Indications.* Chałubiński's clinical method, he explained, was based on a holistic and patient-oriented approach, and was particularly appropriate for the medical practitioner. Chałubiński stressed that disease was a "modified life", and therefore one should always pay attention to the organism as a whole. In contrast, "in the anatomic-pathological approach, we cannot avoid paying less attention to the sick organism, the 'sick life', as Chałubiński would say, and more attention to the sick organ, the sick tissue. To put it tersely, we lose from our sight the patient, and we see only the diseased organ". While Chałubiński and Baranowski saw in each of their patients a suffering fellow-man, in the German school "the patient was replaced by the disease, and thus the physician started to treat his patients as more or less interesting material for their investigations." The unavoidable result was the multiplication of unethical experiments on humans, such as injections of cancer cells or the inoculation of patients with syphilis ([4], pp. 6; 9; 15).

Why was Chałubiński's excellent clinical method almost completely forgotten? Biegański thought that the principal reason for it was the fact that Chałubiński and his students were not interested in medical science, and did not try to make their theories scientific, thus more attractive to the young, scientifically-minded physicians. The *Logic of Medicine,* he explained, was an attempt to systematize Chałubiński's approach, and to make it compatible with the development of medical science. His way of doing this was to affirm that in fact Chałubiński's art of finding therapeutical indications - and therapy in general, if properly understood - was not a practical craft but a fully-developed scientific discipline. Biegański attempted to show that methods inspired by other scientific disciplines can and should be used by physicians. He was particularly interested in the role of classification, of experimental methods, of hypotheses, of statistical methods, and of deductions in medicine. This approach, undoubtedly inspired by Oesterlen's *Medical Logic* (1852) [31], differed from its predecessor in one crucial point. Oesterlen was interested only in medical science, and his book could have been titled "The Logic of General Pathology". He had only very limited faith in therapy, and he affirmed that the most experienced physicians recognize that medicine can propose little help of a positive nature. For him the practical medicine of the future should deal with prevention, not with healing ([31], pp. 435-436).[3] In contrast, Biegański was mainly interested in the theoretical basis of therapy. He tried to identify the common methodological ground of therapy and scientific investigation, and attempted to show that both were governed by the same fundamental laws of logic.

Biegański was obliged to admit that many of the methods developed by other sciences could not - or, at least could not yet - be applied to the practice of medicine. But, he added, even if other scientific disciplines were methodologically more advanced than medicine, this did not mean that practical medicine was qualitatively different from other natural sciences. For example, the physician faced with a patient usually progressed through the formulation of consecutive hypotheses. He first formulated a hypothetical diagnosis after the observation of some of the most obvious symptoms, then he tried to confirm it by further investigation. If the observations failed to confirm the first hypothesis, he formulated a second one, and, if necessary, a third. Bieganski called this "an inductive method". He affirmed that this method, quite similar to the way scientists advance hypotheses, was probably the most frequent way of establishing a diagnosis. But there were other methods as well, which could also be systematized and which were also akin to the way natural scientists worked: 1) the exclusion method (simultaneous formulation of two or more hypotheses, then the elimination of all but one); 2) analogy (the comparison, point by point, of a given case to a known case); 3) the experimental method (e.g., the confirmation of the diagnosis by a positive reaction to a drug known to have a selective impact on the disease).

Biernacki claimed that he was merely systematizing Chałubiński's methods of finding therapeutical indication, and was showing its relationships to the general principles of logic. But in fact the overall approach to therapy of the two thinkers was quite distinct. The search for therapeutic indications always started, for Chałubiński, with a careful observation of isolated symptoms - or in his words, the study of the most proximate "disease moments". The physician should then try to find through a regression process the most remote disease moment which could be responsive to the available therapeutic means. In contrast, Biegański proposed to begin with the global recognition of the nature of a disease, and take this as a starting point for the search for therapeutic indications. For example, in nephritis one must first recognize the nature of the disease: inflammation of the kidney. Theoretically, one should then attempt to act on the causes of this dysfunction. However, in the case of nephritis, the true etiological cause of kidney inflamation (e.g., a previous infectious disease, such as scarlet fever) cannot be treated by the available therapeutic means, and it is equally impossible to act directly on the most important disease moment, i.e., the inflammatory pathology of the kidney. The physician should therefore try to act indirectly on the inflammation by creating conditions which can assist the organism to cure it by its own means (e.g., by administering a diet which facilitates kidney function). This is, however, not sufficient. The physician should also try to discover secondary therapeutic indications, such as treatment of the anemia and the edema often present in nephritis, as well as the symptoms of uremia.

The difference between Chałubiński's and Biegański's approaches may seem insignificant. Both arrive at similar conclusions: the need to treat the most remote accessible *moments* in the causal chain of the disease. But this is not the case. Chałubiński's approach took as its starting point the specific condition of an individual patient, while Biegański's starting point was the recognition and the naming of a disease (in his example, nephritis). Whereas for Chałubiński diagnosis was at best a secondary problem, and was not indispensable for therapeutic decisions, Biegański viewed the right diagnosis - thus medical science - as the very center of his practical method of discovering therapeutic indications ([4], pp. 134-138).[4]

Although Biegański preferred to introduce scientific methods into the practice of medicine, in particular into therapy, he strongly warned against the new tendency to accept uncritically all scientific innovations. The physician, he explained, should be aware of the fact that research in general, and research in medicine in particular, is often hampered by the pre-formed ideas of the individual investigator. He remarked: "The human mind has a property which makes truly objective judgments difficult. Our previously acquired convictions influence the new ones, and give them a specific shape" ([9], pp. 91-92). Prejudice can influence the result of an observation. For example, Haller was unable to observe that mater dura (a membrane surrounding the brain) is a highly sensitive tissue, because he had an *a priori* conviction that all membranes that had fibrous tissue were insensitive. The danger of the influence of preconceived ideas on observation is not unique to medicine, but it is more likely in medicine than in other scientific disciplines. Biegański observes: "Everybody who recommends a new treatment for a given disease always quotes a great number of observations which indicate the unusual effectiveness of the new therapy." Physicians have also a strong tendency to confuse hypotheses and facts. They stick to their opinions, which seem to them beyond any doubt, and they forget that what was true yesterday is often false today. "Theories always change, only our mind remains curiously unchanged, and it firmly believes today, as it firmly believed yesterday, in the creations of its own imagination" ([9], pp. 103-104).

While Biegański's first book was an attempt to adopt the methods of investigation of the natural sciences to the uniquely medical way of thinking, his second book, *General Problems of the Theory of Medical Sciences,* attempted to develop a general scientific theory of diseases. This theory was based on his vision of recent biological knowledge, and more specifically on his ideas on structure and function of the living cell. In the first part of the book, Bieganski summarized his views on determinism in biology, on teleology, on vitalism versus mechanism, on the beginning and the goal of life, on metabolism, heredity, and adaptation. The aim of his long introduction (more than half of the book) was to supply the theoretical basis for the main thesis of the book: disease is essentially a disturbance of vital functions. This concept, Biegański explained, was at the bottom of Chałubinski's clinical theory, but Chałubiński did not make its theoretical basis explicit. Bieganski's goal in his second book was firmly to ground this concept on new biological knowledge, then to found the notion of rational therapy on it. Thus, *General Problems of the Theory of Medical Sciences* was a logical continuation of Biegański's efforts to rehabilitate Chałubiński's clinical emphasis while avoiding its theoretical shortcomings ([5], pp. VI-IX).

Biegański, following Chałubiński, defined disease as a disturbance of vital functions - either an exaggerated stimulation, or the lack of ability to react to stimuli. For him there was no clear-cut distinction between physiological and pathological functions, only a difference in *degree*. Normal function was the reaction of the cell to moderate stimuli, while disease, or modified function, was the reaction of the cell to excessive stimuli. All pathological manifestations could be explained by a disturbance of the vital functions of cells (for him, e.g., "modifications of the inter-molecular cohesion of the biogen"), and death was simply the final disintegration of the vital functions, either at the cellular level or at the level of the whole organism ([5], pp. 247-283).[5]

Disease was defined by Biegański as the reaction of the organism to excessive stimuli. It always had two components: the disease-inducing stimulus and the reacting organism. The physician, whose aim is to reestablish the lost equilibrium, can try to modify both. First, he can try to eliminate, or at least to reduce the harmful stimuli; second, he can try to increase the natural resistance of the organism. Biegański admitted that usually the physician is unable to act directly either on the causes of the disease or on the forces within the organism that work to eliminate them. This does not mean, however, that the physician has only a very limited range of action. In contrast to Biernacki, who viewed symptomatic treatments as a secondary and relatively inefficient method of medical intervention, Bieganski's theory of disease allowed him to view every modification of the physiological functions of the organism as a *rational action*, destined not only to alleviate the suffering of the patient, but to promote cure. Indirect actions such as rest, adequate diet, symptomatic (e.g., antipyretic, emetic, diuretic) medication, either diminish the overstimulation of cells or enhance the action of the defensive forces of the organism. They therefore directly contribute to the re-establishment of functional equilibrium ([5], pp. 293-296).[6]

Biegański strongly opposed the followers of therapeutic nihilism who professed faith in the "curative force" of the organism that was further revived by the discoveries of bacteriology and immunology. This faith, he claimed, is no more rational than the vitalist's faith in *vis naturae*. Disease can be cured only when the disturbed equilibrium of the organism is re-established. But the organism is often unable to establish rapidly the lost equilibrium. Therefore, in every single case of disturbance of the functions of the organism something can and should be done in order to influence the disease process. Every therapy modifies functions by stimulation: either chemical stimulation (drugs), mechanical stimulation (massage, gymnastics, or even surgery), thermal stimulation (hydrotherapy), or electrical stimulation (electrotherapy). The physician is not just "helping nature" but is modifying either the causes of the disease or the reactions to them ([5], pp. 303-304; 295-296).[7]

Bieganski stressed the functional aspects of disease. He therefore was opposed to the metaphor of disease as a "battlefield" between microorganisms and phagocytes made popular by the new science of bacteriology. Such an approach was as mistaken as the interpretation of Darwin's metaphor, "the struggle for life", as a battle between individuals. Disease, Biegański explained, is a modified physiological function, thus a specific type of chemical reaction. Referring to a "struggle" between disease-inducing elements and the organism makes about as much sense as speaking about a "struggle" between two elements in a chemical reaction ([5], pp. 298-302).[8]

Therapy was for Biegański the primary medical science and the source of all medical knowledge. He adopted the definition of the French clinician, Armand Trousseau (1801-1867): *"La médecine, c'est l'art de guérir, elle n'est que cela."* His intention, however, was not to affirm the superiority of the "art of healing" over medical science but to make the two identical by making therapy itself scientific. Therapy should become a rational and deductive science based on formalized logical assumptions and on the achievements of the theoretical medical sciences. It should enjoy the same prestige as the other natural sciences.

In the second, enlarged and modified edition of his *Logic of Medicine* (1908), Biegański further developed his views of the way to transform therapy into an exact science. In the introduction to the second edition, Biegański argued that medicine is a specific branch of the natural sciences. One can thus speak of the "logic of medicine", or the "critique of medical knowledge" as one can speak of the "logic", or "critique of knowledge" (i.e., epistemology) of physics or chemistry ([9], pp. XIX-XV). It is not possible, however, to transform all medical practice into an exact science. The mode of action of many highly efficacious drugs and therapeutic means is still poorly understood, and thus it is impossible to eliminate completely the art (i.e., knowledge acquired by experience) from medical practice. Medicine should therefore be defined simultaneously as a science and an art. But, Biegański hastened to add, art and science are two aspects of the same fundamental activity: the modification of natural phenomena. After all, elements of art do exist even in the most mathematized sciences: the art of presenting the results, the art of planning an experiment, etc. The existence of the art element in practical medicine should not be an obstacle to viewing medicine as a science ([9], pp. 26-29).

There are, however, some particular difficulties in transforming clinical medicine into scientific practice. Medical knowledge is based on clinical observation, and thus Biegański had stressed that the physician's prejudices often modify his observations. This natural tendency to observe only what conforms to one's theoretical presuppositions can explain the persistence over many centuries of inefficient and even dangerous treatments, such as blood-letting. Moreover, one usually perceives new facts through analogy with old, familiar ones, and such analogy can be misleading. Biegański did not think, however, that unbiased observations were entirely impossible. He believed that these distortions of observation could be eliminated, or at least minimized if the observer was aware of their danger. Moreover, he thought that visual impressions were less prone to distortion than other types of sensory impressions. The growing tendency to transform all impressions into quantifiable visual impressions (e.g., the sensation of heat into the height of a mercury column in a thermometer, the taste or smell of secretions into colorimetric chemical reactions) should thus help to transform clinical observations into a more exact science ([9], pp. 42-50).[9]

Another problem in transforming medical practice into scientific propositions is found in the classification of pathological states. The old, functional (i.e., symptomatic) or anatomic pathological classifications were rejected because they did not account for the totality of disease phenomena. Different diseases can have very similiar symptoms (e.g., fever, diarrhea), and some diseases (e.g., many acute infectious diseases) do not produce any important anatomical modifications. The same can be said of the etiological definition of disease adopted during the period of great triumphs in bacteriology. Etiological factors cannot, by themselves, explain the totality of pathological phenomena. They cannot explain *the great differences in individual reactions to similar pathogenic stimuli* (e.g., the differences in pathological changes induced by the same bacterium in different individuals). Bieganski's conclusion was that classifications that depend on a single criterion cannot produce an adequate classification of diseases because diseases do not depend on a single, well-defined cause. Rather, they represent a complex chain of events which appears as a result of important modifications in regulatory mechanisms and which involve the organism as a whole. In order to achieve a rational classification of pathological states one should use a combination of functional, anatomical, and etiological aspects of a disease. For the same reason, over-specialization in medicine, based either on anatomic-pathological divisions (heart diseases, digestive diseases) or on etiological divisions (infectious diseases, tropical diseases) is often more harmful than helpful ([9], pp. 67-78).

Rational classification of diseases, and (even more so) rational therapy, should, Biegański claimed, be based on an adequate understanding of causal chains in pathology. This is not, however, a simple task. While in the physical sciences one can usually isolate simple events and direct causes, such a reductionist approach does not have much meaning when dealing with living organisms. Not only are living organisms highly complex structures, but the very goal of biological understanding is the study of life, which is by definition a complex event. In addition, biology, in contrast to physics and chemistry, deals with closed systems in which the function of the whole is dependent on relationships between the parts.[10] Consequently, Biegański thought that the study of biology, and even more the study of pathology, should be teleologically-oriented. The suspicious attitude towards teleological thinking was, he affirmed, the result of a mechanical transposition to biology of the norms of research in the physical and chemical sciences. But while a teleologically-oriented approach can indeed be an obstacle to physical or chemical understanding, it is indispensable in biology. How otherwise can one understand certain complex phenomena such as biological regulation? And it is even more important in medicine, not only because pathological phenomena are, by their very nature, complex events, but above all because *the therapeutic action of the physician is entirely guided by teleological considerations* ([9], pp. 136-141).

The teleological approach, Biegański affirmed, should facilitate experimental studies of pathological phenomena. He was in favor of such experimental studies, but he was aware of their limits. Only the relatively rare biological phenomena, he explained, that can be isolated and studied in vitro (e.g., enzymatic digestion, neutralization of bacterial toxins by antitoxins) can be examined directly by using material of human origin. In all other cases one is obliged to extrapolate from the results of animal experimentation to human beings. But animals are different from humans, and the transposition of results obtained in animals to humans is far from unproblematic. The discoveries of bacteriology put a temporary end to the hope, expressed by the physiologists (e.g., Claude Bernard in his *Introduction à l'étude de la médecine expérimentale)* of solving all the problems of human pathology by animal experimentation. It was quickly found that disease-inducing microorgansms do not produce the same reactions in humans and in animals, and moreover that the same microorganism can induce different diseases in different animal species. Pathology seems to be quite different from physiology, and experiments with animals can at best help researchers to formulate hypotheses that need to be confirmed by careful clinical experimentation ([9], pp. 152-176).

In addition, experiments with animals, and even more so clinical experiments, reveal highly complex casual chains. Usually only the first and last links of each chain are known, while all the other variables remain unknown. It is not surprising that the results of such experiments are often unreproducible, and that different investigators obtain even contradictory results. One way to overcome this difficulty is to conduct frequent replication of well-controlled experiments. However, this is not always sufficient and/or possible. Another way to overcome this difficulty is to apply statistical methods. The 'numerical method' advocated by Pierre Louis (1787-1872), has had many adversaries, however, from physiologists such as Claude Bernard - who claimed that phenomena which were not completely understood could be studied - to physicians (such as Oesterlen) who affirmed that no valid results had ever been obtained in medicine by employing a statistical approach ([2], vol. I, p. 226; [31], p. 145).

Bieganski agreed with the earlier evaluation of the philosopher of medicine Zygmunt Kramsztyk, who claimed that the statistical method could not be used to prove etiology. This method cannot demonstrate a causal relationship; it can at best reveal a correlation, and thus help to formulate an adequate hypothesis that has to be verified experimentally. But he affirmed that statistical evaluation could be useful for practical purposes because "theory can content itself with conclusions derived from analogies and with hypotheses, while practice must be based on facts or at least on direct generalizations" [28].[11] Biegański strongly disagreed with Kramsztyk's quite pessimistic evaluation of the viability of statistics in medicine. He admitted that statistics were often employed by incompetent physicians, and that many statistical investigations in medicine used either the wrong methodology or inadequate data. He stressed, however, that this fact did not in itself invalidate all statistical approaches to medicine. It only underscores the need for more rigorous utilization of statistical methods. Applying statistical methods, he explained, was the only way to answer the practical question of choice between alternative therapies. It was thus highly important to find the correct methodology for applying statistics in medicine, and to teach physicians how to use this tool ([9], p. 224).

Bieganski did not entirely adopt the point of view of other thinkers of the Polish School (such as Chałubiński and Biernacki) on diagnosis in medicine. He agreed that disease nosologies and other medical classifications do not actually describe natural phenomena. They usually represent non-existing, ideal types derived from facts, not the facts themselves. But, Biegański added, biological classifications, which are also artificial constructions, represent a relationship based on structural analogy and therefore are useful to describe a common descent. In this genealogical sense, pathological phenomena, too, can be classified according to their degree of similitude. Biegański did agree with Chałubiński's and Biernacki's critiques of the confusion of the ideal, schematized "disease types" presented in textbooks of pathology with the real illness suffered by a given patient. For him, also, in clinical practice one does not see "diseases", only sick persons. But in contrast to these two other thinkers, Biegański thought that the classifications of pathological manifestations, their artificial character notwithstanding, are very important, because they play a crucial role in the process of the constitution of medical science and of the transmission of medical knowledge. A diagnosis reflects the existing medical knowledge on a given pathological disorder. It serves, therefore, to link the specific observations of a given physician to the entire body of knowledge of his profession. Biegański strongly opposed Biernacki's view that for practitioners medical classifications represent merely the "bankruptcy of science". Such an affirmation, Biegański claimed, showed that Biernacki did not understand the nature and the purpose of medical classifications. "Science has no reason to blame itself, for it does not establish classes as drawers in which all individual cases should be stuffed, or as a schemata of clinical pictures to which real-life cases should be fitted. Classifications only sum up previous knowledge, and make possible its application to a given case" ([8, 9], pp. 184-189; 86-98; 100-103). On the other hand, if medical classifications reflect the state of medical knowledge at a given moment, they are bound to change along with the entire corpus of knowledge to which they belong. Consequently, Biegański, who was very critical toward the system of classification of diseases of his time, believed that it was bound to change radically in the near future.[12]

The central subject of Biegański's philosophico-medical studies was his attempt to rehabilitate therapy and to place it on a scientific foundation. But how did it happen that such rehabilitation was now necessary? Why had therapy - its long history, its practical importance, and its similitude (Bieganski *dixit*) to other scientific activities notwithstanding - at the end of the 19th-century a much lower intellectual status than biology and theoretical medicine? For Biegański this phenomenon had its roots in recent historical developments. Until the 19th-century, medical theory and medical practice were closely interwoven. Every treatment proposed by physicians, even the most esoteric one, had a firm theoretical basis. For example, when a disease was viewed as the activity of angry spirits, the right way to treat it with it was through magic; when it was viewed as the result of disequilibrium of humors, blood-letting and vomiting were considered the most appropriate treatments. Every therapeutic school, from the iatrochemists to Brown's and Broussais's schools, based their therapeutic activity on their theoretical conception of disease. However, the advent of the anatomic-pathological school in the mid-19th-century broke the tendentious bonds between medical theory and medical practice. When diseases began to be viewed as a sum of pathological modifications of tissues, the ineffectiveness of the majority of drugs and treatments that were unable to influence these modifications became patent. This resulted first in therapeutic nihilism, and then, when this position became too difficult to apply in practice, in the radical separation between the science of medicine and the practical art of healing now disconnected from the growing body of medical knowledge ([9], pp. 231-232; 243-245).[13]

Biegański's historical analysis of the reasons that separated, in the 19th-century, medical science from the art of healing was quite similiar to Biernacki's explanations of the same subject. But while Biernacki proposed to close the gap between medical science and medical practice by elaboration of new therapies grounded on a sound scientific basis (taking serotherapy as a paradigmatic example of such evolution) ([14], pp. 116-117), Biegański advocated the transformation of existing medical practice into an exact science. He concluded his *Logic of Medicine* by proposing to take Chałubiński's *Method of Finding Therapeutic Indications* - which for Bieganski was an attempt at the systematization of the usual bedside behavior of every good physician - as the starting point of this process. Chałubiński's method, affirmed Biegański, pointed the way toward the formulation of a theory of rational therapy. Such a rational approach to therapy, if supplemented with an adequate scientific basis, could permit the future reunification of medical theory and medical practice.

Biegański's main claim was that medical practice, and more particularly therapy, could and should become an exact science, similar in its methods and approach to all the other natural sciences. But this does not mean that he thought that medical practice had no specificity of its own. For him, this specificity was not methodological, but moral. The specific role of the medical profession did not imply creating or utilizing a different, non-scientific logic, but it did signal a need for a specific code of professional behavior. Consequently, Biegański complemented his medical practice with a lifelong interest in medical ethics. This interest culminated in the publication of his *Thoughts and Aphorisms on Medical Ethics* [6]. In the introduction to this book, Biegański explained that the art of healing was composed of two distinct dimensions: factual and ethical. The evolution of the factual aspect of the art of healing depends on the overall development of human knowledge, while the evolution of the ethical aspect depends on the evolution of the ethical ideals of humanity. Until the 19th-century, these two parts were considered as equally important. But the rapid evolution of scientific medicine in the 19th-century affected this equilibrium in favor of science.

Some zealots of the new scientific approach to medicine had attempted recently to substitute science for morals. They claimed that the only truly ethical attitude for a physician is to contribute to the growth of medical knowledge. In this way, they claimed, the physician would ultimately help the sick in a more effective way than a dozen of his well-intended, compassionate, but under-educated colleagues. But, Biegański explained, these enthusiastic advocates of the new "scientific" medicine had forgotten that human illness is far from being a purely scientific problem, and that the role of the physician is defined as "helping the sick" not only with his technical knowledge, but also, in Biegański's words, "through his psychological influence". The latter aspect, Biegański stressed, is of particular importance. One should not forget that the success of a given treatment often depends on the quality of the relationship that obtains between the physician and his patient. In addition, one should also appreciate the limits of therapeutic intervention. In many cases the only help the physician can offer is consolation, and "even if medicine had no single effective treatment able to fight disease, physicians would be necessary to console the sick" ([6], pp. 134-136; 66; 31-32).[14]

Biegański viewed the attitude of the fervent zealots of scientific medicine as belonging to the general 19th-century trend to place unlimited faith in the progress of science. He noted that only at the end of the 19th-century had people begun to realize that rapid technical progress not only did not increase human happiness, but, on the contrary, seemed to deepen the division between the privileged few and the growing impoverished masses. Writing in 1899, Biegański expressed his hope that the tide of blind faith in science was turning at the eve of the new century, and that people would slowly realize that a return to more spiritual and more ethical approaches was needed everywhere, and in particular in medicine ([6], pp. 26-33).

In his *Thoughts and Aphorisms on Medical Ethics,* Biegański pointed to the absence of systematic reflection on medical ethics. When physicians speak about "medical ethics" they usually think solely about the rules which regulate their professional corporate interests, and not about the rules which should regulate their relationships with their patients. But the corporate interests of the medical profession can be - and often are - in conflict with the patient's best interests. In such cases, Biegański affirmed, the duty of the physician to help suffering people is always primary. In the name of this principle Biegański strongly opposed some of the unwritten "ethical rules" of the medical profession, such as the rule forbidding a physician to contradict in public the opinion of another physician, or the recommendation to refuse to care for a patient wishing to change his physician. The patient, he affirmed, should have the right not only to freely choose his physician, but also, when in doubt, to verify the activity of the latter by comparing his opinions with those of other physicians. Biegański strongly criticized the mercantile and corporatist attitude of some of his colleagues. He did not, however, advocate specific legislation in order to regulate ethical behavior in medicine. He had faith in the virtues of professional medical education and of personal example. The essence of his convictions on this subject - and also of its limits - is summarized in his aphorism: "A good physican should above all be a good man."[15]

NOTES

1 Biegański sharply replied to Kramsztyk's criticisms (in *Gazeta Lekarska,* 1897), claiming that Kramsztyk made no effort whatsoever to understand his particular point of view. He was, however, greatly affected by the cold reception of his book by Kramsztyk, whom he considered as the best among Chałubiński's students. In the second edition of *The Logic of Medicine,* Biegański took into consideration many of Kramsztyk's objections ([3], pp. 87-88; 112).

2 Biegański's teleological approach may be related to the existence of longstanding teleological tradition in nineteenth-century German biology, related by Timothy Lenoir in his book *The Strategy of Life: Teleology and Mechanics in Nineteenth-Century German Biology* [29]. I will not, however, deal with the general philosophical aspects of Bieganski's works, but will focus on his earlier, more medically-oriented studies. On Bieganski's philosophy see, e.g. [12, 13, 16, 36]. Biegański's philosophical work at first attracted the attention of professional philosophers in Poland. He published several articles in the leading Polish philosophical journal, *Przeglad Filozoficzny,* and in 1914 he was offered the chair of logic at Cracow University (he declined this offer for health reasons). But his philosophical works were later forgotten. Even Władysław Tartakiewicz (a student of the Lwow philosopher Kazimierz Twardowski and later one of the leading Polish philosophers of his generation), who as a young man edited the posthumous publication of Biegański's *General Ethics,* did not mention Biegański in his monumental *History of Philosophy* (Warsaw, 1968) ([19, 32], p. 16).

3 Oesterlen quoted the expression of the French physician Reveillé Parisé (*La Gazette Médicale,* 1851, **29**). "La médecine est la plus noble des professions et le plus triste des métiers." ("Medicine is the most noble profession and the most depressing occupation.")

4 Bieganski acknowledged, however, that the basis of diagnosis - the classification of diseases - was an artificial division: "classes are always created by our mind. In nature there is no such thing as a class, species, genus: only individuals exist" ([4], p. 23; 43). In the second edition of his book, Biegański's point of view on the art of finding therapeutical indication was closer to Chałubiński's ideas ([9], p. 243).

5 Bieganski often utilized, as Chałubiński did, terms such as "stimulation", "overexcitation", "vital functions", etc. He related his concept of "disturbed function" to Virchow's teachings ([5], p. 243), but such expressions might also reflect the old dualistic vision of disease as overstimulation or understimulation. Indeed, Biernacki's view of disease was probably closer to such a vision than to the modern, biochemical vision of

pathological phenomena such as was advocated, e.g., by Biernacki. In his article on 'The Principal Currents of 19th-Century Medicine' [7], Biegański affirmed that chemical and microchemical studies of the protoplasm represented the future of medical science, but on the other hand he was highly sceptical about the new " 'fantastic' theories concerning the activities of toxins and antitoxins".

6 Biegański's theory can be viewed as an attempt to translate the classical vision of Hippocratic medicine, which defined health as a state of equilibrium between all the "humors" of the body, into the language of the biology of his time.

7 Biegański's point of view can be compared with earlier ideas on rational medicine. Physicians who reacted to excessive, "heroic" treatments on the one hand, and "therapeutic nihilism" on the other, claimed that rational therapy should be based on a clear-cut distinction between the cases in which medical intervention is indeed effective, and those in which medicine can at best offer symptomatic treatment or psychological help (see [15, 23]). This division, implicit also in Biernacki's writings, was rejected by Biegański. For him, every pathological state should be treated and each therapeutic intervention was important, because each "modification of function" affected the global disease process.

8 Biegański claimed that the problem of the origin of anti-microbial substances in the blood was an example of the inadequacy of the "battle metaphor", because in fact it was not at all clear if the anti-toxins were really produced by the infected organism (as those who accepted the "battle metaphor" claimed), or by the infecting microorganisms themselves. The idea that antibacterial substances that appeared in the blood were in fact bacterial products was perfectly acceptable in the 1880's - it was shared by Pasteur, among others. It was, however, clearly outdated in 1897. At that time it was already rejected on an experimental basis [30].

9 Biegański discussed this notion already in the first edition of his *Logic of Medicine*.

10 Biegański was perhaps inspired by Claude Bernard's "organicism": the idea that the organism represents more than the sum of its parts. There is a "directing idea" of each organism, and "what lives, what exists, is the whole" ([2], t. II, pp. 143-147).

11 Kramsztyk [28] affirmed that, while etiology belongs to theoretical medicine, therapy has practical goals, and statistics can be used only to answer practical questions. This idea was not entirely new. Already Pierre Cabanis (1728-1808) had explained that practical knowledge such as agriculture or medicine should be based on probability [22]. On Cabanis' medical philosophy, see, e.g. ([34], pp. 103-109] and [1], pp. 15-22).

12 The idea that biological classifications were constructed by scientists had been discussed by Biegański in the first edition of *The Logic of Medicine* ([4], p. 23).

13 See also [7]. Biegański had a strong interest in the history of medicine, and in 1901 he published the first textbook on the history of medicine in Polish, *Introduction to the History of Medicine* [*Wstęp do historji medycyny*].

14 Biegański's ideas about the importance of the psychological factors in healing are quite similiar to Biernacki's ideas on this subject. Biernacki discussed these ideas in his book, *The Essence and the Limits of Medical Knowledge,* published the same year.

15 W. Biegański, 'Medical Ethics', a lecture given at the meeting of The Association of Polish Physicians in Czestochowa, April 26, 1914 ([6], pp. 122-153). A strong believer in the importance of human factors in healing, Biegański also supported access of women to medical studies. He claimed that women were perhaps unable to create new medical knowledge, but that their access to medicine "will bring some warmth to our cold, over-rational profession" ([6], p. 73).

BIBLIOGRAPHY

[1] Ackerknecht, E.: 1967, *Medicine at the Paris Hospital 1794-1848,* Johns Hopkins University Press, Baltimore.

[2] Bernard, C.: 1938 (1865), *Introduction à l'étude de la médecine expérimentale,* Delgrave, Paris.

[3] Biegańska, M.: 1930, *Władysław Biegański: Life and Work [Władysław Biegański, życie i praca],* Wydawnictwo Kasy Imienia Mianowskiego, Warsaw.

[4] Biegański, W.: 1894, *The Logic of Medicine, or The Principles of General Methodology of the Medical Sciences [Logika medycyny czyli Zasady ogólnej metodologii nauk lekarskich],* E. Wende, Warsaw.

[5] Biegański, W.: 1897, *General Problems of Theory of the Medical Sciences [Zagadnienia ogólne z teoryi nauk lekarskich],* S. Ogelbrandt, Warsaw.

[6] Biegański, W.: 1957 (1899), *Thoughts and Aphorisms on Medical Ethics* [*Myśli i aforyzmy o etyce lekarskiej*], Panstwowy Zaklad Wydawnictw Lekarskich, Warsaw.

[7] Biegański, W.: 1901, 'The Principal Currents of 19th-Century Medicine' ['Główne prądy w medycynie XIX wieku'], *Krytyka Lekarska* **5**, 1-15.

[8] Biegański, W.: 1907, 'Critical Remarks on the Classification of Diseases' ['Uwagi krytyczne nad klasyfikacja chorob'], *Krytyka Lekarska* **11**, 67-76.

[9] Biegański, W.: 1908, *The Logic of Medicine or The Critique of Medical Knowledge [Logika Medycyny czyli Krytyka Poznania Lekarskiego]*, E. Wende, Warsaw.

[10] Biegański, W.: 1909, *Medizinische Logik - Kritik der ärztlichen Erkenntnis* (transl. A. Fabian), Curt Kabitzsch, Würtzburg.

[11] Biegański, W.: *Encyklopedia Powszechna PWN*, vol. I, p. 779.

[12] Biegański, W.: 1910, *Tractatus on Knowledge and Truth [Traktat o poznaniu i prawdzie]*, E. Wende, Warsaw.

[13] Biegański, W.: 1907, 'Analogy and its Meaning in Scientific Research' ['Analogia i jej znaczenie w badaniu naukowym'], *Przeglad Filozoficzny* **10**, 481-488.

[14] Biernacki, E.: 1900, *Chałubiński and the Present Goals of Medical Science*, Wydawnictwo Przegladu Filozoficznego, Warsaw.

[15] Bigelow, J.: 1858, *Brief Expositions of Rational Medicine*, Phillips, Samson and Company, Boston.

[16] Bilikiewicz, T.: 1957, 'Biegański as Philosopher' ['Biegański jako filozof'], *Archiwum Historii Medycyny* **20**, 347-355.

[17] Bilikiewicz, T.: 1929, 'On the German Translation of Biegański's *Logic of Medicine*' ['O niemieckim przekładzie *Logiki medycyny* Wl. Biegańskiego'], *Archiwum Historji i Filozofji Medycyny* **9**, 30-39.

[18] Bilikiewicz, T.: 1929, 'The Relationships Between Władysław Biegański and Prof. Max Neuburger' ['Stosunek Władysława Biegańskiego do prof. Maksa Neuburgera'], *Archiwum Historji i Filozofji Medycyny* **9**, 181-193.

[19] Bilikiewicz, T.: 'The Philosophy of Biegański' ['Poglady filozoficzne Biegańskiego'], in G. Świderski and M. Stański (eds.), *Władysław Bieganski, Physician and Philosopher [Władysław Biegański - lekarz i filozof]*, Poznańskie Towarzystwo Przyjaciół Nauk, Poznań, pp. 127-146.

[20] Bilikiewicz, T.: 1936, 'Władysław Biegański', *Polish Biographical Dictionary*, vol. 2, pp. 29-30.

[21] Borkowski, E.: 1971, 'The Medical Environment in Which Biegański Worked' ['Srodowisko lekarskie w ktorym działał Wladyslaw Biegański'], in G. Swiderski and M. Stanski (eds.), *Władysław Biegański, Physician and Philosopher [Władysław Biegański - lekarz i filozof]*, Poznańskie Towarzystwo Przyjaciół Nauk, Poznań, pp. 83-102.

[22] Cabanis, P.: 1956 (1797), *Du degré de certitude de la médecine*, in *Oeuvres*, Paris, vol. I .

[23] Forbes, J.: 1846, 'Young Physician', *British and Foreign Review* **21**, 262.

[24] Gawrychowski, S.: 1980, *Władysław Biegański - A Physician and a Philosopher [Władysław Biegański - lekarz i filozof]*, Nauka Dla Wszyskich, Krakow.

[25] Gawrychowski, S.: 1981, *The Genesis of the Scientific Achievements of Władysław Biegański [Badania nad genezą naukowych osiągnieć Władysława Biegańskiego]*, Towarzystwo Lekarskie Częstochowskie, Częstochowa.

[26] Krakowiecka, L.: 1971, 'Władysław Biegański as a Physician' ['Władysław Biegański jako lekarz'], in G. Świderski and M. Stanski (eds.),*Władysław Biegański, Physician and Philosopher [Władysław Biegański- lekarz i filozof]*, Poznańskie Towarzystwo Przyjaciół Nauk, Poznań, pp. 103-116.

[27] Kramsztyk, Z.: 1899, 'Władysław Biegański: The Logic of Medicine or the Principles of General Methodology of the Medical Sciences' ['Władysław Biegański. Logika medycyny czyli Zasady ogólnej metodologii nauk lekarskich'], in Z. Kramsztyk, *Critical notes on Medical Subjects [Szkice krytyczne z zakresu medycyny]*, E. Wende, Warsaw, pp. 329-339.

[28] Kramsztyk, Z.: 1899, 'Medical statistics'['Statystyka lekarska'], in *Critical Studies on Medical Subjects [Szkice krytyczne z zakresu medycyny]* Warsaw, pp. 100-109.

[29] Lenoir, T.: 1982, *The Strategy of Life: Teleology and Mechanics in Nineteenth-Century German Biology*, The University of Chicago Press, Chicago.

[30] Lindenmann, J.: 1981, 'Immunology in the 1880's: Two Early Theories', in *The Immune System*, Karger, Basel, pp. 413-422.

[31] Oesterlen, F.: 1855 (1852), *Medical Logic* (transl. G. Whitley), The Sydenham Society Print, London.

[32] Skarzyński, B.: 1957 (1899), 'Preface', in W. Bieganski, *Thoughts and Aphorisms on Medical Ethics ['Mysli i aforyzmy o etyce lekarskiej']*, Panstwowy Zakład Wydawnictw Lekarskich, Warsaw, pp. 3-16.

[33] Stański, M.: 1968, 'Władysław Biegański as a Precursor of Social Medicine' ['Władysław Biegański jako prekursor medycyny społecznej'], *Kwartalnik Historii Nauki i Techniki* **13**, 369-377.

[34] Staum, M.S.: 1980, *Cabanis: Enlightenment and Medical Philosophy in the French Revolution*, Princeton University Press, Princeton.

[35] Stocki, E. and Jasinski, W.J.: 1971, 'The Foreign and Polish Critics of Biegański's *Medical Logic'* ['*Logika medycyny* Władysława Biegańskiego w recencjach zagranicznych i krajowych'], in G. Świderski and M. Stanski (eds.), *Władysław*

Bieganski, Physician and Philosopher (Wladyslaw Biegański Lekarz i Filozof), Poznańskie Towarzystwo Przyjaciół Nauk, Poznań, pp. 235-254.

[36] Ziemski, S.: 1971, 'Wladysław Biegański's Methodology' ['Władysław Biegański jako metodolog'], in G. Świderski and M. Stański (eds.), *Władysław Biegański, Physician and Philosopher*, Poznańskie Towarzystwo Przyjaciół Nauk, Poznań, pp. 179-198.

CHAPTER IV.A

TEXTS OF WLADYSLAW BIEGANSKI
GENERAL PROBLEMS OF THE THEORY OF MEDICAL SCIENCES (1897)
The Concept of Therapy*

In the previous chapters we have analyzed the origins and the development of disease. We have seen that from a biological point of view, disease is not foreign to life. It is a specific manifestation of life, necessary when the cell is facing very strong stimuli. We have said that a function can be defined as a reaction of a living cell to small and medium stimuli, whereas disease, or a modified function, is the reaction of the same live cell to strong and excessive stimuli.

There is no clear-cut borderline between physiological functions and pathological disturbance of functions. There is a difference between a physiological function and a qualitative change in that function, i.e., a strictly pathological function. However, between these extremes, we encounter quantitative changes of functions which provide a transition from physiology to pathology. The entire adaptation of life to stimuli originating in the environment provides a long chain of modifications, the initial link of which is a minimal reaction at the threshold of a stimulus, and the final link is the death of the cell.

We have tried to justify all that in the previous chapters. Now, we will proceed to consider the principles of therapy from the point of view of these general biological principles. Illness is undoubtedly an indication of life, but it is unfavorable for the organism, and consequently the practical task of medicine is to eliminate illness and to control unpropitious manifestations of life.

Therapy was the first of the medical sciences, and all the numerous branches of medicine sprouted from the main trunk of therapy. The need for

---

* pp. 291-300; 303-304.

treating the ill was followed by the development of anatomy, physiology, pathology, pharmacology, pathological anatomy, etc. All the light rays emanating from these sciences converge again in one focus, i.e., therapy. The principles of therapy always result from general views on the essence of illness and its causes. These general causes provide a source and also the final aim of all medical sciences. "La médecine est l'art de guérir, elle n'est que cela", said the excellent French clinician Trousseau. A similar view of the medical sciences has been propagated by the Pole Chałubiński. Today we may observe a certain scorn of the practical tasks of our science. The entire center of gravity of medicine has been transferred to pathology, while the theory of diseases and therapy has been assigned a secondary significance.

My opinion is that one should not adopt such an extreme point of view. Neither should the theoretical search for the essence of pathological disturbances be neglected, nor should the main practical aim of our science be overlooked. Therapy should be elevated to the rank of a strictly rational, deductive science, based on the science of logic and on all the results obtained in the theoretical medical sciences. The principles of the so-defined therapy have been masterfully laid down by Chałubiński in his excellent work: *The Method of Finding Therapeutic Indications.*

Let us now return to our subject. Disease is understood as a reaction of life to excessive stimuli. Thus, every disease has two aspects, i.e., excessive stimuli that have their origin in the environment on the one hand, and the action of human beings upon these stimuli, on the other hand.

If we wish to protect the organism from disease, we may act upon each of these aspects separately. We may either attempt to eliminate pathogenic stimuli as far as possible, or to act on the vitality of the organism and on its resistance. In the first case, we may remove the harmful effect of thermal, mechanical, chemical and contagious stimuli by proper dress, adequate accomodations, regular work, avoidance of improper food, disinfection and other means. In the second case, the resistance of the organism may be increased by skillful gymnastics, by rendering the organism immune to harmful effects by preventive inoculation, etc. In fact, we should act at the same time on both aspects, since the elimination of excessive stimuli alone may turn out to be harmful to the organism, because an organism accustomed to weak stimuli only loses the efficiency of its functions and consequently the equilibrium and the function of the organism may be disturbed when facing a stronger stimulus. The task of hygiene and health care is to indicate and justify in detail the procedures aimed at preventing diseases. We present here only the general principle of prevention, which is so self-evident that it does not need to be discussed in detail.

If we have to deal with an advanced disease, what then are our therapeutic tasks? We have to deal with the same two aspects of disease, namely: pathogenic stimuli and the living organism. Here also we should direct our endeavors to both these aspects. We should, as far as possible, eliminate the stimulus which causes disease, and then expose the organism to the most favorable conditions in order to anulate the modifications caused by the stimulus. The first condition is perhaps the most important one and, consequently, the so-called causal treatment is viewed as the ideal therapy. We act against the cause of disease when, e.g., we mechanically remove a foreign body present in the organism, when, during a deep suppuration, we provide the possibility for the pus to leave the body through a wide cut or when, in the case of food intoxication, we administer a laxative agent.

Wherever the cause of disease cannot be eliminated we may try to weaken or to neutralize it. Sometimes such treatment is successful. We hope to succeed in neutralizing contagious toxins by the injection of an anti-toxic serum, to weaken the action of malaria by administering quinine, the cause of syphilis by administering mercury, etc. But in the overwhelming majority of cases, we are not able to reach the cause of the disease - for either we do not know what the cause is and do not know how to look for it; or knowing the cause we do not know how to counteract it; or finally, because the actual cause or stimulus has already ceased to act and we are seeing its result only. Causal treatment cannot be applied in all the above-mentioned cases.

What should be done then? Nothing else but the focusing of all our attention on the diseased organism and its functions. In the majority of cases, disease is not a local event, but is the disturbance of one function which causes a number of disturbances of other, closely related functions. These intermediate disturbances of functions are sometimes more harmful to the organism than the original, pathological cause. For example, in pneumonia impaired blood circulation enhances the activity of the right ventricle of the heart. This functional disturbance plays an enormous role in the course and elimination of pneumonia, and the majority of fatal cases of pneumonia result from the exhaustion of the activity of the heart and its paralysis. The treatment of this secondary functional disturbance may thus turn out to be very effective.

Let us take another example, e.g., a gastric peptic ulcer. The true cause of this disease is not known to us, and in consequence causal treatment cannot be applied here. One should remember, however, that healing of every ulcer is made difficult by a continuous irritation of the peptic ulcer by the acid gastric juice. For treatment purposes we may therefore attempt to temporarily eliminate the gastric function by restraining from eating. We therefore eliminate the continuous inflow of the irritating harmful stimuli. In this case, we do not act upon a modified function, but upon a normal function, since a modification of normal function turns out to be favorable to the treatment. In the case of nephritis, we also cannot directly act on the cause of the disease. Here the goal of our treatment is to remove all the stimuli that irritate the kidney and, in case of insufficient secretion of water and of decomposition of products by the kidneys, to induce other secretory functions of the body, such as the secretory functions of the skin. When we proceed in this way we do not, however, treat nephritis, but remove the direct threat of this disease and stimulate the repairing process.

Let us take another example: a valvular defect. Here, we cannot eliminate the modifications in the valves that obstruct blood circulation, but our task is mainly to help to develop a balanced circulation of the blood by favoring the hypertrophy of the respective sections of the heart muscle. Such a hypertrophy may develop as a result of an intensified function. We may act in the same direction to stimulate heart action by proper amount of nutrition. In all the cases mentioned above we do not act on the cause of the disease, but on the various functions of the body. By stimulation or inhibition of these functions, we aim at either eliminating the threat resulting in a disturbance of a function that is important to the body, or at inhibiting the stimuli which may increase some already present disturbance of function, or, finally, at developing the most favorable adaptation of a given function to a persisting disturbance. These are the tasks of therapy wherever, for any reason, we cannot directly act on the cause of disease.

Undoubtedly, causal treatment is an ideal therapy; all our scientific research aims at it, and if medicine could ever attain that ideal, the task of therapy would be very much simplified. Today, however, things being as they are and as they will be for many years to come, we must follow secondary means in order to approach as nearly as possible our goal, i. e., the cure of disease. In order that such medical procedures not be a groping in the dark but a conscious approach to a strictly defined aim, we should first perfectly understand, in every case, what is to be fought and why and what can be achieved by our therapeutic means. Hence the most important basis for any therapy must be a skillful search for therapeutic indications. These indications depend not only on the kind of disease, but also on individual indications and, even more important, on the particular condition of a given patient. No routine approach nor any established practice may be pursued; each therapeutic indication should be the result of a specific rational deduction.

One cannot use the same therapeutic means in every case, nor even in every stage of a given disease. A treatment which may be very favorable in one case may turn out to be harmful in another; the same therapeutic means that, for example, was indispensable for a certain patient on a given day can turn out to be superfluous and even harmful to another patient.

The opinion that medicine cannot be learned from books is undoubtedly true for therapy. If one wishes to become a good therapist one should know how to think, to get accustomed to becoming aware of the effects of every therapeutic means, and even of every dose of such a means.

Only the physician who, in the presence of the patient, asks himself the question -- What should and what may be done in a given case? -- is, I claim, a good therapist.

We have purposely discussed at some length the question of indications and the formal side of therapy.

One need not be a clever observer to see that precisely this aspect is deficient in our medicine. The overwhelming majority of physicians treat diseases but not patients, and consequently apply treatment in a routine way. The physician who proceeds by routine is like a soldier who strikes all around with his eyes closed; such a soldier fights against the enemy by striking his friends, too. Therapy taught in medical books is only a set of rules, sometimes even a description of routines, according to which diseases are to be treated. It is not, however, what it should be. Chalubinski was absolutely right in recommending to his students to read handbooks of general pathology and to avoid handbooks of specific pathology and therapy, since the latter are too detailed and thus may do more harm than good to a beginning physician. Chalubinski's scientific knowledge, as some of his critics claim, was perhaps not up-to-date, but he was undoubtedly a clever physician who understood perfectly the tasks of practical medicine.

We have actually deviated from our main subject, i.e., the theory of therapy. We have said that therapy should first be directed against the cause of disease. When the cause is not known or cannot be reached, one should act on the whole organism, i.e., change those functions and stimuli that may turn out to be harmful to the body during a given disease. In selecting our therapeutic procedure we pursue the idea that we must always eliminate the most important and most harmful moments of disease[1] which are susceptible to our treatment.

Now a question arises: What are our medical means and treatments? We can modify the functions of an organism only by stimuli; there is not, and cannot be any other method of consistently modifying a function; all our therapeutic means are such stimuli. We employ various stimuli in therapy, such as chemical stimuli (drugs), mechanical stimuli (massage, physical excercise, certain surgical treatments), thermal stimuli (hydrotherapy, balneotherapy) and finally, electrical stimuli (electrotherapy). By the application of these stimuli we can stimulate or inhibit specific functions. Some of these stimuli have an excitatory effect while others have an inhibitive effect. The inhibitive agents are, e.g., narcotics, hypnotic drugs, bromine derivatives, gentle massage, lukewarm bath, and the positive pole of direct current. Stimulative agents are, e.g., alcohol, caffeine, intensive massage, physical exercise, cold baths, and interrupted electric current.

The effect of these stimuli very often depends on their dose. There are stimuli which in small doses are stimulatory while they are inhibitory when used in large doses. Others conversely are inhibitory in small doses and stimulatory in large doses. These stimuli can act either on all the functions of the organism or, more specifically, only on some selected functions. Thus apomorphine acts only on the vomitory motion of the stomach, fox-glove on the heart, and argot on smooth muscle fibres. These are, however, agents acting simultaneously on several functions. Atropine acts on mucous and sudoriperous glands, on the motion of the pupil and on the heart; pilocarpine acts on the same functions, but in a different direction, etc.

We cannot discuss here the details of the action of drugs, because we only wanted to cite some examples to support our opinion that all medical agents provide stimuli which qualitatively modify the functions of the organism. Our whole medical action involves this qualitative and controlled modification in the functions of the organism.

The direction of that action will depend on the therapeutic goal defined by us for a given patient.

An enormous role in understanding the tasks of therapy will be played by understanding the nature of a disease. The idea that the disease represents a struggle of the organism against pathogenic stimuli is as old as Hippocratic medicine. The frequently encountered phenomena of self-healing of the organism were interpreted in the light of his idea: the active victory of the organism. It was assumed that the organism is able to overcome the disease and to free itself from it thanks to the presence of an inherent force that was called *vis medicatrix naturae* by ancient medicine. That *vis medicatrix* represented a form of life force, and like the latter it also constituted some indefinite, almost mystical element.

In the past century, science has rejected the concept of life force. It has retained, however, the concept of the healing force of the organism. This remarkable fact testifies to the recent neglect of therapy. The concept that nature heals disease is so well established today that it has become almost an axiom of our medical art. This concept is, however, erroneous in principle and it contradicts our contemporary biological knowledge. If a disease is a disturbance in a function induced by the action of a pathogenic stimulus, it is senseless to speak here of any kind of struggle, just as it is senseless to express in this way our understanding of a physiological function. Nobody will claim that digestion is the struggle of the organism against food; nobody should therefore claim that an inflammation represents such a struggle. Both physiological and pathological functions represent various reactions of a living cell to the action of a given stimulus. The only difference between a physiological and a pathological function is the force of this stimulus.

A disease is therefore not a struggle, but a reaction of the organism to stimuli, and the difference between struggle and reaction is enormous. The concept of struggle is one of independent purposeful action, whereas reaction is only subjected to physical and chemical laws. Chemical reactions are well-known. It would, however, be a great *licentia poetica* to start speaking of struggle in chemistry. We would not object to the word 'struggle' itself, were it not given too literary a meaning.

The metaphor of struggle has long been introduced in biology. The struggle for existence in the teachings of Darwin, however, does not at all represent a struggle in the common sense of the word, but is a metaphor which expresses the fact that under similar conditions a stronger and more efficient individual will have more chance of surviving than a weaker one.

We would not object to a metaphorical sense of the word struggle in the understanding of disease. This is, however, not the case. Disease is usually viewed as a real struggle and event, as a regular battle waged by an army of phagocytes against bacteria. Such an idea is decidedly erroneous.

We repeat once again that disease is only the reaction of a cell to an excessive or unusual stimulus. Now the question arises whether that reaction is purposeful or not, and whether a change in the function of a cell involves the purpose of counteracting that stimulus. We know that physiological functions of the organism are purposeful measures; do the pathological functions also represent similarly purposeful measures?

The physiological functions are the result of division of labor, itself a consequence of phylogenetic development. The functions of life are mutually and purposefully interrelated as a result of the differentiation of the cells of the organism. If the physiological functions operate within correct limits, their purpose remains well fulfilled. In the case of a disturbance of a function, the physiological purpose is hampered. However, if a function does not fulfill its physiological purpose, it may fulfill another purpose and the entire disturbance may only aim at eliminating a pathological stimulus. This is the opinion of many pathologists who view pathogenic symptoms such as inflammation and high temperature as purposeful functions aiming at separating pathogenic causes from the organism. The question arises, however, whether we have any irrefutable evidence that the organism is capable by itself of eliminating the cause of disease. We know, for example, that a sick organism left to itself often recovers. Did self-healing occur thanks to an active participation of the organism, or did the organism play a passive role and the recovery occur because of the disappearance of the stimulus? This question is not easy to answer. In the case of inflammation we see that the organism plays an active role and removes the dead tissue. This active participation, however, is limited to the elimination of the effects of disease and dead tissue, and is followed by the regeneration of the latter.

There is however no organism able to eliminate a cause that is an active stimulus of a given disease. The stimulus usually belongs to the environment but not to the organism and the latter cannot affect it directly. We should make a clear distinction between a pathogenic stimulus and the effects of the action of this stimulus on the body. The body can actively affect only the consequences of a stimulus, but it does not exert a distinctive effect on the pathogenic stimulus itself. And, since the latter is the cause of disease and since the disease will continue as long as the stimulus continues to exert its effects, we cannot say that the organism heals diseases. The organism can only eliminate the effects of disease, but not the disease itself, i.e., the pathogenic stimulus.

What follows from this? If the organism does not heal itself, but diseases appear and disappear as a direct consequence of pathogenic stimuli, a physician is not an agent who helps nature, but takes, or should take an active part in therapy by eliminating pathogenic stimuli and thereby healing diseases.

When the cause of the disease cannot be reached, the physician does not help nature, but by changing at will the functions of the organism he either indirectly affects the cause of the disease itself, or he modifies the reactions of the organism in order to obtain a recovery.

NOTE

1 "Moment of disease", a term first used by Chałubiński, means everything resulting from disease and adversely affecting the organism. Strictly speaking, it is a logical term covering both functions and stimuli. See Chałubiński: *The Method of Finding Therapeutic Indications.*

WLADYSLAW BIEGANSKI
THOUGHTS AND APHORISMS ON MEDICAL ETHICS (1899)*

Medicine has for a long time distinguished medical science from the art of healing. The first is subjected to general laws of progress in science, and its modifications are related to the general trends prevailing in the field of human knowledge. The second, i.e., the art of healing, is divided into a practical part and an ethical part. The practical part of the art of healing is also related to scientific achievements, and its progress is closely dependent on the development of the medical sciences. The ethical part bears no direct relation to science and depends more on emotions than on reason, and its modifications are closely related to the evolution of the general ethical principles of the human race.

The beginnings of medical ethics reach back to ancient times. From the moment a certain number of people specialized in the science of healing, laws must have been laid down binding doctors among themselves and defining their relationships with persons who seek their assistance. Social life has its own needs, and these social needs lay down laws which at first are customary but are later put down as concise rules and constitute written laws. In the field of medical ethics the first written law was the oath of Hippocrates. This oath for the first time clearly expressed the idea that the medical profession is a specific corporation whose members should be bound by an appropriate and special education, and that an oath should be sworn by everyone joining the medical profession.

The Hippocratic oath also includes a certain number of regulations of the physician's conduct with patients. It mentions that a physician should not be shy but firm, and should not be conceited but rather aware of his knowledge. The Hippocratic oath can be considered a system of medical ethics laid down as clearly and concisely as everything else written by that author.

I am not attempting to write a history of medical ethics. I only wish to stress the fact that all the prominent representatives of our science have never

---

* pp. 25-32; 133-139.

forgotten medical ethics. If we neglect the Middle Ages, we may notice in the more recent times the example of Ambroise Paré who even dedicated one work, Canons, to medical ethics. In this work, written in rhymed couplets, he pronounced many excellent thoughts on this subject. Next, we notice Sydenham who, although he did not write a specific treatise on medical ethics, in all his works continuously exhorted doctors to ethical conduct. Sydenham was the author of that sublime thought that the physician should be "*clementiae divinae minister*", "*medicus, non autem formularum praescriptor*". We also encounter the beautiful personality of Sylvius who masterfully combined an excellent teaching and enormous practice with a humane and cordial attitude towards his patients. In short, as rightly pointed out by Haeser, the history of our science reveals that in the past great physicians have always been great philanthropists.

Only in the 19th-century, or at least in its second half, can one observe a distinct exception to this general rule. We have had, and still have, many prominent physicians, surgeons, and specialists who have left, and still leave for us excellent works, but only a few among them have also left behind them the memory of a great philanthropist.

I do not claim that at present there are no philanthropists among the doctors. However, they should be looked for not among the famous doctors and outstanding university professors, but among the multitude of practitioners of no importance to the sciences. Such a philanthropist often dedicates his whole life to the well-being of his fellow-men and in the end dies unsung, without even an obituary in a medical journal, because he has made no contribution to medical science.

An impartial investigator will notice that the relations between medicine and ethics which have developed at the present time are a logical result of the entire trend of modern medicine. Only in the 19th-century was medicine proud to adopt the name of experimental science. Scientific medicine concerned itself only with the scientific goals of our art. It affirmed that only a scientific knowledge, laboriously derived from experiments, can restore health to the ill, which is the only thing for which the latter ask. Wherever the physician cannot restore health to the ill, he is at least able to tell him his diagnosis and to predict his future fate. However, today, sentiments and mawkishness do not enter within the scope of medical science, and, moreover, they are said to be harmful to a physician by robbing him of the impartiality and the detached attitude which are necessary for the identification of diseases and for looking for the means to fight them. The fate of an individual man no longer means anything when compared with the goals of science. If as a result of bold surgery unlucky patients die by the hundreds, no harm has been done if these failures can improve surgical technique. If in inoculating syphilis one injects a serious disease in healthy people, no harm has been done if in this manner the question of the contagiousness of syphilis is settled once and for all.

This is how fanaticism was created in science, which subordinated human life to scientific objectives. This fanaticism has been openly propagated by the most excellent universities and laboratories. And all the visitors and listeners have become imbibed with, if not scientific fanaticism, then contempt for the feelings and the fate of the individual. A fanatical lawyer might be ready to cry out that the whole world might perish so long as justice be done; the fanatical adherents of scientific medicine likewise cried out "*pereat mundus fiat... scientia medicinae*".

We have to admit that these extreme and ruthless views in medicine resulted from the general intellectual climate of the 19th-century. This was the century of the rise of the natural sciences and of inventions, and its main slogan was "science is power". In the 19th-century, the cult of science invaded even spheres which, by their very nature, were entirely foreign to the spirit of science, such as religion and morals. Let us remember here Auguste Comte's dreams of positive religion and of churches in which the cult of science would be practiced, and in which the statues and pictures of the saints would be replaced by those of Newton, Bacon, Galileo, Aristotle, and others. Let us also remember Strauss' dreams of a new religion, and the so-called "free communes" in Germany in which the cult of science was pursued in a ridiculous and pretentious manner. In the 19th-century, science made claims of reigning over the entire mental life of the human being. It was considered to be a source not only of material success but also of happiness and morality, and was expected to eliminate human destitution and all offences.

These noble but fanatical dreams were pursued by the intelligentsia during a great part of the 19th-century. Only at the very end of that century did the idea dawn that these dreams would never become true. The new inventions and the fabulous mastering of the forces of nature had brought about a deep division among people; on the one hand, unbounded riches were concentrated in the hand of few, and on the other, there was a growing number of homeless proletariat. Poverty had not disappeared but rather increased more and more rapidly, threatening the future with an awful spectre. The spread of education had not moralized the masses, but on the contrary demoralized them. The inevitable struggle for existence which cannot be halted, the pursuit of wealth, the cult of force that stemmed directly from the cult of science, and the disappearance of all noble ideas, all had contributed to the expansion of selfish impulses, which were hampered only by the fear of sanctions.

The happiness that science was to bring to the human race had disappeared somewhere. What remained was exorbitant expectations, and the pursuit of increasingly greater comfort, both brought about by technical progress. The unhampered greed for pleasure became in fact a heavy burden on the lives of people, disturbed them, and gave them no satisfaction, without which a state of happiness cannot be reached.

For this reason, a reaction to the previous trend began to set in throughout the world. An increasingly greater number of persons became convinced that science could not make the hope that had been pinned on it come true. The conviction that the human soul consists not of reason alone but also of feelings which are not directly affected by thought, was increasingly brought to the fore.

As a consequence, a distinct turning point could be seen in all fields. In science there appeared a generalizing philosophical trend which aimed at replacing pure empiricism. While in philosophy a distinct turn towards idealism was made, in the arts there started a search for new approaches which took the form of the impressionist and symbolist movements. The hesitations in the religious area were reflected by an undeniable increase of the force of Christianity on the one hand, and by the pursuit of something new and unattainable, expressed as interest in buddhism, spiritism, etc., on the other hand. In economics the principles of the welfare state increasingly overcame those of free competition. Finally, ethics, which had been almost entirely forgotten, again focused general interest. Whereas in the middle of our century the theory of knowledge was viewed as the most important among all the philosophical problems, today ethics is viewed as the most crucial philosophical question. It is unhesitatingly, although sometimes queerly, discussed by poets - e.g., the Scandinavian school and Tolstoy - and it also has become the favorite problem of the majority of philosophers. In short, whatever the field of human thought we take up, everywhere we see change of trends.

These modified trends have not all taken a uniform shape. The notions have not crystallized yet, no common ideal has been found, and no distinct goal has been determined to be aimed at. There is consensus on one point only: we cannot pursue the previous direction. The human race is now like a cart which for many years has been driven along the road to perfection, but which has suddenly been stopped and turned away by the drivers with all their might, because they have noticed an abyss on the way, and now, slowed down, they wander freely until a new direction can be found.

Such changes have occurred, and are still occurring, in the general direction of human thought. Medicine, too, cannot continue its previous trend, but must change its direction. I shall not discuss in detail the possible changes in the scientific orientation of medicine, because I have already mentioned them elsewhere. Here I shall draw your attention only to those changes which must occur in the practical direction of medicine and in the ethical tasks of our art. The characteristic feature of practical life is that it lags behind theory. Theory, as we have already said, has modified its direction, but practical life continues through inertia to advance along the previous road. Some time will probably elapse before a turn will occur here too. Such a turn must, however, occur. The question now arises what is the direction in which it will proceed. In our previous works we have already stated that the center of gravity of medical science is shifting from loose facts to generalizations, from organ to function, and from anatomical changes to disturbances of functions. It is becoming increasingly probable that a disease will be viewed in the future not as a change occurring in organs, but as a sick, functionally modified life. Thus, the individual patient who almost completely disappeared in the previous anatomical approach to disease, must regain his rights. The physician should see not the diseased organ alone, but the diseased organism, the sick person. Undoubtedly, such a shift in the scientific point of view would affect the moral principles of our science.

First of all, the now commonly adopted principle that scientific knowledge is all that counts in the art of healing, would be shaken. As a matter of fact, we have a very meagre stock of absolutely certain therapeutic facts. Therefore, success in healing does not always dependpon the scientific indications of therapy but, in the overwhelming majority of cases, on the psychic effect a physician exerts on his patient. The recognition of this fact must become increasingly obvious and must exert a great influence on our practical behavior.

The physician is able to exert his psychic influence and to heal through suggestion only if a bond of sympathy exists between him and his patient. In the future, therefore, empathy with the sick will regain the importance that it had at the very beginning of our science, and that it lost only as a result of the dogmatic turn to scientific medicine.

The division of labor in modern society has led to the formation of a number of professions, each of which meets a specific need and satisfies a particular aim. Different aims entail different obligations, or at least arrange these obligations in a different way. Discipline and obedience to one's superiors are a cardinal duty of the military profession, but it is not indispensable in other professions. A priest is obliged to tell the truth to the penitent, but if a physician did the same to the patient it would not be so praiseworthy. A merchant is bound to behave with particular courtesy and show ultimate indulgence to the whims of the buyer, but an army officer behaving in the same manner toward the soldier would be totally out of place. Thus, what is applicable to one profession aiming to achieve a given goal is not indispensable, and may be superfluous or even harmful for reaching a different goal. For this reason one should consider in detail the goals and obligations related to particular professions, i.e., in short, one should formulate a specific professional ethics.

As already mentioned, the most important problem facing every system of ethics is the designation of its main goal, since the specific obligations derive from the final goal. Thus, in discussing professional ethics we must consider the question of what is the principal goal of the medical profession.

The answer to this question is not difficult, since the self-evident and obvious idea, the first which comes to mind, is that the aim of the medical profession is to heal the sick. It is clear that if therapy is recognized as the highest aim of professional activities, this is not to say that the goal is this or that method of treatment. Abstracting from all particular therapeutic means and therapeutic methods, we can define this goal, broadly, as assistance to patients. Consequently, we may define that aim more propertly as assisting sick people. Such is undoubtedly the final aim of our profession. I need not, I dare say, prove its truth. But I must add here that the highest aim is always represented in its perfect and ideal form, meaning not just any kind of assistance to ill people, but the most perfect and the best assistance possible. Aim always expresses desire, and the latter is closely related to feeling and liking. We thus desire and aim at what we like the best and what is perfect. Having established that the highest aim of the medical profession is the ideal of perfect assistance to sick people, we can easily lay down the general principles of the ethical aspects of medical practice, and consequently the professional obligations of the physician.

Since, as we have already mentioned, we define as right all that leads to the intended aim, and wrong all that deviates from that aim, the professional act of a physician that aims at, and as far as possible attains the ideal of perfect assistance to the sick, is defined by us as a right one, and that which contradicts that ideal is defined as a wrong one.

This principle is the only one which should be decisive in the evaluation of professional activities. All other considerations should be subordinated to and be in agreement with it. We must add here that in ethical appraisal one should evaluate not only acts, but also their intentions. The consequence of a given act can be positive and can lead to the highest aim indicated by us; if, however, this act stems only from the pursuit of purely personal interests, it will not be considered by us as moral or ethical. If, for example, a physician sent to attend a very rich patient takes care of him with the utmost devotion, solely in the hope of obtaining a high fee, such a deed cannot be called right in the ethical sense. It will be either ethically neutral, if the physician has enough free time, or ethically wrong, if he neglects other not-so-rich patients in favor of the rich one.

It follows from the above that ethical appraisal is very difficult, and one should be cautious in pronouncing it, since we cannot always penetrate the secret intricacies of the soul of another person, and know the motives that were driving it.

When giving an opinion on professional conduct we should also employ the highest ethical principle that, when applied to our profession, can be stated: treat your patient as you wish to be treated yourself. If our conduct conforms to that principle, we may call it right, but if it does not we should call it wrong.

All professional duties can be deduced from the ideal of the highest goal. As mentioned above, these duties are requirements and a requirement is nothing else than a condition necessary to reach a given goal, i.e., a condition without which that goal cannot be reached. Duties may be characterized as primary truths, and hence a duty is beyond logical argument. If one acknowledges, however, the legitimacy of the final goal, one should also acknowledge the legitimacy of its requirements. Thus, if we consider that the highest goal of the medical profession is to give perfect assistance to the sick, we also must acknowledge the importance of all the conditions which make that goal attainable. These conditions are precisely the professional duties.

These duties were determined in the following way:

The highest aim is to help the patient without any reservation. Since disease is a misfortune which can affect anybody, the first duty of a physician is to assist in the same perfect way all human beings without any difference and without any regard to whether the patient is rich or poor, whether he is our personal friend or our enemy and, finally, without any regard to his race, nationality, or religion. The only discrimination among patients a physician is authorized to make depends on the degree of the danger of a given disease and the urgency of the need for help.

Next, since the help should be perfect and thus given as early as possible, a physician must rush to every call of a patient, irrespective of the time of day or night and irrespective of the distance. The ideal of perfect help requires, in addition, that the physician should be familiar with up-to-date medical science and should follow new medical tendencies and achievements. The neglect of that obligation not only impairs the personal value of the physician, but also constitutes an offence against his ethical duties.

Here I give only a few examples to explain the relation and the close association between the duties of a physician, taking in consideration his highest aim. I have no intention of enumerating and justifying all the ethical standards of the medical profession, since they are well-known and common. I say known, although they are never fully applied. No physician is always able to fulfill the duties of his profession. This should not surprise us, if we notice that the highest goal of the medical profession is an ideal and its requirements, i.e., the duties, are ideal, too. The essence of an ideal is that it cannot be completely reached and is only an example to follow. Following an ideal is already a praiseworthy deed. The aim of ethical ideals is precisely to develop striving for perfection.

The obligations to fellow physicians, i.e., the interprofessional relations among physicians, can also be deduced from the assumed goal of the medical profession. The highest goal mentioned by us requires of a physician that each time he has any doubt in identification of a disease or has insufficient skill in performing the treatment, he should seek the assistance of other physicians. A physician should never refuse when called for consultation, and the only aim of a consultation should be the well-being of the patient. Having in mind the well-being of the patient and taking into consideration the fact that confidence in the physician's medical skill is a very important factor in therapy, the physician asked for a consultation or called to the patient who has already been treated by another physician should not discredit the latter and should express his views with moderation in case of difference. For that same reason the physician should avoid shocking manifestations of public disagreement and envy among physicians, should refrain from advertisement which humiliates the medical profession, etc. In short, if we consider the highest goal of our profession and if we are able to subordinate all our personal motives to that goal, we shall perhaps be able to reach unanimous professional co-operation.

I should like to clearly stress here that friendly relations among colleagues are not a goal in themselves, but are the requirement of another, higher goal. Nobody would claim that the goal of our profession is to maintain good relations among physicians. On the other hand, it is noteworthy that physicians very frequently identify the idea of professional ethics with that of good relations among colleagues, and when ethics is discussed in our circles it is mainly in the sense of corporative ethics. However, correct and friendly relations between fellow physicians do not testify by any means to the high level of their ethical culture. Let us imagine a situation in which all the physicians of a given town agreed to support one another; the relations among them would then be highly harmonious in spite of the fact that such an agreement, if it were not faithful to the higher ideals of the medical profession, might lead to highly unethical acts. Superfluous and expensive medical consultations might be required. Treatment which actually requires only one physician might be performed with other physicians assisting, in order to provide mutual support and in order to increase the profits of other colleagues. In short, disgusting exploitation can take place under the guise of unanimous actions.

Conversely, bad relations among colleagues do not always testify to a low level of professional ethics. The physicians of a given locality may not know one another, nor maintain social relations, nor even notice one another. If, however, they manage to overcome their personal animosities, participate together in medical consultations when necessary and do not refuse assistance to one another, we cannot deny them a certain ethical culture even if we cannot describe their relations as ideal.

Ethical culture involves being concerned with the common good and subordinating personal motives and interests to a higher, general aim. The exclusive or predominant identification of correct interprofessional relations with professional ethics leads frequently to a false appraisal of the actions of physicians. An act which is allegedly good from the corporate point of view can in fact be considered unethical from the point of view of right professional ethics.

Let us take for example a problem which is frequently raised both in practice and in theory: how a physician should behave when called to a patient who is being treated by another colleague. According to the principles of corporate solidarity, a physician should refuse his assistance to the patient, and should require instead a consultation with the previous physician. Can this approach be reconciled with the principles of proper professional ethics, which views as its highest goal and only authoritative criterion the well-being of the patient? Can we define as right an action which set limits upon the free will of a patient and his family and prevents him from exercising any kind of control over the physician?

"You should continue to be treated by the physician whom you first called, even if you have lost confidence in him and you doubt that you are being well treated." This is the formula of corporate deontology translated into the patient's language. Such a restriction is in fact entirely unjustified. We should accord the patient the right both to select a physician and to verify his actions, and the patient can exercise the latter right only by comparing the opinion of several physicians.

Joint consultations are not always helpful in building up the confidence of the patient. The patient may suspect that they take place under the influence of the first physician in charge of the case and that corporate ethics would not allow the physicians called for consulations to voice independent opinions.

Therefore, even when we disagree with the patient's preconceived ideas, we have no ethical right to impede the patient's will. The desire for good relations among physicians is not a sufficient reason for imposing constraints on the patient.

We have already mentioned that a clash of opinions during a medical consultation should take place without outside witnesses, and that the results of a consultation should be communicated to the patient and his family in a cordial form. These requirements are not intended to maintain friendly relations among colleagues, but are dictated by the higher goal of the well-being of the patient. We do not acquaint him with the difference of opinions which occurred during a consultation because we wish to maintain the patient's confidence and preserve his equanimity of mind. The same higher aim, however, may in some circumstances require a distinct presentation (to the patient and his family) of disagreements that occurred during the consultation. Let us assume that we are called to a consultation on a patient who has been wrongly treated by the consulting physician, i.e., the physician has wrongly identified the disease and has not administered the therapy which is considered as necessary and recommended by medical science. During the consultation, we convince the patient to change the treatment.

What would happen if this colleague were a conceited and obstinate man who did not wish to recognize our arguments? What should we do? If, in order to preserve correct relations among colleagues, we kept secret our true opinion and, in agreement with good interprofessional manners, presented our view to the patient and his family in a gentle, hesitant way in order not to offend the consulting physician, we would behave wrongly and in an unethical way. The highest aim of our profession requires us to pronounce our true conviction openly and bluntly, without any regard to the possibility of breaking good relations with a colleague, since ethics requires that the consideration of good interprofessional relations should be subordinated to the highest goal, i.e., the well-being of the patient.

Supreme lex medicorum - salus aegrati.

From the above-mentioned remarks it follows that solidarity cannot be a goal in itself, and that it puts an obligation on us only if it is not in contradiction with the highest goal of our profession: the well-being of the patient. If the comradeship leads to a deviation from that goal and is based on selfish personal and corporative interests, it loses all its obligatory, ethical significance.

WLADYSLAW BIEGANSKI
THE LOGIC OF MEDICINE OR THE CRITIQUE OF MEDICAL KNOWLEDGE (1908)*

The question arises whether the logic of medicine has all the features of a separate discipline and can be isolated both from other medical disciplines and from philosophical disciplines. Every branch of science becomes an independent discipline if its subject matter is clearly separated from other related subjects and neighboring scientific fields, and if in addition it is sufficiently broad. The subject matter of medical logic is clearly and distinctly differentiated from the subject matter of all the other medical disciplines in that the former deals with the formal side of medical cognition and the latter with its content and essence. While other medical sciences explain *what* we know, logic explains *how* we know it. It could, therefore, be said that medical logic is engaged in the study of methods of medical knowledge, whereas other medical disciplines are engaged in the study of its essence.

The above-mentioned definition is, however, not entirely precise. Every medical discipline specifies not only its results, but also the method of attaining knowledge. If medical logic were engaged only in the study of research methods, it would infringe on the scope of other sciences. This is not so, because medical logic is not concerned with methodological details, but only with general principles. Consequently, medical logic could be called a general methodology of the medical sciences.

But this definition is not sufficient. Medical logic provides not only the general principles of the methods employed in medical cognition, but also estimates their values with respect to the final aim of every cognition, namely with respect to truth. Such an estimation of value is called a critique; hence medical logic is not a descriptive general methodology, but a critical one.

Allowing for this last-mentioned and perhaps most important point under discussion, we also could call medical logic a critique of the methods of medical knowledge or, more briefly, a critique of medical knowledge. The latter

* pp. XIII-XV; 24-29.

knowledge or, more briefly, a critique of medical knowledge. The latter abbreviation is, strictly speaking, even more to the point, since medical logic does not only deal with methods of medical knowledge. Every cognition is based on certain fundamental ideas which provide a foundation for cognition and a basis on which, using appropriate methods, the edifice of science can be erected. Thus, for example, the idea of three-dimensional indefinite space provides such a basis for geometry, the idea of number for arithmetic, that of cause for the natural sciences, that of disease and therapy for medicine, etc.

A critical consideration of the general principles of research methods does not in itself exhaust the entire formal side of cognition, since medical logic must also take into consideration a critical point of view on the general concepts of our science. Hence, the more general phrase "critique of medical knowledge" is more to the point and more appropriate than the more detailed phrase "critique of the methods of medical knowledge". This comment provides sufficient reason for separating the subject matter of medical logic from those of other medical disciplines.

Every specific science dogmatically accepts certain concepts and certain detailed research methods and employs them in constructing the entire field of cognition. Science is not engaged in the analysis of the genesis of its assumptions or its research methods. These research methods are usually viewed as generally accepted practices, and their value is measured by the results obtained by their application. Consequently, if a certain concept or a certain method gives satisfactory results when applied in practice, then the concept or the method is viewed as adequate, true and correct by the scientist. However, the philosophical theory of science, to which medical logic belongs, takes another point of view and subjects those views of the specific sciences to analysis and to criticism. The starting point of a specific science and its end point (scientific concepts are sometimes such end points) are the subject matter of investigations of the theory of science.

Let us now consider the limits of medical logic and its relationship to the general philosophical sciences. We have mentioned that medical logic is in fact a special branch of the theory of science, which investigates the most general assumptions made by all sciences and the methods employed by them. Specific sciences, however, modify those general assumptions and methods, and give them an individual character which depends on their particular research subject. The idea of causality in contemporary physics differs in many respects from that in contemporary biology, and experimental methods employed in the exact sciences are not entirely the same as the experimental methods employed in biological research. Methods and assumptions adapt themselves to the subject matter of a given science and are the source of the differences occurring among them.

There is still another problem to be discussed: the fact that specific sciences may have particular assumptions and methods, unknown to other sciences. For example, the general theory of science takes into consideration exclusively or at least mainly the general properties of assumptions and methods which are common to all sciences, but it is unable to take into consideration all the modifications to which they are subjected in specific sciences. This task belongs to the theories of specific sciences, to which medical logic also belongs. The more the characteristic features of the subject of research or the cognitive aims of a given science or group of specific sciences differs from other sciences, the more indispensable is the need for a separate consideration of this science. This is the case with medicine. The assumptions and the methods of physics, mechanics, astronomy or other similar sciences can be studied even in fine detail, using the general theory of science, but the assumptions and methods of medicine, in particular when combined with medical practice, necessarily require a detailed and individual consideration. In this case a specialization is necessary. We can therefore do without any separate logic of physics, chemistry or similar theoretical sciences, but we cannot evaluate the assumptions and methods of medicine from the general standpoint of the theory of science. Consequently, medical logic becomes an increasingly independent branch of science, and sooner or later it must become an entirely independent discipline.

Medical logic is therefore the philosophical theory of medicine in the general sense of that word. It covers both the medical sciences and medical practice. The philosophical theory of science has for a long time been divided into two parts: the so-called theory of cognition, and logic in the strict sense of the word. The first deals with the general concepts of the sciences, the second with the general methods of cognition. Because the term "logic" has such a restricted meaning, the term "medical logic" adopted by us might seem inappropriate, since our "medical logic" deals not only with the methods but also with the concepts of medical knowledge. However, because the theory of knowledge has for a long time been included in philosophy and in logic, this term is very frequently employed not in its restricted but in its general meaning. Consequently, my opinion is that it would not be erroneous to assign to it a wide meaning in the present work. After all, this term may be accepted because it is very concise; in addition, this use already has a tradition, since the first work by Oesterlen on that subject bears the title *Medical Logic*.

We choose therefore to maintain the name of the first edition in the title of the present work, adding to it its synonym: *Critique of Medical Knowledge*.

Medicine and medical disciplines

Today under the general term 'medicine', we understand a set of a dozen or so separate disciplines which directly or indirectly aim at providing help for the sick. The disciplines that try to reach that goal directly are called practical or, even better, applied medical disciplines. Included among them are pathology, therapy, surgery, obstetrics, clinical pharmacology, etc. The goal of other medical disciplines, called theoretical disciplines, is the study of the conditions of life and disease, and the study of therapeutic means. By such study they contribute indirectly to the science of healing. These theoretical medical disciplines include anatomy, physiology, pathological anatomy, experimental pathology, experimental pharmacology, etc. Such a system of medical disciplines interrelated by a common, though more or less distant goal, is in fact a recent achievement, a consequence of many centuries of the evolution of medicine.

From the Renaissance on, various theoretical medical disciplines were formed, the goal of which was to investigate in an impartial way, devoid of any practical or utilitarian aim, various phenomena found in healthy or ill organisms. Their great expansion has, in all probability, indirectly affected the principles of medical practice. That effect did not, however, manifest itself immediately. Consequently, in the 17th and 18th-centuries the evolution of the theoretical disciplines was accompanied by the development of various pathological systems aiming at the explanation of the methods of medical practice. Those systems had all the features of the older practical science. Only in the middle of the 19th-century, when the progress in the theoretical sciences had become evident, did the pathological systems collapse, and medical practice become transformed from a practical discipline to an applied one, or more properly speaking into a number of applied scientific disciplines.

For us the term "applied scientific discipline"[1] means a discipline which applies the results obtained in the theoretical sciences to certain practical goals. It might be argued that such an application should not be considered as a science and we should not speak of applied sciences, but of the application of theoretical science to practice. This, however, is not true. Practice has its specific problems distinct from those in which theoretical science is engaged. These problems may be solved on the basis of the data supplied by the theoretical sciences, but the search for the solution is the subject of a specific kind of scientific research. Were it not for the needs of practical application, neither such problems nor their solutions would exist. The application of auscultation to diagnosis described by Laënnec was a practical application of the theoretical sciences. It was only a deduction derived from comparison between clinical observation and the anatomo-pathological modifications. However, this application, or this pure deduction, created a number of new questions concerning pathological conditions, and led to the development of the concept of respiratory or bronchial murmur, of consonating rales, of valvular heart murmur, etc. These problems are entirely foreign to a theoretical science, since they appear only when its results are applied to medical practice. By solving them, we introduce a new element into the previously pure deduction and pure application, an element that imparts a character of scientific research to the entire subject matter. In such a way a theoretical science leads to the development of an applied science. Such an applied science of physical diagnostics was developed by Skoda through the introduction of experimental study of auscultory and percussible phenomena into the deductive study of Laënnec.

A detailed examination of the chief applied medical sciences - such as medical diagnostics, surgery and clinical therapy - may show us that they are based on three entirely different principles:

1- the application through deduction of the findings of the theoretical sciences to practice;

2- the detailed description of medical acts - e.g., various surgical treatments, methods of auscultation and percussion, examination of secretions, etc.;
3- the theoretical explanation of results - the pathological conditions leading to the generation of resonance, or to audible heart murmur, the justification of aseptic therapy of wounds, etc.

The above-mentioned practical medical fields are applied sciences in virtue of the first principle; they are "art", by virtue of the second principle; and finally, by virtue of the third principle, they are a theoretical science.

There is a trend in medicine today to attempt to found entire medical practice on the theoretical sciences and thus to transform it into an applied science. But this goal cannot be completely attained. First, because the entire content of these sciences and every practical application cannot be explained by facts and generalizations derived from the theoretical sciences; not everything in practice results from the application of the theoretical sciences. As a consequence, in medical practice there exist empirical truths which cannot be subordinated to general theoretical truths. We do not know why mercury is helpful in syphilis, or iron in chlorosis, why morphine eases pain or why chloroform anesthetizes, etc. All those generalizations were derived from empirical observations and they have remained in applied science, although we do not know how to deduce them from theoretical truths. Hence, our applied sciences are not exactly "applied sciences" in the true and full meaning of the word, because they contain a quite substantial amount of empiricism. This fact has even induced a certain deviation into the history of medicine. At the beginning of the period of scientific medicine, when the theoretical sciences were recognized as being central to medicine, everything that could not be proven by scientific methods was viewed as erroneous or illusory. As a consequence, all empirical knowledge was excluded from practical medicine and what was left was only applied science in the strict sense of the term. This happened during the period of therapeutic nihilism propounded by the Vienna school in the middle of the 19th-century. It was an error which was soon eliminated, when the power of facts overwhelmed scientific dogmatism.

Besides empirical knowledge, the applied medical sciences also contain a considerable element of art, which cannot be entirely reduced to scientific principles. A substantial part of medical practice, the so-called "medical art", involves manual dexterity, training of the eye and ear for observation during examination, etc. The art element would always remain outside the scientific domain. In the old medical science this element of art provided the main basis for medical practice, and although the main part of today's practical performance has scientific foundations, these foundations cannot entirely exclude the art component. Consequently, Magnus [2] and the entire recent scientific trend in medicine were mistaken when they claimed that practical medicine should be based only on theoretical science, and that present-day medicine was not an art but solely a science. This opinion reflects the same dogmatism as that of the Vienna school, which condemned all the achievements of therapy if they did not result from, and were not accountable for, scientific principles.

"It is not true - says Kramsztyk [1] - that medicine, or treatment results directly from knowledge, that is, from the simple sum of scientific information. In order to apply this knowledge, to examine patients using all the methods and instruments available, to take care of patients, to use appropriately medical means, and above all, to use efficiently one's hands as well as tools in so many different functions, a specific skill is required. A physician acquainted with practical work will know how to use each method and how to fulfill his functions in the best way. Such practical skill is by no means less essential for the physician than theoretical knowledge. A student who has not acquired adequate knowledge of anatomy, physiology, and pathology, cannot be a good physician. But the one who knows little about the methods of treatment will not be a good physician either. In addition to proficiency in scientific theories, the physician must train his eye and his hand, but also, albeit it is not often mentioned, he should bend his will and his emotions to the needs of his profession.

We see, therefore, that medicine involves not only elements of science, but also those of art. Nusbaum [3] is right in saying that "medicine is one of the youngest sciences but one of the oldest arts, and consequently is both a science and an art. Once it was mainly an art, and in some periods it was mainly a science, but it should be both an art and a science." And, no doubt, it is so. The renunciation of the term "art" by contemporary scientific medicine constitutes in fact a misunderstanding. In order to eliminate this misunderstanding it is necessary to define precisely what should be considered an art and what a science. Art is performance and science is understanding or comprehension. Science answers the question what should be, and art - how it should be done. Consequently, art cannot exist without a certain amount of theoretical knowledge and, conversely, there is no science without a certain addition of art.

The difference between science and art is essentially not qualitative but quantitative. In some human activities the role of the theoretical element is limited while the method of performance is crucial. This is the case for all kinds of crafts, which can be divided into higher and lower depending on the amount of theory involved.

On the other hand, there are activities in which theoretical explanation occupies the foremost position and the method of performance recedes into the background. Here belong, among other things, the theoretical sciences. However, even the most theoretical and abstract science involves an element of art, for example in its methods of presentation and arrangement of arguments. If science is defined as a system of ideas and opinions, then we must consider the method of systematizing these ideas and opinions as an art. Any lecture in mathematics or any other purely theoretical science is and should be judged as an "art object". We have, therefore, clear, complete and vivid lectures, as well as those that are obscure, heavy, chaotic, lengthy, etc. In other sciences, the element of art plays an even more important role. This is the case, for example, in the experimental sciences, since the performance of any experiment is an art in the full sense of the word. There are even "born artists" among the experimentalists, who are able to realize the art of experimentation to an unheard-of perfection. All the above leads to the conclusion that both art and science are, as Nusbaum has explained, two aspects of the same thing, namely, the action of mankind on phenomena.

These comments on the relationships between art and science prove clearly that medicine cannot be only a science, but is and must be a combination of science and art. Consequently, medical cognition involves two separate tasks, namely, scientific tasks and practical tasks. However, the present doctrine of scientific medicine neglects the practical tasks of medicine and reduces medicine solely to its scientific tasks. The consequence is the appearance of numerous, erroneous trends and inconsistencies that were so well, although perhaps somewhat exaggeratedly, described and characterized by Biernacki.

We shall deal with those problems in subsequent chapters. At the end of this chapter let us mention the impact of the reduction of medicine to medical science. Considered from such a restricted point of view, medical logic is only the logic of medical science and the critique of scientific knowledge, and it is described as such in the works of Oesterlen and Claude Bernard and even in the recent work by Magnus. However, as early as the time of Chalubinski, Polish authors raised and developed in detail the question of the significance of the practical tasks of medicine for the study of medical knowledge.

The present work aims as far as possible at combining both these trends and attempts to present an accurate and complete critique of theoretical, i.e., scientific, and practical medical knowledge.

NOTE

1 Detailed comments on the theoretical and applied sciences can be found in the two first chapters of Biernacki's book, *Principles of Medical Knowledge*.

BIBLIOGRAPHY

[1] Magnus, H.: 1904, *Kritik des medizinischen Erkenntnis*.

[2] Kramsztyk, Z.: 1895, 'Is Medicine an Art or a Science', *Gazeta Lekarska*.

[3] Nusbaum, H.: 1895, 'Is Medicine a Science or an Art?, *Gazeta Lekarska*.

CHAPTER V

ZYGMUNT KRAMSZTYK AND THE CRITICAL EVALUATION OF MEDICAL PRACTICE

Zygmunt Kramsztyk (1848-1920) was born in Warsaw, a son of an orthodox Jewish family. His father, a rabbi, supported better integration of Jews in Polish society, while maintaining their religious and cultural identity. He actively participated in the 1861-1863 insurrection in Poland and was jailed by the Russian government. All his (male) children were active in Warsaw intellectual life, and all of them actively fought for the assimilation of Jews in Polish society. Zygmunt Kramsztyk was not only a physician and philosopher of medicine; he was also a medical publisher, being co-owner and co-editor of *The Medical Journal [Gazeta Lekarska]* of Warsaw, and later editor of *Medical Critique*. He was also active in the foundation of the Warsaw Zoo and of the Warsaw Society of Nature Lovers, and, in addition, he wrote essays on Polish literature [24].

Kramsztyk, who was slightly older than other philosophers of medicine of his generation, studied medicine in Warsaw in the late 1860's and early 1870's, and had the opportunity to attend Chalubinski's lectures. Kramsztyk was considered as Chalubinski's best student, and a disciple of Chalubinski's clinical tradition. Later he specialized in ophthalmology, first in Warsaw, and later in Vienna and Berlin. This was clearly a fortuitous choice. Ophthalmology was one of the few disciplines in which Polish physicians were able to keep pace with the rapidly moving front of medical research. This specialty was relatively less affected by the "scientific revolution" in medicine in the second half of the 19th-century, because ophthalmologic practice (unlike the understanding of the physical basis of vision, which was strongly influenced by the progress of physics and physiology) was primarily based on experience and on manual dexterity. In the mid-19th-century, Wiktor Szokalski (1811-1890), a student of the German ophthalmologist Sichel, created a well-known school of ophthalmology in Poland, and some of the Polish ophthalmologists became famous abroad [3, 4, 27, 28, 29]. Kramsztyk was Szokalski's student and followed his professional tradition.

During his entire life Kramsztyk remained a medical practitioner. He was never tempted by fundamental research or by reflections on general philosophical topics. He was very interested in the practice of ophthalmology, and he quickly became a successful and highly-esteemed specialist in this field. Nearly all the concrete examples cited in his writing are taken from the his medical specialty, and the few which deal with questions of internal medicine are also clearly influenced by his personal experience as a physician. Kramsztyk was probably the most practice-oriented among the Polish "bedside philosophers". From the mid-1880's, Kramsztyk regularly published in the Polish medical press (mostly in *The Medical Journal)* articles which discussed selected aspects of medical activity. Kramsztyk's original reflections on medicine have their roots in his ability to observe and evaluate his own experience and standard medical practice ofhistime, and in his life-long interest in the history of medicine. This last interest made him aware of the ever-changing nature of medical knowledge, and enabled him to maintain a critical distance from the latest fashions in medicine. Many of the ideas developed later by other thinkers of the Polish School appeared first, albeit often in a sketchy form, in Kramsztyk's writings.[1] However, Kramsztyk's most important contribution to the Polish philosophy of medicine was undoubtedly his creation of the journal *Medical Critique*. He owned this journal and edited it practically single-handed for eleven years. Kramsztyk's aim in founding *Medical Critique* was to create a forum for a critical examination of medical knowledge and practice. But it also became the principal site for the crystallization of the thought of the Polish School of the Philosophy of Medicine [8].

Kramsztyk's philosophy of medicine was strongly affected by his professional position as a successful medical practitioner, enthusiastic about his daily work and fascinated by the evolution of his medical specialty. Thus the dilemma - "medicine: art or science?" - central to the reflections of Biernacki and Bieganski was of not much importance to Kramsztyk. For him the solution was simple: medicine is an art (i.e., for him, a craft). It is practical knowledge, not a science. Although medicine is based on scientific knowledge, it is above all a profession, albeit, Kramsztyk hastened to add, the most difficult and the most demanding of the professions. According to Kramsztyk, every human action depends on knowledge. On the other hand, knowing depends on doing. Even writing or teaching depends on a certain level of practical knowledge. Scientific observation, and even more so scientific experimentation thus contain an important element of art [7].[2] This interdependence of knowing and doing did not mean for Kramsztyk the blurring of the limits between the two. While the scientist asks "Why?", and aims at understanding the relationships between natural phenomena, the practician asks "How?", and tries to discover the best way to attain his goal. Moreover, the human mind is in general not flexible enough to be trained simultaneously and with equal efficiency for contemplation and for action. As a rule (and, Kramsztyk added, the exceptions to this rule are very rare), men become either investigation-centered or action-centered. These two attitudes imply a very different frame of mind. A scientific investigator should always doubt, hesitate, and keep a highly critical attitude. But a hesitating attitude would render action-centered decision-making difficult. Unlike the scientist, the man of action, even if aware of the shortcomings of his knowledge, should have sufficient confidence in it to be able to make quick and firm decisions. For him, knowledge is only the starting point of his work, not its goal ([7], pp. 117-118).

Kramsztyk refuted the claim that in the 19th-century medicine became a true science. For him, this claim made no sense whatsoever. Medicine, he affirmed, is by definition an art, and thus cannot be miraculously transformed into a science "as if such transformation of one thing into an entirely different thing was at all possible" ([7], p. 120). He did not understand why physicians disliked being called craftsmen. An honest examination of the daily activities of an average physician would undoubtedly reveal the overwhelming importance of the practical aspects of his profession. Every physician knows that he usually obtains the best practical results when he consciously applies his craftmanship to alleviate the sufferings of his patients. Medicine, Kramsztyk added, is partially based on scientific knowledge. But this is equally true for other technical applications of knowledge, and this feature does not transform them into a science. The aims of science and of techniques are very different. An engineer is judged by the quality of his concrete achievements, not by his knowledge of fundamental physics. It is amazing, therefore, added

Kramsztyk, that so many physicians are judged mainly on the quality of their scientific publications, not on their practical ability to help the sick.

For Kramsztyk, the modern tendency to give priority to the scientific aspects of medicine clearly had a negative impact on this profession. The young physician, destined to a small-town practice, leaves medical school with his head full of anatomy and histology, but often without even rudimentary knowledge of the basic techniques of his profession. Worse, in medical school he was taught that his true vocation is to be a scientist. He therefore views the practice of medicine as a secondary or even an inferior activity. As a consequence, the medical faculties produce an impressive number of frustrated pseudo-scientists. On the other hand, those physicians who happen to be quite satisfied with their daily work are pushed to publish in order to improve their reputation and professional standing. "I published quite a lot of papers several years ago, but now that my practice is sufficiently developed, I do not need to publish any more" a well-known physician once said with disarming honesty ([7], p. 125). For Kramsztyk such "publications" are bound to be worthless. The hypertrophy of the medical press, filled with a huge number of uninspired papers, made him suspect many of the authors of those studies of publishing for the same reasons as the physician quoted above, although they were usually less willing to admit this fact openly. The effort to make the practice of medicine "scientific" thus creates frustrated physicians on the one hand and poor scientists on the other. There is only one remedy for this situation: to separate science and healing. A scientist needs all his time and energy to produce good professional science, while a practicing physician needs time to take care of his patients. A physician can, of course, study experimental biology (or astronomy, or history, or art) as a serious interest, but he should not confuse this activity with his true vocation: healing the sick ([7], pp. 122-129).

Kramsztyk believed that medicine is a profession. It is, however, a very difficult profession which demands a great deal of learning, thought, and responsibility. There is a great difference between a good physician and a poor one in their ability to help their patients. Unlike Biernacki, Kramsztyk did not think that the therapeutic weapons of physicians are inefficient, and that the principal difference between a good and a poor physician is the difference in their ability to give psychological support to their patients. His own specialty, ophthalmology, taught him otherwise. In the late 19th-century, the practice of ophthalmology was indeed less stressful and less frustrating than many other medical specialties (e.g., pediatrics), because the ophthalmologist seldom dealt with life-threatening conditions. On the other hand, ophthalmologists, who in many cases were really able to help their patients, were in all probability less inclined to therapeutic nihilism. They were able to correct many sight defects by the simple, but nevertheless highly scientific production of lenses, and other sight defects by surgical means. They also had at their disposal several drugs (such as athropine) which could produce controlled and measurable effects on the eye. Kramsztyk did agree, however, that even in ophthalmology the number of therapeutic means the activity of which was fully understood (surgery excepted), was still very limited. A rational, fully understood therapy is the ultimate goal of science, but this goal was not yet in sight, and Kramsztyk was not absolutely certain that it would be attained one day ([9], pp. 189-196).

Meanwhile, Kramsztyk thought that it was mistaken and it might be even dangerous to behave as if the goal of rational therapy was already realized. But this was precisely what many physicians did. They looked for scientific, i.e., causal rationalizations for each of their moves, and preferred to stick to this pseudo-rational way of thinking in order to avoid the necessity of recognizing the limits of their knowledge. For example, one of Kramsztyk's colleagues, a highly respectable and reputed physician, explained in a medical consultation that his patient, suffering from tabes (tertiary syphilis), manifested weakness of the central nervous system. A direct stimulation of the debilitated central nervous system was, however, impossible. One could thus attempt to stimulate the peripheral nervous system through the stimulation of the skin. The logical conclusion of the physician was to recommend that the patient attend a health resort using mineral water with skin-stimulating properties. This recommendation was presented not as a result of the personal experience of the physician - e.g., knowledge that such a water cure had helped other patients suffering from tabes - but as an inevitable result of a precise and rational equation.

The truth, according to Kramsztyk, was quite different. The physician knew what the disease was, and what served as a possible treatment. But he needed a "rational" connection between the two. In some cases, the finding of a simple and rational explanation for a treatment was difficult, because the "distance" between the disease and the proposed treatment was too great. It is not easy to make a direct connection between mineral water and tabes, and intermediary stages are needed in order to construct a bridge between the two. The physician thus passed from water cure to skin excitation, from the skin to the peripheral nervous system, from the peripheral to the central nervous system, and finally from the central nervous system to tabes. The "rational physician" forgot, however, to mention that the final effect sought by him was the irritation of the skin. Scraping the skin with a hard brush should have been effective too ([9], pp. 193-196).

The short moment of decision during which the physician chooses the appropriate treatment for his patient is, Kramsztyk explained, influenced by several factors: the theoretical knowledge and the practical experience of the physician, his compassion for his patient, and considerations relative to the physician's own professional status ([9], p. 189). Such complex and often unconscious motivations behind a decision are not specific to medicine. They characterize decisionmaking in every practical profession. But unlike other professionals, physicians try to present their decisions as *rational,* i.e., motivated solely by their science. Some physicians, more cynical than others, are aware of the fact that the rationality of their explanations sometimes serves principally to make an impression on their patients. For example, Kramsztyk quoted another of his colleagues, who after hearing some therapeutic propositions asked with an ironic smile, "well, and now how are you going to give a rational explanation for this decision?" But such an overtly cynical approach is the exception, not the rule. The majority of physicians sincerely believe that they understand the action of the drugs they prescribe. They refuse to admit that their knowledge is far from complete, and they also try to forget that the so-called "rational therapy" was particularly powerful in the period when lack of scientific knowledge allowed the development of all-encompassing medical systems ([9], p. 194; 192).

Kramsztyk was far from being hostile to rationality in medicine. He only strongly objected to the uncritical faith in theoretical suppositions in healing, and to the confusion, frequent in medicine, between hypotheses and facts ([10], p. 186). This confusion between facts and hypotheses was amplified by the frequent mistake of viewing the schematic textbook representations of disease as faithful descriptions of real-life pathological states. But, Kramsztyk explained, textbook representations of pathological modifications induced by a given disease are, as a rule, didactic simplifications. They thus include not only observations of pathological states but also the presuppositions that one holds about them. For example, the drawings in pathology atlases represent the synthesis of a great number of individual cases, and moreover this synthesis is influenced by the image of the disease at the time the picture was drawn. For this reason such pictures often become quickly outdated. Kramsztyk was aware of the fact that one cannot avoid schematic textbook descriptions of pathological states, because such descriptions are indispensable for the transmission of knowledge. But he stressed that one should be very careful not to confuse them with real-life pathological states, and above all not to use uncritically the textbook representations to support what is believed to be a rational therapy ([11], pp. 55-58).

Kramsztyk advocated the limitation of schematic representations of pathological states and their replacement with minute and precise descriptions of concrete clinical cases. But he was aware of the fact that clinical observations do depend on previous knowledge, and that even the most direct observations are influenced by pre-conceived ideas. His analysis of this phenomenon was much more detailed than the analysis proposed by Bieganski. "In the process of observation" - Kramsztyk explained - "the investigator is never completely passive. His mind is not a blank sheet, but contains many general ideas and many pieces of information that are unconsciously transferred to the observed facts." Those pre-conceived ideas deeply modify the way of viewing the external world: "a man does not see new facts with the prejudice-free eyes of a newborn child, but he perceives them with the eyes of his mind, which contains many bits of information, theories and of pre-conceived ideas, and he is bound to see the world in the light of his theories and pre-conceived ideas" ([8], p. 42; [12]). The observer, who views nature with the 'eyes of his mind', "notices above all those phenomena which are consistent with his previous knowledge and overlooks all the others. He may even observe non-existing phenomena if they are necessary to confirm his views" [17]. This is particularly true for medicine. In medicine, by definition all the observed phenomena are very complex. While there exist simple physical and chemical facts, there is no such thing as a simple biological fact:

> "life is an extremely complex process, and when studying life, in addition to the phenomena the interrelations of which we want to expose, there always appear numerous other phenomena related to and affecting the phenomena investigated by us. No clinical fact will ever be as firm and as certain as a simple physical fact.... Due to the great complexity of the subjects clinical facts can be bent much more easily than the physical ones. This is why there are so many general theories in pathology, and why these theories so often break down" [12].

The complexity of pathological phenomena and the specific nature of medical knowledge make unbiased, theory-free clinical observations particularly difficult. Physicians have thus a particularly strong tendency to make selective observations: "the physician's attention is usually directed only towards the phenomena that he has been trained to see, those with which he is familiar, and those which are the most frequent. This is a usual feature of the human mind: we are able to perceive only familar phenomena, because they are the most present in our mind and because we have for them a ready-made name and a ready-made theory" [18]. On the other hand, physicians have the tendency to confuse theory and facts. Kramsztyk analysed in detail two contradictory views of the genesis of glaucoma, both in fact containing theory-informed views, but both presented as based on uncontested "clinical facts" [12].

Kramsztyk believed that while one should be aware of the difficulties besetting clinical observation, this difficulty should not diminish the importance of such observations in the formation of medical knowledge. Observations, Kramsztyk explained, are a fundamental tool in the progress of medical knowledge. They should, however, be carried out as carefully as possible, and their authors should be aware of the pitfalls of subjectivity. One way to limit those pitfalls is to multiply the number of independent observers. "Collective clinical observations" can greatly improve the viability of this method and thus increase its contribution to pathology [19]. However, in synthetizing many individual observations an additional difficulty emerges: the opacity and time-dependency of medical terms. In medicine there is no such thing as a direct transcription of immediate sensory impressions. Every observation made by a physician is perceived through a filter of ready-made concepts, and is immediately translated into medical terms. This translation is a source of difficulty in comparing medical descriptions. The meaning of a given medical term can be obscure, and can differ for different observers: "we are more familiar with the terms than with their meaning" and sometimes "we do not adapt the terms to fit the phenomena, but seek an adequate content for a familiar term". Moreover, when science changes, the same term can completely change its meaning, and then "different persons use the same term with very different meanings.... Such terms with varying meaning are even more annoying in historical investigations" [20].

Kramsztyk, like the other members of the Polish School, stressed the individual character of every illness and the difficulty of making generalizations in medicine. He explained that even the best understood eye diseases, such as trachoma, in fact represent a large spectrum of disease pictures, include many borderline cases, and reflect a great range of intrinsic variability [13]. The individual character of each clinical case was the principal reason for Kramsztyk's strong objection to the use of statistical methods in medicine. Statistics, he explained, encouraged an over-simplified view of diseases and discouraged the observation of individual differences between patients and the adaptation of treatment to the needs of each individual case ([10], p. 185; [6], p. 110). Statistical reasoning cannot supply causal explanations and is thus useless in the determination of the etiology of pathological disorders. It is true, admitted Kramsztyk, that theoretically statistical methods can be very useful in evaluating the efffectiveness of a given treatment, but in practice the impossibility of making the test conditions truly uniform, and the strong tendency of physicians to observe only those results which they wish to observe, make statistical evaluations of the effectiveness of treatments totally useless ([6], pp. 107-109).

The complexity and individual character of each illness account also for the fact that it is not possible to make precise predictions in medicine. Medicine, unlike astronomy or physics, does not measure identity, but only similarity, and is not grounded on one overall and encompassing theory but on a series of separate generalizations. Kramsztyk, to be sure, did not think that in the natural sciences it is possible to make absolutely precise predictions. Even in astronomy, he explained, accurate predictions are possible only because one observes a single phenomenon, the movement of heavenly bodies, and this isolated phenomenon is observed only for a very short time relative to its duration. "We would be more or less in the position of an astronomer if we were able to observe a given phenomenon in human life for many thousands of seconds, and then were asked to predict its fate during the next minute" ([5], p. 152; 137). In addition, the physician is, by definition, always dealing with highly complex phenomena. Even a physicist, who is working in the framework of a well-established theory, often predicts the phenomena he is observing in the artificial setting of the laboratory: he cannot predict the exact pattern of behavior of a handful of feathers thrown in the wind. How can we expect that the physician would be able to predict the evolution of a pathological state, a situation much more complex than the behaviour of a handful of feathers? ([5], p. 147; 151-152). Unfortunately, physicians have a tendency to forget the unpredictability of living organisms. Therapeutic evaluations are often made on the basis of a single reaction of a given drug or treatment, observed in the laboratory. Often physicians do not take into consideration the fact that their ideas on the functioning of this drug or treatment represent at best a partial truth. Moreover, Kramsztyk stressed that even when the physician can adequately evaluate the physiological effects of a given drug (emetic, narcotic, etc.) he cannot be sure if the induced effect contributes critically to the recovery of his patient ([5], pp. 153-155).

Kramsztyk was acutely aware of how little is known, and how much is unknown in medicine. He therefore appreciated the point of view of medical skeptics. But he strongly disapproved of this point of view. In many cases, he affirmed, the physician cannot only give psychological support to his patient, but also actively help him to recover his health or at least slow down the disease process ([10], pp. 175-176). But he had to be aware of the two principal dangers besetting the medical profession: the routine application of treatments, and the blind following of the latest fashion. Those two dangers, quite different at first sight, are in fact, Kramsztyk affirmed, two sides of the same coin. Both result from an over-confidence in authority, from lack of critical thinking, and from an inability to learn from one's experience. A critical attitude towards innovations in medical science, coupled with the willingness to look at every new patient with fresh eyes and to view every illness as a unique case, were for Kramsztyk the preconditions for the progress of medicine. In Poland, they were also the prerequisite for overcoming the handicap of provinciality, lack of originality, and over-confidence in foreign achievements [16].

Kramsztyk decided to found a specific journal that would publish original reflections on medical knowledge and medical practice. His aim was to develop the capacity of critical thinking of Polish physicians, and thus to improve the level of Polish medicine. *Medical Critique,* he explained, should critically evaluate present developments in medicine. However, in order to achieve this goal, one should also take into consideration the history of medicine. Only the historical perspective permits a dynamic view of science. "For a scientist, science is immobile; for an historian, it is a rapidly flowing stream. For the former, science *exists* - for the latter it *is becoming*" [15]. An historical perspective was the best way to prevent what Kramsztyk called "the modern scholastics" - the excessive dependence of physicians on established authority. The principal difference between the new scholastics and the traditional scholastics was that the idol of the new scholastic was not the most ancient authority, but the newest one. It was usually based on admiration of the latest results of science, but ignorance of its history. But "only the history of science, the history of changing views on the same phenomena, can shed the right kind of light on present ideas and protect us from their uncritical over-estimation" ([8], p. 43).

Kramsztyk published *Medical Critique* for 11 years (1897-1907). The termination of its publication has usually been explained by political and economic difficulties during the period following the revolution in Russia and Poland in 1905 [26] .Although economic difficulties probably hastened the end of *Medical Critique,* this was perhaps not the only reason for Kramsztyk's decision to discontinue publication. Other elements - the strong opposition to Kramsztyk's medico-philosophical ideas among Polish physicians accompanied by the degradation of Polish-Jewish relationships in Poland - probably played an important role in this decision.

In 1907, Kramsztyk felt that his view of medicine was losing ground. He thought that he had failed to convince his colleagues to abandon the blind following of the latest medical novelty, and to develop a critical, philosophically- and historically-oriented reflection on medicine. In the last issue of *Medical Critique,* Kramsztyk bitterly complained about the blindness of those physicians, who see the latest development in science as the definitive one, the only one to which one should pay attention, and for whom science has just started today, at best yesterday, and the past is not worth thinking about. But, Kramsztyk affirmed, the cult of the present narrows the scope of the mind, offends the dignity of human thought, and, moreover, is harmful to science itself, because science can only acquire its proper meaning from an historical perspective on changing scientific ideas. And, he added, "the struggle against the worship of the present is indispensable, but it is a very difficult task, and does not bring any satisfaction.... It leaves one alone, not only without followers, but even without a small number of faithful companions - nothing but mute harp cords and deaf listeners" [23].

Kramsztyk's bitterness had perhaps other causes as well. At that time the Warsaw Medical Society for the first time refused the admission of Jewish members. As a sign of protest Kramsztyk resigned from the Society and founded a new one, the Society for Social Medicine. He also ended his collaboration with *The Medical Journal.* In the same period Kramsztyk, who was deeply interested in Polish literature, published a book in which he proposed a new interpretation of the famous epic, *Pan Tadeusz,* written by the national poet, Adam Mickiewicz. This novel interpretation gave a central place to the Jewish musician Yankiel, and stressed the elements of Polish-Jewish friendship in the plot. Kramsztyk's interpretation was strongly criticized and ridiculed by the Polish press. The experience was a shocking one for Kramsztyk, who all his life fought for the assimilation of Jews into Polish society [24].

When he founded *Medical Critique,* Kramsztyk's aim was to contribute to the development of Polish medicine [16]. He also fought for the purity of the Polish language in medical publications, and kept a special section on langage in his journal. Kramsztyk strongly believed that "the national ideal is transmitted not through the blood, but through the spirit."[3] Some Poles clearly thought otherwise, and their views may have been important in Kramsztyk's decision to cease publication of *Medical Critique.* Kramsztyk himself did not explain the background of this decision to the readers of his journal; in December, 1907, he only published a brief statement: "This issue of *Medical Critique* is the last one. The journal will not appear any longer. Farewell, and cordial thanks to all its collaborators and readers. Zygmunt Kramsztyk" [22].

NOTES

1 For example, Kramsztyk was the first to explain that observations in medicine were dependent on the a priori ideas of the observer, and that every biological experiment was based on the underlying (although often non-explicit) assumption that the living organism has a purposeful design [2]. He was also the first to discuss the limits of the physician's understanding of the action of drugs and the importance of psychological factors in healing [5] to analyze the problems of the application of statistics to medicine [6], and to discuss whether medicine is a science or an art [7].

2 Kramsztyk's article, written in 1895, preceded Biegański's similar views on the same subject in the second edition of his *The Logic of Medicine* ([1], pp. 26-29).

3 This statement was made in an obituary note for the histologist, H. Hoyer [21]. Hoyer's father was a German: he studied in Germany but later he chose to live in Warsaw, and continued to work there even after the 1861-63 insurrection, when Russian repression made the continuation of his scientific studies difficult.

BIBLIOGRAPHY

[1] Biegański, W.: 1908, *The Logic of Medicine, or the Critique of Medical Knowledge [Logika medycyny czyli krytyka poznania lekarskiego]*, E. Wende, Warsaw.

[2] Kramsztyk, Z.: 1899 (1897), 'The Goal as a Directing Line of Research' ['Cel jako wskazówka badania'], in Z. Kramsztyk, *Critical Notes on Medical Subjects [Szkice krytyczne z zakresu medycyny]*, E. Wende, Warsaw, pp. 89-95.

[3] Kramsztyk, Z.: 1899 (1884), 'Celebration of the 50th Anniversary of the Medical Activity of Professor Szokalski' ['Mowa dla uczczenia pięćdziesięcioletniej rocznicy działalnosci lekarskiej Professora W. Szokalskiego'], in Z. Kramsztyk, *Critical Notes on Medical Subjects [Szkice krytyczne z zakresu medycyny]*, E. Wende, Warsaw, pp. 237-247.

[4] Kramsztyk, Z.: 1899 (1896), 'Wiktor Szokalski, Posthumous Memoirs' ['Wiktor Szokalski, wspomnienia pośmiertelne'], in Z. Kramsztyk, *Critical Notes on Medical Subjects [Szkice krytyczne z zakresu medycyny]*, E. Wende, Warsaw, pp. 248-251.

[5] Kramsztyk, Z.: 1899 (1893), 'Medical Prognosis' ['Rokowanie Lekarskie'], in Z. Kramsztyk, *Critical Notes on Medical Subjects [Szkice krytyczne z zakresu medycyny]*, E. Wende, Warsaw, pp. 130-167.

[6] Kramsztyk, Z.: 1899, 'Medical Statistics' ['Statystyka lekarska'], in Z. Kramsztyk, *Critical Notes on Medical Subjects [Szkice krytyczne z zakresu medycyny]*, E. Wende, Warsaw, pp.100-111.

[7] Kramsztyk, Z.: 1899 (1895), 'Is Medicine an Art or a Science?' ['Czy medycyna nauką jest czy sztuką?'], in Z. Kramsztyk, *Critical Notes on Medical Subjects [Szkice krytyczne z zakresu medycyny]*, E. Wende, Warsaw, pp. 110-119.

[8] Kramsztyk, Z.: 1899 (1897), 'Project of Medical Critique' ['Prospekt Krytyki Lekarskiej'], in Z. Kramsztyk, *Critical Notes on Medical Subjects [Szkice krytyczne z zakresu medycyny]*, E. Wende, Warsaw, pp. 42-45.

[9] Kramsztyk, Z.: 1899 (1887), 'Rational Treatment' ['Racyonalne leczenie'], in Z. Kramsztyk, *Critical Notes on Medical Subjects [Szkice krytyczne z zakresu medycyny]*, E. Wende, Warsaw, pp. 189-196.

[10] Kramsztyk, Z.: 1899 (1887), 'Certainty in Treatment' ['Pewność w leczeniu'], in Z. Kramsztyk, *Critical Notes on Medical Subjects [Szkice krytyczne z zakresu medycyny]*, E. Wende, Warsaw, pp. 168-188.

[11] Kramsztyk, Z.: 1899 (1884), 'On Clinical Descriptions' ['O kazuistyce klinicznej'], in Z. Kramsztyk, *Critical Notes on Medical Subjects [Szkice krytyczne z zakresu medycyny]*, E. Wende, Warsaw, pp. 46-65.

[12] Kramsztyk, Z.: 1899 (1898), 'A Clinical Fact' ['Fakt kliniczny'], in Z. Kramsztyk,*Critical Notes on Medical Subjects [Szkice krytyczne z zakresu medycyny]*, E. Wende, Warsaw, pp. 66-71.

[13] Kramsztyk, Z.: 1899 (1898), 'The Difficulties of Clinical Studies of Trachoma' ['O trudnościach klinicznego badania jaglicy'], in Z. Kramsztyk, *Critical Notes on Medical Subjects [Szkice krytyczne z zakresu medycyny]*, E. Wende, Warsaw, pp. 72-88.

[14] Kramsztyk, Z.: 1899 (1894), 'Medical Experience' ['Doświadczenie Lekarskie'], *Gazeta Lekarska*, in Z. Kramsztyk, *Critical Notes on Medical Subjects [Szkice krytyczne z zakresu medycyny]*, E. Wende, Warsaw, pp.197-207.

[15] Kramsztyk, Z.: 1899, 'On the Importance of Historical Knowledge' ['O znaczeniu wiedzy historycznej'], *Krytyka Lekarska* **3**, 253-256; 253.

[16] Kramsztyk, Z.: 1899 (1898), 'In Foreign Clothes' ['W obcej szacie'], in Z. Kramsztyk, *Critical Notes on Medical Subjects [Szkice krytyczne z zakresu medycyny]*, E. Wende, Warsaw, pp.17-20.

[17] Kramsztyk, Z.: 1906, 'Difficulties in the Investigation of Chronic Diseases' ['Trudności w badaniu chorób przewlekłych'], *Krytyka Lekarska* **10**, 149-157.

[18] Kramsztyk, 'Z.: 1906, Clinical Symptoms' ['Objawy kliniczne'], *Krytyka Lekarska* **10**, 40-44.

[19] Kramsztyk, Z.: 1903, 'The Participation of the Will in Scientific Investigations' ['Udział woli w badaniach naukowych'], *Krytyka Lekarska* **7**, 240-247.

[20] Kramsztyk, Z.: 1907, 'Defects of Terminology and of Scientific Classifications in Ophthalmology' ['Wady mianownictwa i układu naukowego w okulistyce'], *Krytyka Lekarska* **11**, 115-124.

[21] Kramsztyk, Z.: 1907, 'Professor Henryk Hoyer - a Posthumous Memoir' ['Professor Henryk Hoyer - wspomnienia pośmiertelne'], *Krytyka Lekarska* **11**, 99-103.

[22] Kramsztyk, Z.: 1907, 'Farewell' ['Pożegnanie'], *Krytyka Lekarska* **11**, 194.

[23] Kramsztyk, Z.: 1907, 'On Being Up-to-date' ['Aktualność'], *Krytyka Lekarska* **11**, 163-165.

[24] Krakowiecka, L.: 1970, 'Zygmunt Kramsztyk', *Polish Biographical Dictionary*, vol. 15(1), pp. 137-139.

[25] Kus, M.: 1966, 'Polish Medical Sciences in the Years 1864-1914' ['Nauki medyczne w Polsce w okresie 1864-1914'], *Studia i Materiały z Dziejow Nauki Polskiej* seria B, **10**, 121-135.

[26] Ostrowska, T.: 1970, 'The Dynamics of Development of the Polish Medical Press in the 19th-Century' ['Dynamika procesu rozwojowego polskiego czasopiśmiennictwa lekarskiego w XIX stuleciu'], *Studia i Materiały z Dziejow Nauki Polskiej* seria B, **14**, 33-59.

[27] Skulimowski, M.: 1966, 'Polish Medical Sciences in the Years 1795-1864', *Studia i Materiały z Dziejow Nauki Polskie* seria B, **10**, 105-112.

[28] Słowiecki, M.: 1981, *History of Polish Science*, Interpress, Warsaw, pp. 192-196.

[29] Szokalski, W.K.: 1883, *Dictionary of Polish Physicians*, Warsaw, pp. 497-502.

CHAPTER V.A

TEXTS OF ZYGMUNT KRAMSZTYK 'RATIONAL TREATMENT'

(*Gazeta Lekarska*, 1887)

A man who walks alone will not go astray only if the road in front of him is sunlit, only when he can see. The sunlight of human actions is reason; when a man can fully cover with his mind all he has to do, he will not hesitate and will obtain the best possible results. No wonder that a doctor should want to have such a perfect guide when fulfilling his difficult and complicated tasks. To treat patients in a way exclusively based on reason is a goal to ward which our art is tending, and which it is slowly but inevitably approaching.

In the physician's conduct there is a certain moment of reflexion and hesitation, a moment when examination of the patient has been completed and a decision is to be pronounced. Many reasons contribute to the physician's decision: his scientific knowledge, his personal experience, his compassion for the patient's plight and also considerations relative to the physician's professional reputation. This process is mostly unconscious. If we were able, however, to slow down and to extend the moment of decision, there would be many things to read from it. The final decision is often preceded by much reflection and hesitation, but, after all, such hesitation is the prerequisite of every practical trade, an attribute of any essential step in life. To eliminate this hesitation and reflexion, to always have a single clearly laid-out path - what an attractive perspective for a physician, what a simplification of his task!

To rely entirely on his mind, the physician must know and understand the whole case; he must cover all its details and must link them all with ties of natural inevitability. He must also, in the same way, understand the therapeutic means and their effects on a healthy and a sick body. If a physician had mastered in such a way the subject of his work, he would never hesitate, his proceedings would result almost exclusively from a logical equation, and such an equation would be as reliable as a mathematical equation and would never lead him astray. The aspiration to rational treatment would be fulfilled if the treatments were indeed based on and resulted from a precise equation.

A truly rational treatment would require that we be able to express the physical condition of the patient or his disease in a mathematical formula, in which the difference between this formula and another formula expressing good health or a normal condition would indicate to us the appropriate treatment, also expressed as a mathematical formula.

This distant and desirable goal has already been reached in a very minute part of medical knowledge. There really exists a branch of medicine in which the physician can actually calculate the therapeutic means: the optics of the eye and its refraction defects. The correct condition of the eye has its specific mathematical formula. It is quite easy, or at least possible, to examine the organ of sight of a given person and express its condition in a numerical formula. The difference between the two values is actually the formula for the therapeutic means sought, i.e., the refraction index of the lens which would return the eye to normal vision.

However, even in the case in which the treatment is to a very large extent governed by the mind, there are additional factors, too. Even if optical defects have been studied in detail, there is no single possible decision concerning the treatment. There are still many factors which need to be determined before a suitable optical device may be recommended for the patient. Sometimes eye glasses, even if well-adjusted to the indexed refraction of the eye, are not tolerated well by the patient. Therefore, even in this optimal case, the treatment is not entirely rational, because a wholly rational conduct would be infallible. The doctor, using a perfectly rational treatment, should be able to calculate as closely as possible how much he can help a given patient, and the result of his calculation must always be accurate.

If this is the case with optics, what can be said of the vast realms of medical science, concerning all other diseases, and all the therapies available for patients. Many important modifications must be introduced in our knowledge of physics and chemistry, in our concept of health and illness, and in our knowledge about therapeutic means, before we can attain a level of understanding that will allow the human mind to match disease and treatment logically.

An engineer or a mechanic, who is able to solve the majority of his problems with infallible, simple calculation, also needs, in addition to his scientific knowledge, a practical sense and concrete experience. Often a proficient foreman can manage a factory better than the most learned chemist. Although many nowadays have taken a fancy to including medicine among the exact sciences, it is still far from ranking among well-codified practical tasks, for there is nowhere in nature a subject which is so complicated as medicine.

Truly rational treatment is a goal medical science aims at, but this goal is yet so distant that it seems doubtful we will ever reach it. Only our awareness of the scope and penetration of scientific knowledge, or our conviction that what is wished for so much must some day come true, can make us believe that that distant goal will be attained one day.

If a single, small error is present in a mathematical formula, or if only one minute detail remains unknown or is neglected, the result not only might but certainly will be wrong. Anyone using logic in a case that is not ripe for such reasoning will certainly go astray. He may be likened to a man walking in the mountains on uneven ground in very poor moonlight and who relies exclusively on his visual impressions. When he sees a moonlit stone, he directs his steps towards it, but a dark abyss, unnoticed by him, opens just under his feet, and he perishes before reaching the stone.

Medicine has experienced several periods resembling such firm steps in uncertain light. When we proceed rationally, when our line of conduct derives from sound reasoning, it does not at all mean that we are wise, but only that we are convinced we are wise. A line of conduct can be completely wrong and still be rational. If someone who has false concepts of life and on the effects of medicine is deeply convinced that he cannot be wrong, he will proceed firmly according to his knowledge; we shall have to admit that his behavior is rational. We can disagree with his concepts, and consequently refuse to behave as he does, but from his point of view his line of conduct is rational. He cannot see that the theory he holds is not faithful to nature, only to his mind; he would project it on the world in the same way a deformed eye projects its own deviations on the outside world and believes in its subjective reality. Such a man is a fanatic.

The concept of rational treatment is not at all a modern invention. It was in all probability more frequent in ancient times. When human knowledge was more limited and people were unable to become aware of many questions, it was easier to develop strong convictions. When madness was an act of the devil and when people believed that they had highly efficient means to fight the devil, such a line of conduct was rational. It was a rational fanatism, and perhaps in all cases rationalism is, in a way, a fanatical approach.

The fanatical doctors who were perfectly sure that blood was the seat of all diseases did not hesitate to bleed patients in order to reduce the quantity of illness. And the fact that in many cases blood-letting was not effective could not shake the fanatical belief of the doctors; it was only considered as proof that blood-letting had not been sufficient. So they bled the patient the next day, and the day after the next, and every day until the patient fainted. Often such fanaticism was clearly fatal to patients, but it might be justified subjectively, since the doctors were acting in perfectly good faith, and they were sincerely and strongly convinced they were doing the right thing.

No matter how much a man wants to be resistant to theories introduced from the outside, no matter how much he is convinced that his mind is open only to the observation of real phenomena and to his own impressions, everyone has his personal philosophy or outlook on life, even if he is not always conscious of this fact, and everyone is to some extent a fanatic.

We believe, however, that a period of destructive fanatism could not easily be repeated nowadays. Today we are less eager to trust theories, or at least we do not believe in them strongly enough to harm our patients for the sake of such beliefs. Before having turned their theoretical reasoning into therapeutic means, physicians first test them experimentally. Today we have, however, another kind of rationalism, different from the old one. This is an insincere, apparent rationalism, the goal of which is to make us believe in the rationality of the doctor's line of conduct.

Years ago I was asked to consult together with two other colleagues, in order to give advice on a tabetic patient. The most respectable of us said: "The nervous system of our patient is ruined - it needs stimulation. But since we cannot penetrate into its very center, we will have to send the impulse through peripheral nerves; the skin nerves have to be stimulated. Let us advise a spa treatment at Ciechocinek for the patient." And so it was decided. I was deeply depressed. If it had been I who had advised the patient to go to Ciechocinek, it would have been because a treatment in Ciechocinek seemed to be helpful for other patients. The advice of my learned colleague sounded, however, like an absolute necessity. Ciechocinek was the result of a precise equation.

Fortunately enough, another colleague of mine improved my mood. This time we were only two at the consultation. I said something quite indifferent, which I do not even remember now, but at which my colleague smiled sarcastically and asked : "Well, now, how are you going to give a rational explanation for this decision, my dear colleague?" So he was rational, too! Since then, I have become slightly suspicious about rational methods of treatment.

Many times I have met colleagues who challenged my inclination for this kind of rationalism, and finally I have discovered its secret. This secret consists in a kind of mental tunnel. The rational doctor starts his action in the same way as any average mortal. First of all, he goes to his medicine chest, picks up a bottle and vanishes in the tunnel. Then he reappears at the other end with the bottle in hand. Anyone seeing, or rather listening to him, is wondering from where the hell the chap got castor oil in the first place? But he just carried it through the tunnel. In other words, being in possession of a medicine, the doctor searches for a detail of the case that could be connected (in a satisfactory manner) with what he has already got in his hand. The rational man travels both ways: from the therapeutic indication to the symptom, and back again. But the first part of the journey is concealed from the world and here is where all the magic originates.

For example, in some inflammations a cold compress was usually applied: "My dear colleague, the blood vessels are dilated. When we apply an ice compress, they will contract and the patient will recover." How clear and simple! However, a couple of years later this method of treatment has changed. It has been found by that time that hot poultice yields better results. Hot poultice? This is good too! "My dear colleague. Hot poultice is indispensable for our patient; the exudate is hard - it must be softened to enhance suppuration." But sometimes the explanation is not so easy. There are cases in which the disease is so remote from the therapeutic means to be recommended that linking the two becomes a problem. The distance between tabes and Ciechocinek is too long. There is no way of covering it in one jump. Therefore, an intermediate station has to be selected - say, skin stimulation. First, the spinal chord is tied to the skin, and, well... from the skin to Ciechocinek the road is straight enough. But is the affair not transparent enough? After all, you do not need to go to Ciechocinek to irritate the patient's skin. A simple brush will do.

Very often the term "rational doctor" is a compliment, but sometimes it is an undeserved one.

No doubt, rational treatment also has its positive aspects, and it has played an important role in science. Every theory calls for new methods of treatment. And although not all of them survive and many of them are used in a way which is different from the one for which they were initially designed, some of them contribute to the art of medicine. If we knew the history of all the therapeutic means, if we knew the effects they have had attached to them, it would probably become apparent that in the majority of cases useful therapies were developed thanks to theoretical reasoning. Rationalism is undoubtedly a source of many innovations, although very often these innovations have an erroneous starting point.

In fact, no doctor can totally refuse reasoning. He uses it most readily in cases in which his knowledge is clear, simple, and indubitable. His line of action should be an adequate combination of reason and of experience. However, in cases in which his experience leads him to be suspicious about the conclusions of his reasoning, his experience and not his reasoning should indicate the right decision.

The word "rational" has different meanings, and therefore misunderstanding may easily occur. For example, some people oppose rational treatment to thoughtless or incoherent treatment, as if rational should be opposed to stupid. But, in fact, this term refers to keeping entirely to formal logic, and is opposed to experience. Or, at least this is the meaning of "rational" I have employed in this work.

ZYGMUNT KRAMSZTYK
'IS MEDICINE AN ART OR A SCIENCE?'

(Gazeta lekarska, 1895)

Here, in the Edessa mountains, dwell the men of brain.
They count the stars and weigh each tiny drop of shower,
But I shall ever seek their help in vain.
There is much wisdom in their minds, but little power.

Years ago I visited a well-known opthalmology clinic, the head of which was a very experienced doctor of remarkable reputation. His assistant, who has since then taken the position and fame from his master, was at the time occupied almost exclusively with studies of the pathological anatomy of the eye.

On morning rounds the assistant presented us with a new patient suffering from a creeping corneal ulcer. It took him half an hour to describe in detail all the theories which explain the relations between the ulcer and the pus collected in the eye, after which he went to another patient. When the head of the clinic arrived, he was also led to the same patient. But this time the speech prelection was short: "A! Hypopionkeratitis - but why does the patient still have no cataplasm?" He had drawn our attention to the only detail his assistant had forgotten.

Problems may be considered from various points of view. And in fact, people of different trades looking at the same object would see its different aspects. Walking through the woods a botanist will attach his attention to moss and mushrooms, a forester will estimate diameters and heights of trees and the possible profits that could be obtained from them, a painter will take interest in the forms, colors, lights, and shadows, a hunter will seek for the game, and a child will only look for berries.

What elevates me to the heavens will not move them by an inch... they watch the sky like astronomers or like wolves do - the look of a shepherd is different from that of a lover or a poet.

Science and art seem to be the two components of a chemical reaction. It is through our senses that the impressions penetrate from the outer world to our minds and are processed, arranged, and converted into our knowledge. The knowledge is transformed by our will into actions and thus acts again on the outer world.

Art and science are bound together like two parts of a chemical reaction which are inevitably connected with each other, and cannot exist one without the other. First of all, each action depends on the knowledge attained by us before: art is the daughter of science. If we undertake a reasonable action in order to attain a predetermined goal, we have to know beforehand something about the subject and about the procedure. Thus, if we want to cure a patient, we must know the nature of the illness and the effect of the therapeutic means at our disposal. Our information and the notions we use may be true or not, but still we must have them, believe in them, and proceed accordingly. Otherwise, our actions will not be an art; they will neither be conscious nor reasonable. They might be likened to a thoughtless hand weaving, which is neither purposeful nor reasonable action and does not lead to any distinct goal. Each physician, like any man of action, should know the goal of his action and the procedure itself; he should therefore have appropriate scientific knowledge.

But science also cannot exist without art -- without purposeful acts. As long as a scholar limits himself to passive observation and to contemplation, he can do without any action. However, the consciousness enclosed within the mind of a scholar is not yet a science. In order to become a science it needs to be delivered to others. The ability to arrange knowledge in a certain order, to put it into words, is an art for which both talent and experience are required. A lecture or a scientific paper is a work of art, and science cannot exist without this art. Moreover, research work is not limited to passive watching, thinking, or contemplating. The different aspects of the observed material must be collected and examined as early as the observation stage. Scientific experiments, which frequently require scientific abilities and specific skill, should be considered a very fine art.

Thus art and science, although theoretically separable from one another, are closely connected in every scientific study and in every human act. In practice, therefore, they cannot be separated into two distinct domains of the human mind. Any attempt to separate completely these two realms should be considered as sterile and useless.

However, the existence of the question "science versus art" is related to the subjective aspects of this problem, and to the way it is viewed by the workers in different realms of human activity. Whatever means are used by a scientist to solve his problems, one question always sticks in his mind: Why? His action is aimed at one goal: to find the relationships between the phenomena and to trace down their causes. Whatever knowledge is sought by a practician on a given subject, he has always a distinct goal before him: How can this be done? What action will it take to reach a given goal?

The human mind will turn to one or the other direction: either it will investigate the phenomena, or follow some useful kind of activity; either a man of reflection or a man of action will develop. There is probably no human mind so flexible as to be able to pass smoothly from meditation to action. No human mind is able, after the execution of an important action, to immediately forget it, together with its consequences, and to start thinking about the causes of phenomena. And after all, treatment of a patient should be considered as such an important action - first of all by the physician himself. Even if such flexible minds might exist, they would be highly exceptional. No one can expect a normal human mind to be able to do this trick: "There is only one spark in a man, and once, in his youth, it is fired." Inborn talent and long exercise can develop a mind to turn to one or the other direction: it can become either investigative or active, but not both.

Reflection does not lead to more reliable and effective action. Sometimes the opposite is true. Large horizons and the prediction of all connections and all possible effects of a given action can hamper the acting hand. The task of a scientist is to see everything, to avoid jumping to conclusions, to have always a tendency to doubt, and to avoid certainty before all the doubts on a given subject have vanished from his mind. Doubts and hesitation are therefore indispensable features of a scientist. But for the man of action, knowledge is not the goal of his work, but the starting point of his action, its bedrock. He must have no doubts, he must be brave in action, he must have confidence in his knowledge, in spite of all its shortcomings. The man of action must very often make quick decisions; he must select the most probable possibility and reject or inhibit other information and all his doubts, which may otherwise inhibit his action.

If a single goal, or one supreme task were to be selected for humanity, one might view such a task to be understanding the world - i.e., doing science. If the most supreme happiness of a human being was to be named - it would be the peacefulness of the scholar. Great achievements in science provide for the greatest and most imperishable, and thus the most desirable fame.

But science is not the only need of humanity, and it is not the field that should be cultivated by the greatest number of workers. Humanity aims at knowing the world, but its other needs have to be met, too. Whatever position one takes in the world, one's first and most important duty should be not to gain and possess as much knowledge as possible, but to fulfill one's obligations as best one can. After all, we do not ask the gardener to have a deep knowledge of botany, or to know all the up-to-date discoveries in this field, but to provide us with the best vegetables, the sweetest fruits, and the most beautiful and aromatic flowers.

To devote oneself to study one needs proper working conditions. The trade of a doctor provides no such conditions. A professional scholar who is dedicated to science may use his after-work hours for recreation. A doctor may dedicate to scientific studies only hours taken from his recreation time.

The first indispensable condition for research work is peacefulness of mind. No duty should be more important for a scientific worker than his research; his peaceful contemplation must not be driven from his mind by any other problems. A physician cannot enjoy such peacefulness of mind, which is the basis for research and for scientific meditation. The more responsibility for other people that results from the profession one is engaged in, the more one is disturbed and one's mind occupied with the duties of that trade. The profession of physician is one of the most difficult from this point of view. Only a commander of troops during a war has greater responsibility. A feeling of discomfort which might spoil the night's rest, grounded in difficult, risky surgery; an irregular case of disease, constant suspense, mostly incomplete and seldom perfect results of treatment; unexpected complications, reproaches from patients, depressing albeit often unjustified remorse, unclear feelings of guilt: all of these are implanted in the everyday life of a physician and reflect his usual thoughts and feelings. Satisfaction following a successful treatment, even if it is a hundred times more frequent than the mishaps, never removes the grief and sorrow that follow a single unfortunate case.

A physician can devote to science only a few very exceptional moments of absolute freedom of mind, moments which are so scarce in his profession. One should take this fact into consideration when judging scientific papers published by physicians.

In the presence of the burden of duty and responsibility for every act, all peaceful scientific contemplation must be inhibited, compromised by the weight of action. At the moment of surgery, the incision the doctor is about to make will have, and must seem to him to have, more importance than all scientific theories together.

When medicine is ranked among the arts, doctors often take this definition as an insult both for the profession and for them personally. It can be heard sometimes that the "art period" of medicine is over and that in the process of its improvement, medicine has become a true science, as if such a transformation of one thing into an entirely different thing were at all possible.

The application of scientific knowledge is not a scientific investigation, and it is not true that the physician's professional performance is exclusively the consequence of the sum of his scientific knowledge.

A physician should first honestly analyze what he has done in his professional life, consider the source of the most important, the most certain benefits he has been able to bring to his patients, estimate how much scientific knowledge he puts into his daily actions, forget all the remarks he has become used to making, before stating that his trade is an art or a science. To restore sight dimmed by cataract, to restore breath in the case of a choked larynx, to free an intestinal loop, to deliver a baby which could not be born in a natural way -- all these simple manual operations are art, aren't they? But even if we do not act with hand or knife, but use therapeutic means, when we separate the insertions within the iris with atropine instead of with a knife, when we purge the stomach and the bowels with drugs, when we kill noxious bacteria with chemicals and not with a red-hot iron, we are also men of action, we are also dealing with an art.

Medicine is not one of those liberal arts which pursue novelty and oddity. In medical action usually we have to be as close as possible to well-known and fixed rules. The doctors do not like to acknowledge it, and are irritated when medicine is ranked among the crafts or is called such, but, in fact, usually the best results are obtained when the doctor's action is almost automatic, when his work is so familiar and diligently performed that it becomes a craft. It is certainly, however, a craft of the highest rank, because the subject of the doctor's action is the human body, which is the most precious property of a man, and because the most precious treasure of a man - his life and health - is at stake.

Medicine is really the most difficult craft of all. Its subject is a hundredfold more complicated than the subject of any other craft, and much thought must often be exercised before a physician can arrive at a given conclusion and decide about an action. The physician's acts are also different from those accomplished in other crafts, because the physician must be far more involved in what he is doing than any other craftsman, and because he always deals with pain and misery. This moral aspect of the doctor's activity should be carefully taken into consideration.

"How can anybody ask if medicine is a science? Must it not be learned?" Recently we have read this question again. Science consists in explanation of the world. It is a rational attempt at understanding all its phenomena and arranging them into an indispensable, causal order. But man has to learn how to carry out each specific action. Music or painting are surely arts, but how long do the great men of these arts have to exercise before they are able to act? Even to sweep the street efficiently and quickly - even this simple art - has to be learned. It is not a question of skill only, but also of knowledge. Each action of man, each art, is the application of certain properties of bodies or natural phenomena to meet the requirements of man. It is therefore also the application of science. Many advantages for mankind result from truths reached by science, which aims at idealistic, unselfish goals. In addition, research on many problems is carried out mainly for practical purposes. Certainly, the more sophisticated a human action, the higher its level, the more extensive knowledge it requires. Research in organic chemistry has provided many dyestuffs for industry. The modern dyer must therefore be well-acquainted with the respective area of chemistry and must have thorough good experience with chemical processes - but dyeing is an art, while chemistry is a science. Electrical energy is at our service, but an electrical engineer is different from a scientist. An engineer must know mathematics, mechanics, and many other subjects, but to build a bridge is not a scientific task. A physician must have extensive knowledge of the body and its functions, as well as of their deviations, but this edifice of knowledge is not erected by physicians.

If we look at the history of the great scientific discoveries in anatomy, histology, physiology, chemistry, and pathology, we would inevitably find that all these must be credited to men who devoted their lives to scientific research, and not to medical practice. The clinical arts, that is, methods of examination, application of treatment, and surgery, are practiced by physicians, and although in some cases the first concepts are not invented by physicians, only they are able to elaborate the clinical methods and details of treatment. The physicians who refine the art of medicine are men of special merit. Such a man, in addition to the contribution made by his good, reasonable, and diligent fulfilment of his work, provides humanity with new weapons to fight disease. Such a physician is a man of much higher, double merit. But other physicians, too, if only they carry on their profession in a decent way, deserve the greatest esteem, for they faithfully fulfill their task.

Often young physicians evaluate their elder and well-known colleagues, but only one criterion is applied in such evaluations: what has he done for science, what papers has he published, where has his name been quoted? But if he has no publications, if his name has not been quoted in the textbooks, he is immediately perceived as a man of no merit. It looks as if the value of a physician, his value to society, depends only on his published scientific works. But what are we here for? For what reason does society call upon and support us? After all, whenever a physician gives efficient advice, whenever a patient following this advice recovers, his life is saved, or he escapes remaining infirm, the doctor has surely fulfilled his obligation toward humanity and society. Healing is the foundation of our activity and our most important task. Publications or scientific work, although they might add to the physician's fame, are a different kind of contribution and are of secondary value to the profession. One can be an outstanding doctor of the highest merit and deserving the highest regard, without producing any publications. Of two doctors it is not always the one who is well-known as a scientist who is also a better, more brilliant doctor, or who has higher merit than his colleague whose name is not known to the public.

Our power is in action and in healing. When we go so far as to practically dispose of our essential activities, we voluntarily depreciate the value of our trade. A brilliant surgeon who gained his fame by successful medical advice and brilliantly performed surgery is openly depreciating his own value by saying that it is not the number of surgeries performed, but laboratory research which makes a good surgeon. If that were so, perhaps the value of a good bacteriologist or histologist should be measured by the number of surgeries he has performed.

What an aberrant view! Is it not evident that the more surgeries a surgeon has performed and, what is more important, the greater the skill with which they were performed, the better physician he is and the better his life-mission is fulfilled? Not only is he not obliged to have a laboratory to carry out scientific research, but I do not even think that to carry out such work would be of value for him. His time is occupied by something else, and his mind wanders in a different sphere. By contrast, when a detailed examination of a tumor, a bacterium or a secretion is necessary, and when, first of all, a scientific opinion is required on what has been found, there is no doubt that this can be done much better by a scholar who has a laboratory and who is devoted to scientific research. After all, no human profession is an isolated unit with no connection with the other professions; everybody must seek help from others. No beautiful piece of furniture can be made by one man: the carpenter, the turner, the carver, the upholsterer, to say nothing of those who prepared the material, all had to take pains to produce a single object. In pure science too, not every scholar is equally skillful at every method of investigation, and often one scientist seeks help from another, more experienced one, as happened recently with the investigation of argon. Why then should a doctor, in order to solve his problems, need to personally carry out scientific research?

There is something insincere, something hypocritical about asking a physician to be a scientist or at least to be known as such. If such an aspiration were sincere, it would often result in an unpleasant and strenuous duality. A physician is attracted by pure science or by research work; this he considers to be the most suitable and pleasant occupation for himself, but the harshness of life has obliged him to stay in a profession which, in comparison with his beloved scientific contemplation, must seem unpleasant if not abominable. On the other hand, other physicians who happened to become scientists often yearn for medical practice because it leads to a higher income. Thus, we have half-physicians ready to change into scientists, or at least claiming so, and half-scientists who would readily devote their time and mind to practice. Would it not be more reasonable, and far better for them and for medical practice and science as well, if everyone did his best to keep to his true occupation? The person called by fate to be a physician should have a feeling for his profession, he should come to like it because the profession of physician is sufficiently absorbing to fill the life of a man without having to forego scientific research.

Often the yearning for scientific work is insincere, and scientific papers are written by doctors for reasons that are very different from the search for knowledge. Sometimes papers are published because they establish fame, and fame, although it is not the fame of a practitioner, can establish a good practice. I remember a very popular doctor once saying to me with completely ingenuous sincerity: "I published many papers some years ago, but since my medical practice has developed I see no reason to continue". It is evident that no scientific value can be expected from papers based on such reasoning and aiming at such goals. Our abundant, vast, almost unlimited medical literature very often raises the suspicion that many of the physicians who published there had the same idea as had that sincere colleague of mine.

Reasonably enough, many doctors turn to different intellectual tasks in addition to their professional occupations. A man should not limit his mind to one closed domain and live like a machine. Each such additional interest will extend the limits of one's mind, will allow one to view one's own profession from a new angle, and will certainly relieve one's mind by drawing the attention away from the harsh aspects of the medical profession. There are some well-known physicians who take an interest in astronomy, history, or natural history, to say nothing at those, quite numerous, who seek solace in literature or in other fine arts. Biological research is chosen by many doctors as a hobby or a secondary occupation. Any such secondary occupation is admirable if it does not interfere with one's primary occupation. But all such occupations are not indispensable to the physician's work.

A doctor must have a vast knowledge of biology, but this knowledge is not sufficient to be a physician. It is not true that medicine, or therapy, stems directly from this knowledge, that is, from the simple sum of scientific information. In order to apply this knowledge, to examine patients using all the methods and instruments available, to take care of patients, to use appropriately medical means, and above all to use efficiently one's hands as well as tools in so many different functions, a specific skill is required. A physician acquainted with practical work will know how to use each method and how to fulfill his functions in the best way. Such practical skill is by no means less essential for the physician than theoretical knowledge. A student who has not acquired an adequate knowledge of anatomy, of physiology, or of pathology, cannot be a good physician. But the one who knows little about methods of treatment will not be a good physician, either.

In addition to proficiency in scientific theories, the physician should train his eye and hand; but also, albeit it is not often mentioned, he should bend his will and emotions to the needs of his profession. Medical education must develop both these aspects harmoniously. But in contemporary medical education the center of interest has totally deviated towards theory, while practice is greatly neglected. The medical student learns too much, much too much scientific theory, and too little of what will be indispensable for him from the first day in an independent position, in order to cope with the practical tasks of his profession.

How many physicians have never, during their studies, applied a catheter, or made a surgical incision, or turned up a lid, or used a speculum to look into the larynx, or examine the eyeground? Inexact description of epithelium of a given organ can deprive the student of a doctor's degree, but nobody asks the student if he can introduce a catheter into the body, although knowing how to do the latter is in fact much more important for the patient.

A young doctor settles in the provinces, his head filled with epithelia but his hands lacking the simplest medical experience. He can truly say, following Goethe: "I need just what I do not know, but what I know - I cannot make use of." In the majority of cases, much of the scientific knowledge quickly evaporates from the memory of the physicians - the best proof that it was not required for their professional tasks. How many things are taught at school just in order to pass an examination and then forgotten? Out of each hundred medical postgraduates, only one or two, or, in the first few years, at best three, will devote themselves to scientific research. It may seem that the curricula of medical schools are made for these few students only, while the others are impelled to adapt themselves to the needs of these few future investigators.

The experience and skill required in the medical profession are gained gradually, usually by the physician's own efforts. He will, however, always suffer from essential deficiencies. He is self-taught and therefore the therapeutic functions are often performed carelessly by him. I am not referring here to the few postgraduates who have accepted positions in hospitals and have obtained the title of specialist, but to the vast number of physicians living in the provinces. If these physicians had been sufficiently prepared by the universities, there would be less cripples and less untimely deaths in the world.

We do not propose to train physicians who have the skill to automatically perform one or another manual function, neither do we propose to produce one-term-trained specialists who look as if they were manufactured in a factory. We do appreciate a doctor who properly fulfills his professional activity, who has a good understanding of this activity, and who seeks to learn everything that will help him to improve it. A physician should be familiar with biology in general, not only with the biological knowledge which is directly related to medicine. But the interest in science should not overshadow his profession. Science should not be presented to him as a leading light, the one and only way of life, for this life takes a different route and he must follow a different star.

When leaving the university, the doctor must not only extend his knowledge and learn the practical skills that are most essential to him, but he must also challenge the ideas that have been imposed on him: "Medicine is a science and research work is its basic task and the source of its glory, while practice is a secondary task or at best a less noble one." This is what a young doctor has in mind when he enters the medical world, and right away he is asked by society to do something entirely different. The young doctor then experiences an intrinsic rupture, a conflict, a struggle between the schoolbred tendency and the social requirement. Sometimes this conflict does not vanish for a lifetime, and the doctor feels himself a stranger to his profession, carrying out his work under compulsion, not being able to devote time to scientific research because he has been caught in the arms of practice, and all his time has been taken away from him: "They have disjointed my wings, they have dislocated them so that I cannot any longer soar upwards."

When, during an inauguration ceremony of a new medical faculty, the university professor indicates to his young students their tasks in life, he suggests that science and research work should be considered as their goal. There is not a single word about the treatment of patients, about ways of dealing with patients, about the days, from dusk to dawn, spent with patients -- no indication that these are the very things for which people will ask, and that these tasks will leave the medical faculty no time for biological studies.

Only a modest dowry will the student carry on his journey to his profession, the hardest profession of all, should he take to heart the "exciting" and "beautiful" speech of his professor!

ZYGMUNT KRAMSZTYK
'A CLINICAL FACT'
(*Krytyka Lekarska,* 1898)

A fact is the most important and essential element of science. Not only is it the foundation on which science is built, but it is the only source from which science flows, and in which it has already been contained. Theories do not grow independently, but they are drawn from facts, aren't they? Even the most abstract theory, such as, for example, the cosmogonic theory of Moses, is based on facts, and facts are included within it. When explaining how the Earth, the firmament, all the creatures and Man came into being, Moses was familiar with all these entities and with the fact that they existed, to say nothing of the psychological facts included in his theory -- that is, the human concept of will and of aim. His cosmogonic theory was actually deduced from the facts he knew.

But what is a science exactly, and what does it aim at? At a description of all phenomena in the world by the smallest possible number of statements and rules, or perhaps even by one ultimate basic law. To fulfill this function, science must first explain the nature of the phenomena, or the "facts", that are to be compiled, processed, and reduced to a minimal number.

Nature can be explained in two ways: either by description, by presenting existing entities and phenomena, or by searching for the links between phenomena, whether causal or intentional, by gathering facts into more and more extensive groups and thus arriving at more and more general facts. The latter method of explanation of nature is the true way of science, but it must be preceded by a simple description.

These two lines of scientific study correspond to two types of facts of a different character: one is a descriptive fact, the other is a relation, or a fact of perception, or a connection between phenomena. A descriptive fact demonstrates solely the existence of an object or a phenomenon without offering any explanation of its meaning or its connection with other phenomena. Such a new fact, if perceived and described, does not explain natural problems, but on the contrary makes them more extensive, since it raises new problems in the future, however, new facts may very likely be helpful in the explanation of already-known but still unexplained facts. There are many facts of an exclusively descriptive nature, e.g., in the anatomy of the brain. Many very complicated facts had been collected in this area before any explanation could be given of their meaning and the connections involved.

It may seem that a simple observation of a fact or a description of phenomena which has not been noticed previously is in general accidental. A traveller may have noticed a plant which had not been known before, only because the land it was growing on had not been visited yet by botanists. A physician may have observed and described an unusual monstrosity, while his only contribution to science is the incident by which he has come across this monstrosity. But the incident is of less consequence than is usually thought, and its importance is continually decreasing, because people who observe nature are better prepared for their task and, as the number of known phenomena has been steadily increasing, incidental discovery of new phenomena becomes less and less probable. In order to make such an incidental discovery, one must first be well-informed how far science has advanced; one must therefore have considerable knowledge. Second, usually facts are unknown not because nobody has encountered them but because they are hidden, or because they are inaccessible to direct observation. To perceive such facts new methods must be invented, or at least one must acquire and become familiar with numerous difficult and complicated methods of scientific research.

In the long run, an important proportion of the research which is carried out in physiological laboratories aims at discovering descriptive or anatomical relations. When we stimulate a nerve and observe its effects on various organs, it is only in order to find the route of this nerve or its anatomy. The conventional tools of pure anatomical research, that is, the eye and the knife, or even a microscope, were not sufficient to determine the route of such thin and long, winding and splitting nerves; a physiological reaction had to be called on for assistance.

Prospects for many discoveries appear as soon as a new test method is employed in the research work. As long as the method is new, anybody who becomes familiar with it and employs it competently will meet new facts and unfailingly make new discoveries. But soon enough, due to the great number of experimenters, the mine will be exhausted, and all the observations possible will be completed. The new method will stop being a Fairy whose good grace can easily bring fame to a discoverer, unless he knows very little, so little that he does not know what has been discovered already, and views everything noticed by him as new and important.

Newly-printed old discoveries do not bring any glory. In the scientific press we often meet, however, descriptions of facts that are well-known to the author, as can be gathered from the long lists of earlier papers on the same topic reflected in the bibliography. The author has simply seen a known fact, and then published it - but what for? A great proportion of the articles published in medical journals are sophistry, revealing nothing new. Reading the weekly publications or even thicker journals we arrive again and again at the conclusion that we have found nothing new. The authors have probably been stimulated to write by the "writing fever" that is so popular nowadays. The journals must be happy to receive such articles since they have assumed the obligation to print a given quantity of paper at specified intervals, and without a daily portion of old sophistry such an obligation cannot be fulfilled.

However, from time to time, even without new testing methods, new observations occur, since to make an observation one needs not only an adequate method, but also remarkable talent for seeing things. People are usually dependent on what their masters have turned their attention to, not only in their judgments but also in the use of their senses. To see something that has not been noticed before and to show it to others, one must be able to look at the world through one's own eyes. One must possess independence of mind, a very rare quality and the basic feature of genius. But since such minds seldom occur, descriptions of the new and important aspects of commonly known phenomena are extremely rare.

Sharp minds have, however, much wider opportunity in recording facts of the second order, where one looks not for the existence of phenomena, but for relations between known phenomena. So far, a very low proportion of these facts is known in the same detail as the descriptive facts. To see that the two phenomena are always associated helps to better understand their nature and to make our knowledge of their nature more straightforward. I may even say that material progress in science might consist mainly in perception of relations between phenomena.

However, if such facts are to be integrated into the structure of science, they must possess certain features. The facts must be simple enough and must cover only those phenomena and those factors which are taken into consideration. The Magdeburg hemispheres have proved that air is heavy and that it applies pressure; lack of air inside, and presence of air outside the sphere, were consistent in all the experiments, and no factors which were impossible to remove interfered or made the conclusions more difficult. The fact that the products of the burning process have greater weight than the materials burned is also simple. It can be said that physics and chemistry are founded on such simple and clear-cut facts; it follows that because these are most reliable facts, the chemical and physical theories seem so impressive and unshakeable. It is sometimes surprising how much information science can gradually read and produce from a single fact.

Unfortunately, such simple facts do not and cannot exist either in pathology or in biology. Life is a very complex process. For this reason, when studying life, in addition to the phenomena, the interrelations of which we want to expose, there always appear numerous other phenomena related to and affecting the phenomena investigated by us. This is the reason why no clinical fact will ever be as firm and certain as a simple physical fact. Even an experiment has only a limited capacity to simplify the phenomena of life. If we attempt to remove completely all the auxilliary phenomena beside the phenomenon that we are investigating, life itself would have to be destroyed. However, clinical facts can be more or less intricate and more or less certain. The theories should therefore be built up only on the simplest and most certain facts. In practice it is, however, very often the other way round: a scientist will generalize a new, complex, and inexplicable fact into a theory which has the intention of covering a large range of phenomena, or even the whole of pathology. Due to the great complexity of the subject, clinical facts can be "bent" much more easily than physical ones: this is why general theories are encountered so frequently in pathology, and why these theories so often break down.

It is not true that we are able to see facts impartially. Any deduction contains and follows an induction, and deduction plays a very important role in induction. Man does not see new facts with the unprejudiced eyes of a newborn child, but he perceives them with the eyes of his mind which contains many bits of information, theories, and pre-conceived ideas, and he is bound to see the world in the light of his theories and pre-conceived ideas.

Let me give you a quite recent, but characteristic example. Last year two scholars, Bitzos and Abadie, published in *Archives d'Ophtalmologie* two new, but entirely different theories of glaucoma. Today every theory of glaucoma must also cover iridectomy and must explain the way this operation removes the symptoms of the disease. One of the theories claimed that iridectomy can be effective due to the cutting through the iris plexus nerve, while the other conditioned the positive effect on the removal of the major part of the secretory area. Consequently, one author affirmed that iridectomy will be successful only when it has covered the middle part of the iris, in which the plexus nerve is present, and that a positive effect would be obtained even if we limited our action to cutting through the iris. This was presented as an unquestionable clinical fact. For the other author, the observation that iridectomy is conditioned by, and is the more effective (the greater the section of the iris removed, whether at the infrairis or in the circumferential area) is a clinical fact. At least one of the authors must be wrong, although both of them express their opinions in perfectly good faith. However, iridectomy has been performed so frequently, and for so many years, that in all probability all its variants and effects are well-known already. It may be concluded that clinical facts are difficult to interpret, that they can easily be misinterpreted, and that they are perceived by us through the prism of theories. Numerous similar examples can be noted by anyone who reads medical books or listens to doctors' discourses.

The investigators should not be blamed for such lack of precision in science. It is the consequence of the intricacy of the phenomena with which we cope. We have to create theories and hypotheses because without them we would be completely lost in the unpenetrable forest of clinical facts. But if the hypotheses are to tolerably satisfy critical minds, if they are to be a little bit sounder and closer to the truth, they should be based on simple facts, or at least on certain and stable ones.

ZYGMUNT KRAMSZTYK
'ON BEING-UP TO DATE'

(*Krytyka Lekarska,* 1907)

The more distant the object I am looking at is , the wider the space I can cover with my sight. From the Garluch mountaintop or from a boat in the middle of the ocean, I have the impression of covering the whole hemisphere with my sight; the immense dome of the sky meets the equally extensive area of earth all along the horizon; I can see mountains, seas, fields, forests, towns, hamlets, lakes, rivers, roads.... I can see all these at once, I can match the details, observe the borders, understand relations and interactions. But if an object is located somewhere between my eye and this immense view, the closer the object, the larger the part of my view that will be blocked out. Finally, if the object has been placed just in front of my eyes, a finger, a tiny leaf or a small rod will be sufficient to hide all the remaining part of the world, and this object will become the one and only thing I am able to see.

Usually one remains only for a short time on top of a mountain, but closely located objects may be found every day in front of one's eyes. Therefore, a person has only infrequent occasions for admiring breathtaking, vast spaces; instead, he may every day look at his fingers or a tiny rod and lose everything else from sight. This tiny blind, this fleeting object of momentary notice is the novelty.

A man lives for the present moment, for novelty. He can readily forget that it is passing by, that it is but a small, swift wave in an immense flood of waters. This small wave completely fills his mind, and ties him down. The life of a man is occupied with today's pain, today's joy, today's hunger and worry and hope. He forgets yesterday and does not have enough time to contemplate tomorrow.

Novelty is the fame of a vagabond dancer, a new athlete in the circus, a railway accident, a newly-elected high ranking official, a jest in the newspaper. A great majority of people do not speak and cannot even think about anything else but such timely events.

Novelty in medicine is the dry cupping method announced loudly by some doctor, or a new salicylic preparation, or a new therapeutic method boisterously presented at the latest conference but doubted by its promoter. Doctors, like other people - or, perhaps more than other people - are always looking for an occupation for their minds or for their debates, to introduce slight modifications into the medical routine, and above of all to provide something suitable to print.

All these things are novelties, and their expression and their banner is the latest issue of the journal which publishes short reports. Nobody today discusses yesterday's curiosity; everybody neglects everything from the past; everybody neglects history, but no one looks into the future or has a premonitory feeling about it. The latest scientific discoveries seem to be definitive ones, and, at the same time, the first ones worthy of attention. True science has begun today, or yesterday, or at most in the last century, and everything else is viewed as out-of-date, nothing worth our attention, just child-like phantasies. There is practically no large view before us, for we have only a clear picture of things that are just in front of our eyes. The closer the horizon is to a man's sight, the more readily he is convinced that he knows everything, because a tiny object that is blocking the world from his eyes seems to him to be the whole universe. He does not know, nor does he even have an intuitive feeling, that far beyond this little object there exists a world of immense, infinitely various and enduring things.

Novelties not only limit the expanse of the mind and insult its dignity, but also cause very grave harm to science. In the history of ever-changing scientific notions, science assumes its proper form, that of an everchanging stream. If, however, it is cut off from history in its momentary, present form, it looks fixed, hardened, coagulated, even if sometimes it may have the look of a beautiful statue. A look into the future, an aspiration or craving for something better, reveals our limits and indolence, and it shows us from afar a more splendid science and a more complete happiness which need and deserve to be fought for, which may lend wings to the desires, which may stimulate our work and efforts. An extensive look embracing the whole of science demonstrates the size and the limits of the essential scientific notions and their interrelations. It makes a uniform, magnificent tree of science out of tiny, scattered grains of observations. Once in a while a great event or an extensive scientific or social change tears up the curtain and shows us for a short moment the vast worlds of life and science, but the view must soon be veiled and the mind must return to its small backyard. We may be likened to conches hiding in the marshes: hardly once in a year, when they are pushed outside by a storm, they show up on the surface of turgid water, open their mouths, sigh to heaven, and return to their grave.

A spark and a moment - novelty - will always attract and chain people down. To open the human mind, to extend, to animate, to refine it, to enhance it and to bend its route to new roads, to make people happy with the greatest happiness, it is indispensable to develop broad views, to view science from a large perspective, to unveil history and - even if only by dreams or yearning - to look to a better future.

The struggle against the worship of the present is indispensable, but it is a very difficult task and does not bring any satisfaction. For it is difficult to attract people to vast spaces. It leaves one alone, not only without followers, but even without a small number of faithful companions, nothing but mute chords on the harp and deaf listeners.

CHAPTER VI

FROM *MEDICAL CRITIQUE* TO THE *ARCHIVES OF THE HISTORY AND PHILOSOPHY OF MEDICINE*: THE INSTITUTIONALIZATION OF THE POLISH SCHOOL OF THE PHILOSOPHY OF MEDICINE

Medical Critique, although having survived for only 11 years, existed during the ascendancy of the Polish School of the Philosophy of Medicine. In this period, Biernacki published *The Essence and Limits of Medical Knowledge* (1899), *Chalubinski and the Present Goals of Medicine* (1900), *The Principles of Medical Understanding* (1902), and *What is Disease?* (1905), Kramsztyk published his *Critical Notes on Medical Subjects* (1899), and Biegański *The Logic of Medicine or The Principles of General Methodology of The Medical Sciences* (1894), *General Problems of Theory of The Medical Sciences* (1897), *Thoughts and Aphorisms on Medical Ethics* (1899) and *The Logic of Medicine or The Critique of Medical Knowledge* (1908). This intensive activity was, in all probabilility, closely related to the existence of a specific journal dealing with philosophico-medical subjects and permitting the exchange of ideas, information and the development of debates on the critical evaluation of medical knowledge.

In an article celebrating the first year of *Medical Critique,* the histologist and philosopher of medicine, Henryk Hoyer, stressed the pedagogical aims of the journal. Its goal was redefined as teaching physicians the importance of critical thought in medicine and bringing them elements of philosophical knowledge [12]. Indeed, *Medical Critique* published a number of philosophically-oriented essays, mostly dealing with the philosophy of the natural sciences. The two principal authors active in this movement were Bieganski and Hoyer. Biegański published several essays in *Medical Critique:* 'The Main Orientations in 19th-Century Medicine' (1901); 'Neo-vitalism in Modern Biology' (1904); 'The Teleology of Pathological Phenomena' (1905); 'The Concept of Causality in Biology' (1906). The second author of philosophically-oriented articles, Henryk Hoyer (1834-1907), was above all a well-known physiologist and histologist who struggled to maintain his scientific research in Poland in the harsh conditions that followed the 1863 insurrection. Later in his life, because of the deterioration of his eye-sight, he was obliged to reduce his scientific activity, and he became interested in the philosophy of medicine - or, rather, in the philosophy of the biological sciences.[1] Hoyer's essay on 'The Fundamental Ideas of Science' opened the first issue of *Medical Critique* (1897). Later, in the same journal, he published his studies on 'The Importance of Historical Knowledge' (1899), on 'Scientific Laboratories' (1903), and on 'Theory and Practice' (1906). Although Hoyer is usually listed along with other thinkers of the second generation of the Polish School, his philosophical ideas were quite different from those of Biegański, Biernacki, and Kramsztyk. All his philosophical works were characterized by a strong belief in the importance of the fundamental sciences to the progress of medicine. Hoyer was aware of the complexity of living organisms, but he viewed this fact merely as an obstacle to the rapid progress of scientific biology and medicine. A representative of the generation of medical scientists trained during the "golden age" of German physiology and histology, he strongly affirmed his faith in the future of scientific medicine.[2]

Medico-philosophical subjects were not the only topic discussed in *Medical Critique*. The journal had seven main divisions: 1) theoretical reflections on biological and medical subjects, 2) history of medicine, 3) the education of physicians and problems of the medical profession, 4) biographies of physicians, 5) critical evaluation of medical articles and publications, 6) notes about medical language (an important subject in occupied Poland, in which the struggle for the maintenance of the Polish langage was one of the principal forms of national resistance to foreign occupation), and 7) medical news and current problems. The journal introduced from time to time other subjects, too: thus a new division, "Public and Hospital Hygiene" was introduced in 1900, but it disappeared the year after. The last two divisions (medical language and medical news) were usualy discussed in short notes. Many of the longer articles dealt with the critical evaluation of scientific publications, and with theoretical aspects of biology and medicine. Classic studies in the history of medicine and biographical notes of known physicians were less numerous than more theoretical works. The most frequently discussed subject was, however, the concrete problems of the medical profession. Articles dealing with this subject were classified either in the first division (if they were more theoretically-oriented) or in the second division (if they were more practically-oriented), and they often occupied a central place in both those sections.

The predominance of articles discussing the concrete conditions of medical practice in Poland gave *Medical Critique* a clearly social orientation. For example, one series of articles published around 1900 dealt with the problem of medical education, and criticized the over-emphasis on theoretical subjects in the medical curriculum and the lack of adequate practical preparation of future physicians. Several other articles denounced the over-mercantilism of the medical profession and the lack of honesty of some of its members. Others discussed the neglected state of many hospitals, the lack of adequate therapeutic means, and the difficulties of coping with the poverty of most of the patients. Another much-debated topic was the reasons for the relative backwardness of Polish medicine and the ways to remedy this situation. These articles did not hesitate to strongly criticize the lack of originality and the provincialism of Polish medicine and bio-medical research [8, 14]. Clearly, *Medical Critique,* far from being a purely theoretical journal, reflected the whole range of preoccupations of practicing physicians in Poland at the end of the 19th and the beginning of the 20th-century.

Medical Critique ceased to appear in 1907. The *Lwow Medical Weekly, [Lwowski Tygodnik Lekarski],* which had on its editoral board Biegański, Kramsztyk, Szumowski (who was starting out as a philosopher and historian of medicine) and the philosopher Kazimierz Twardowki, the founder of the Lwow School of Philosophy (Adjukiewicz, Kotarbinski, Lukasiewcz, etc.) at first was intended to partially replace *Medical Critique* [60]. But in the end the role of philosophico-medical reflection in the *Lwow Medical Weekly* was very restricted. Only from time to time (mostly in the years 1910-1911) did this journal publish a supplement with a few theoretically-oriented articles. In fact, between 1908 and 1924 the Polish School had no publication of its own. When Adam Wrzosek founded in 1924 the *Archives of the History and Philosophy of Medicine* he hoped to create such a publication again. Wrzosek clearly viewed the new journal as a continuation of the tradition of *Medical Critique.* In his presentation of the *Archives* project he stressed the continuity of the tradition of Polish philosophy of medicine - from Dworzaczek and

Chałubiński, through the studies of Biernacki, Biegański, Kramsztyk, Hoyer and Nusbaum - and explained that the ideas discussed in *Medical Critique* had never lost their relevance [53]. But the analysis of the writings of the third generation of the Polish School, which appeared in the *Archives,* shows a gradual abandonment of the specificy practice-oriented tradition of the Polish School in favor of more traditional approaches to the philosophy and history of medicine. What follows is the story of the failure of the effort to revive the spirit of *Medical Critique.*

Henryk Nusbaum (1849-1937) was the only one among the founders of *Medical Critique* who was a member of the editorial board of *Archives of the History and Philosophy of Medicine.* Nusbaum was born in a family of Jewish intelligentsia which, like Kramsztyk's family, actively supported the assimilation of Jews in Poland. His father, a journalist and publisher, participated in the 1861-63 insurrection and later fought for the integration of Jews in Polish society. Nusbaum studied medicine in Warsaw, then in Dorpat, and was trained in physiology and in physiological chemistry in Vienna, Bern, and Paris. In Warsaw he worked in the physiology laboratory headed by Felix Nawrocki. He failed, however, to obtain a university position and was obliged to maintain a private practice as a neurologist, sharing his time between his patients and his philosophical interests. Nusbaum became active in the movement for the assimilation of Jews, and for this reason he was opposed by both Poles and Jews. He finally decided in 1918 to convert to Catholicism, a decision that may have helped him later to obtain a university position. In 1920, Nusbaum (at the age of 71) was appointed lecturer of philosophy of medicine at Warsaw University and in 1923 he became an honorary professor of history of medicine.[3]

Nusbaum was the symbol of the continuity between the second and third generation of Polish philosophers of medicine. He was not, however, a typical representative of his generation. He did share some of the ideas developed by Biegański, Biernacki, and Kramsztyk, such as the the need to recognize the complexity and the individuality of disease phenomena and the impossibility of creating a natural classification of diseases. But his philosophical reflections were less oriented toward understanding the nature of medical practice, and more toward the study of medical science, viewed by him as part of the biological sciences and subject to the same strict causal laws as the other natural sciences [18].

Nusbaum published in the *Archives* a series of articles on the philosophy of medicine that centered around the familiar question: Medicine - art or science? This subject had already been discussed by him in 1895 in several articles published in the *Medical Journal [Gazeta Lekarska]*. In those early works Nusbaum affirmed that medicine is at the same time science *and* art, or, rather that it is an applied science. He explained that while the aim of fundamental science is the search for truth, and there is but one truth, applied science (i.e., technology) proposes less perfect results and offers more than one solution to a given problem. For Nusbaum, therapy is an applied science and not an art (i.e., craft) because every therapy is in fact an experiment with unknown results. Only if the rules of therapy were entirely codified and its results predictable, could it become an art [16].[4] In a series of articles published in the *Archives* thirty years later, Nusbaum slightly modified his original concept by affirming that one should distinguish three components of medicine: fundamental science, applied science, and art. As a fundamental science medicine belongs to biological studies. It is a very young science because it was born as a result of the progress of physics and chemistry in the 19th-century. As an applied science, medicine deals with the study of the consequences of diseases and of therapy. Finally, as an art medicine deals with the prevention and alleviation of human suffering. This model of medicine as having three (instead of two) separate aspects - art, applied science and science - reflected, in all probability, the phenomenon already visible in the 1920's, of the growing separation between clinical and fundamental biomedical research. Nusbaum argued that it was difficult, if not impossible, to delineate the limits between fundamental and applied science, and between applied science and art. Each aspect influenced the others and was influenced by them. Nusbaum's text strongly suggested, however, that he did not consider these three aspects as equally important. Fundamental research shaped *in fine* the other two aspects of medicine. Although Nusbaum recognized the specificity of medicine, he considered that progress in this discipline was dependent upon the future development of the reductionist, physico-chemical approaches to the study of pathological phenomena [19, 20].

According to Nusbaum, physicians need to study philosophy in order to better understand the phenomenon of life. This idea was first explained in 1898 in his article, 'The Importance of Philosophical Training for Physicians' [17]. It later became the center of his most important philosophical book: *Philosophy of Medicine* (1926). The title of the book does not adequately reflect its content. The book focuses on the history of the evolution of the concept of life from antiquity to the 1920's. Only at its end is there a chapter on medicine as an applied science and as art, in which Nusbaum reiterates the arguments developed in his earlier articles. He also added a brief chapter on the physician's social tasks. Nusbaum's conclusion was that medicine is above all a science whose main goal is the understanding of the phenomenon of life. Such a vision of the essence of medicine is very different from the one advocated by the mainstream thinkers of the Polish School. It is thus not surprising that in this book Nusbaum did not mention any of the other Polish philosophers of medicine, nor did he allude to their works [21].

After the First World War Nusbaum clearly separated himself from the tradition of the Polish School. This was not the case for the second representative of the older generation of Polish physicians who was active in the *Archives,* Stanisław Trzebiński (1861-1930), a professor of history and philosophy of medicine at Vilnaus University [58]. Trzebiński's views were closer to the ideas of his generation of philosophers of medicine. Although Trzebinski did not write in *Medical Critique* (he was at that time busy with his medical practice and only later in his life did he start writing articles on philosophico-medical subjects), he regarded himself as a follower of this tradition. He published several articles in the *Archives* that discussed the thought of the previous generation of philosophers of medicine (Biegański, Chałubiński). In these articles he attempted to study the genesis of these thinkers' ideas and to critically evaluate these ideas in the light of the new developments in medicine and biological science [40, 41, 42].

Trzebiński had a particular interest in both rationality and superstition in medicine. He refuted the view which opposed the rationality of modern scientific medicine, based on the latest acquisitions of science, to the irrationality of earlier medicine. He argued that every therapy was perfectly rational in its time. But when the dogma on which it was based was rejected, the same therapy (e.g., bloodletting) became irrational. In medicine,

Trzebinski explained, what is recognized as absurd is frequently just something that is in contradiction to the dominant opinions of a given period. An idea that was clearly recognized as absurd, may appear much less so when medical theories are replaced by others. For example, the idea of using dried material from plague bubos to prevent infection, first developed through a Paracelsian analogy of healing with a substance that is similiar to the disease itself, was later viewed as utterly absurd until it was partially validated by the theory of immunity. Thus, Trzebiński affirmed, it is impossible to state in an unequivocal way that a given treatment is absurd. One can only make statements about its practical value. For example, Broussais's "system" was finally abandoned because the excessive bleedings that he prescribed were harmful to the patients. In contrast, a similar system - homeopathy - was never abandoned by physicians because, Trzebiński explained, homeopathic treatment is never harmful, particularly if administered together with other therapeutic means, and it can even be helpful due to its positive psychological influence on patients [39].

Trzebiński argued that excessive faith in the rationality (i.e., for him, ideology) of a given time can, and often does, induce errors in observation. For example, in the Middle Ages preconceived notions concerning the human body strongly influenced the anatomical drawings of that time. Trzebiński also recommended not rejecting immediately as "non-scientific" popular observations and so-called "superstitions", for some of these superstitions are later found to contain elements of truth [38]. Finally he criticized the excesses of *rational therapies*. Physicians who are uncritical followers of scientific medicine aspire to a fully rational therapy. But they forget that the complexity of pathological phenomena makes this ideal impossible to realize, at least in the near future. Physicians often behave, however, as if this ideal were possible to realize, and have a tendency to confuse *rationality* with *rationalism* - a blind faith in what they consider to be rational medicine. They forget, therefore, to pay attention to the role of irrational factors in illness - an attitude that Trzebiński qualified as being quite irrational. It is very important, he explained, to take into consideration the role of non-rational factors in healing. Physicians who disregard the therapeutic role of the patient-doctor relationship often complain that patients who are not satisfied by their rational therapy, claim to be cured by some uneducated healer. They do not realize that this healer is in all probability able to influence some psychological aspects of the patient's illness which are neglected by the physician. Thus, one can say that from a certain point of view the behavior of the healer is more rational than the behavior of the physician [37].

Trzebiński, who died in 1930, had only a limited influence on the development of the philosophy of medicine in Poland between the two World Wars. The key person in this movement was undoubltedly Adam Wrzosek (1875-1965). Wrzosek was active in the transition between the second and third generations of Polish philosophers of medicine, and later played a central role in the institutionalization of the philosophy and history of medicine in Poland.

Wrzosek studied medicine in Kiev, then in Berlin (1898), and from 1906 taught pathology at the University of Krakow. He also studied anthropology and specialized in physical anthropology. At the same time he developed an interest in the history and philosophy of medicine and published his first articles on these subjects in *Medical Critique.* He was, among other things, the author of the first detailed biography of Jedrzej Śniadecki [49]. After the proclamation of the independence of Poland his position as head of the department of higher education in the Ministry of Education allowed him to make a decisive contribution to the creation of five chairs in history and philosophy of medicine in schools of medicine [52]. He himself later occupied one of the newly-created chairs, that of Poznan (1921). He founded there the *Archives of the History and Philosophy of Medicine* (1924). Wrzosek also centralized the activities of the Polish Society of Amateurs in History of Medicine, a federation of several local societies (Warsaw, Krakow, Vilnaus, Lwow, Poznan), which held annual or bi-annual meetings and published its announcements in the *Archives.* A highly prolific writer, he published approximately 470 articles during the years 1898-1964 [22]. Besides his work in the field of history and philosophy of medicine, Wrzosek was also active in anthropology. He founded (1926) and edited the first Polish journal of anthropology, and contributed numerous anthropological studies. He attempted to reconcile his interest in anthropology with his work bearing on the history of medicine by advocating the study of the contribution of the racial factor to scientific creativity [50].[5] The central place occupied by Wrzosek in the institutionalization of the history and philosophy of medicine in Poland enabled him to influence these two disciplines and to obtain a wide hearing for his rather peculiar views.

Wrzosek started his career as an historian and philosopher of medicine: in the articles he published in *Medical Critique* (1899-1900), Wrzosek was faithful to the general principles advocated by this journal, and, at that time, he showed interest in the concrete problems of medical practice. For example, he strongly criticized the tendency to train future physicians by showing them "interesting" pathological phenomena, while forgetting to teach them how to deal with frequently found, "uninteresting" cases. He also criticized the attitude of physicians who sought "universal" treatments and forgot that they were not dealing with disease - an abstract concept - but with sick individuals. He strongly denounced unethical clinical experimentation (such as Neisser's injections of extracts of Treponema syphilis into healthy men), and the recent rush to publish, which transformed laboratories into factories for scientific papers and encouraged premature publication (e.g., Koch's tuberculine). The essays Wrzosek published when he was 25 also showed concern for social issues. For him the medicine of the future would be social medicine, and its main preoccupation would be the prevention of disease [44, 45]. In another article Wrzosek discussed the problem of healers. He disagreed with Biernacki's opinion that medical sects and healers could not be eliminated because there would always be patients whom official medicine was incapable of helping, and who would thus turn to the "miracle makers". Wrzosek affirmed, in contrast, that the clients of healers in Poland were not the disappointed patients of classical doctors, but rather uneducated peasants and workers who went to the healer because he was cheaper, and because he knew how to speak their language. He therefore advocated better education, low-cost medicine, and more devoted country doctors as remedies against quacks, medical sectarians, and healers [46].

Later, however, Wrzosek abandoned his interest in the practice of medicine and centered his reflections on the history of medicine and on general philosophical considerations. The beginning of this trend can already be seen in an article, in 1901, dealing with the relationships between philosophy and medicine. Wrzosek first noted that many among the well-known philosophers were physicians. He tried to explain this affinity between philosophy and medicine by the fact that philosophical ideas often influenced the development of concepts of disease and methods of healing. On the other hand, physicians' ideas on psychic phenomena, human will, etc., were of interest to philosophers. The trend to link medicine and philosophy was particularly important in the 19th-century. Famous physicians such as Virchow, Helmholtz, Büchner, Wundt, Haeckel, and Lotze could also be classified as philosophers. For Wrzosek, however, the physician-philosophers were above all those physicians who contributed to the evolution of theoretical reflections on biology, not those who dealt with the theoretical aspects of medical practice. Consequently, he viewed the theoretician of biology, Jedrzej Sniadecki, as the most important Polish physician-philosopher. Wrzosek refuted the claim that the time of physician-philosophers was now over, due to the growing specialization in medicine. The over-specialization, Wrzosek affirmed, is perhaps not the result of the growth of medical knowledge, but on the contrary reflects the limits of physicians' knowledge. It is not to be excluded, therefore, that the future development of medical science would limit the number of specialties, would increase the fundamental unity of this discipline, and would see the return of the physician-philosopher [47, 48].

Later Wrzosek published in the *Archives* numerous biographical studies of either well-known Polish physicians or Polish philosophers of medicine (Chałubiński, Biegański, Biernacki, Trzebiński). When he discussed an original subject, e.g., the definition of disease, Wrzosek first recapitulated the ideas of all his precursors (Sniadecki, Chałubiński, Hoyer, Biegański, Biernacki), and then developed his own definition, a variation on the definition proposed by Biegański in his *Logic of Medicine* [54].[6] Wrzosek stated first that the definition of disease reflected the state of medical knowledge in a given period. Thus the definition of disease in antiquity was based on the notion of equilibrium of the organism, in the 19th-century on an anatomic-pathological perspective, and in the 20th-century, thanks to the rapid evolution of bacteriology, diseases were defined by their etiology. But Wrzosek, perhaps fearful of the relativistic implications of this affirmation, hastened to add that "Polish philosophers of medicine avoided the shortcoming of such a unilateral view" and developed a more complete view of disease - which was less dependent on the ideas of a given time [54].

The central place attributed by Wrzosek to the achievements of Polish physicians in general, and of Polish philosophers of medicine in particular, was quite typical of his later writings. Such a tendency can be observed in the writings of other collaborators of *Archives* as well. It is true that the philosophers of medicine of the earlier generation (Kramsztyk, Biegański), who wrote under foreign occupation, often formulated national aspirations. They affirmed that one of their goals was to contribute to the evolution of original and independent medical thought in Poland. But they stressed the necessity of the development of such thought precisely because they were keenly aware of the serious shortcomings of Polish medicine in their time. Such critical evaluations of the present situation of Polish medicine, one of the principal subjects of numerous articles in *Medical Critique,* were practically absent from *Archives.* For the authors who wrote in *Medical Critique* (particularly for Kramsztyk) the history of medicine was important not because it could serve to glorify the past, but because an historical perspective was an indispensable condition for a critical evaluation of the present. In contrast, the great majority of historical articles published in *Archives* (in particular in the 1930's) were hagiographic and nostalgic evocations of the past, with no critical dimension whatsoever.

Wrzosek's other preoccupation, i.e., the role of the Catholic religion in the development of science, was also typical of the new tendency of *Archives.* Three of its most important collaborators (Wrzosek, Szumowski and Bilikiewicz) developed a pro-Catholic and anti-materialist point of view. Such religious preoccupations were alien to the preoccupations of the earlier generation of Polish philosophers of medicine who separated their religious faith (or the absence of it) from philosophico-medical issues, and never discussed religion in their publications. Wrzosek tried to correct this omission by carefully enumerating all the well-known Polish physicians and philosophers of medicine who were practicing Catholics (Śniadecki, Dietl, Chałubiński, Biegański), affirming that "There was no single eminent physician in the history of Poland who was an atheist, pantheist or materialist, because all our great physicians, like all our great political leaders, poets, and thinkers, were impregnated with deep religiosity in the spirit of the Catholic faith" [55].[7]

Another important collaborator of *Archiwum,* Władysław Szumowski (1875-1954), also strongly supported an anti-materialist point of view. He also represented the new tendency among the *Archiwum* group - the abandonment of specific interest in medical practice in favor of a return to professional history and professional philosophy [26]. Szumowski studied medicine in Warsaw, then physiology in Friburg (with Arthus), and prepared himself for a research career. His career was cut short, however, by tuberculosis. As the disease forced him into inactivity, Szumowski started to study history and philosophy at Lwow University. He wrote a thesis in philosophy ('On the Theory of the Passions in the Writings of Descartes and Malebranche') under the direction of K. Twardowski (1907). He also studied (under the direction of the historian Finkel) the life of the Polish physician Krupinski, active in the late 18th century. In 1908, Szumowski began to teach history of medicine at Lwow University (it was the first full-time course in the history of medicine in a Polish University); he continued his teaching until 1920, while maintaining a part-time medical practice. In 1920, he was appointed to the Chair of History and Philosophy of Medicine at Krakow University. Szumowski thus had formal training in medicine and in history and philosophy. Later he strongly insisted on the necessity of such multi-disciplinary training for each aspiring historian or philosopher of medicine. For him, professional training in the methods of humanistic investigation was an indispensable condition for the future development of the history and philosophy of medicine in Poland.

In 1919, Szumowski, together with Wrzosek, played a decisive role in the establishment of chairs of philosophy and history of medicine in various medical schools in Poland. Szumowski always believed that the history of medicine should be philosophically oriented, and explained this point of view in his *magnum opus,* a textbook of the history of medicine entitled *The History of Medicine from a Philosophical Point of View* [32]. He was the originator of the idea of combining the teaching of the history and philosophy of medicine in a single chair. This idea was proposed by him for the first time in articles published in 1919 and 1920 [28, 29]. Szumowski also played an important role in the diffusion of ideas of the Polish school of philosophy of medicine. His affinity for the study of the history and philosophy of medicine was inspired by Biegański's writings, then by the works of Chałubiński and Biernacki. From 1917 on, he attempted to popularize these already half-forgotten works. In 1917, he published an article entitled 'A Few Words on the Polish Medico-Philosophical School' [27], the first work which stated that such a school existed, and which gave it a name. He also presented papers on the Polish

School at the International Congresses of History and of History of Science [33] [36], and exposed his readers to the ideas of the Polish school of philosophy of medicine in his books *The History of Medicine from a Philosophical Point of View and Philosophy of Medicine* [35].

In his historical writing Szumowski advocated the idea of the importance of religion for the evolution of medical science. He agreed with other philosophers of the Polish School that medicine cannot be simply classified as a science, because complete explanations of pathological phenomena, or even of biological phenomena, were not yet in sight. He thought, however, that medicine had a strong scientific component, and that this component did not appear, as Biernacki or Nusbaum affirmed only in the 19th-century, but already existed in ancient Greek medicine and had never been absent from Western medicine. Medicine was thus not only an ancient art, but also one of the oldest sciences [31]. Szumowski claimed that the progress of medical science had its roots in two orientations: theological determinism and mechanical materialism. Both contributed to the evolution of medical knowledge. But, Szumowski added, 19th-century medicine was unfortunately over-dominated by materialist currents. He therefore observed with satisfaction the return in the 1930's and '40's to the more spiritual and anti-materialist traditions in medicine [34, 35].[8]

Szumowski often stressed the importance of adequate training in history and in philosophy for the professionalization of history and philosophy of medicine [29]. The youngest among the permanent collaborators of *Archives*, Tadeusz Bilikiewicz (born in 1901), reflected the new tendency to abandon reflection on medical practice in favor of professional research in the history of medicine.[9] Bilikewicz studied medicine and psychiatry in Lwow, then philosophy in Krakow. He published, in 1928, his first philosophical work *The Problem of Life in the Light of the Principles of Comparative Psychology*, which represented a dualist approach (separation between body and soul) to the problem of life [2]. In 1926-1928, Bilikiewicz went to Switzerland to complete his training in psychiatry. There he met Henry Sigerist, one of the leading historians of medicine in the period between the two World Wars, who in 1928 became head of the Institute of the History of Medicine in Leipzig. Sigerist helped Bilikiewicz to obtain a scholarship from the Rockefeller Foundation in order to study the history of science. Bilikiewicz worked, in 1930, in Leipzig (collaborating there with Sigerist, Sudhoff, and Temkin), and in 1932 in Paris. He prepared, under Sigerist's direction, a thesis on *Embryology in the Baroque and Rococo Periods* [4], in which he argued that there was a correlation between the socio-cultural development of a given society and the evolution of scientific views that prevailed in it (e.g., Bilikiewicz affirmed that the description of spermatozoids as autonomous, freely moving entities was related to the decline of absolutism, and the position of "ovulists", who attributed the principal role in embryogenesis to the egg, was related to the growing role of women in French court society). Bilikiewicz claimed that Sigerist proposed that he stay in Germany, but he preferred to accept Szumowski's offer to participate in the teaching of the history and philosophy of medicine at Krakow University. When in Poland Bilikiewicz did not continue the direction suggested in his thesis. In his later studies he did not follow the study of the relationships between the development of medical knowledge and socio-cultural factors, and he worked primarily on biographical studies of Polish physicians in the 16th and 17th-centuries.[10] Bilikiewicz was unable to obtain a university position as historian of medicine in Poland, and he thus worked as a psychiatrist [7]. Only in 1945 was he successful in being appointed to a chair of history of medicine at Gdansk University, where he organized an Institute of the History of Medicine.[11]

During its fifteen years of existence (1924-1938), the *Archives of History and Philosophy of Medicine* published 134 articles on the history of medicine, 26 articles on sources of history of medicine, 26 original memoirs and letters of known physicians, 2 translations of classical works in medicine, 17 articles on the history of biology, and 27 works in the philosophy of medicine [59].[12] The *Archives'* content was thus clearly dominated by the history of medicine. From 1930 on, one can observe an intensification of studies in the history of Polish medicine, and in particular a large number of biographies of famous Polish doctors; at the same time, philosophical reflections on medical practice disappeared.[13] The most striking difference between *Medical Critique* and the *Archives of the History and Philosophy of Medicine*, was, however, not the proportion of articles on the history of medicine, but the virtual absence of a critical spirit in the *Archives*. In contrast to *Medical Critique*, the *Archives* - as signaled by the absence of the word 'critique' in its name - practically never published articles which critically evaluated medical knowledge, or the daily behavior of physicians, or the social conditions in which they were obliged to work, or the ethical aspects of physicians' professional practice. Although when founded the *Archives* was meant to replace *Medical Critique,* the former gradually became only one more journal reflecting the traditional (i.e., erudite, descriptive, and non-critical) history of medicine.

NOTES

1 Hoyer studied medicine in Breslau, then physiology and histology in Berlin (he was Virchow's student). He worked in Breslau with Reinchardt, and in 1860 received an appointment as professor of histology at the Warsaw Medico-Chirurgical Academy. On Hoyer's difficulties to maintain an adequate level of scientific research after 1863, see [9, 15, 57].

2 Hoyer was not, however, a mechanist, and he affirmed that the existing physico-chemical concepts were not sufficient to understand all the phenomena of life. He was an advocate of the energetic theories advanced by German physicists at the end of the 19th-century, and he thought that those theories could better explain biological phenomena [11]. Hoyer's faith in the progress of science, quite alien to the way of thinking of other philosophers of the Polish School of his generation, may be illustrated by the fact that he viewed the history of medicine solely as "the most effective way of preventing the return to false orientations, fortunately now left behind us" [13].

3 Nusbaum taught only philosophy of medicine, not history of medicine, because Franciszek Giedroyc (1860-1944), who was nominated in 1921 to the chair of history and philosophy of medicine at the Warsaw University, decided to dedicate himself exclusively to historical and biographical studies, and separated the teaching of history from the teaching of philosophy [23].

4 Kramsztyk's article on this subject was published in the same year and in the same journal.

5 Wrzosek affirmed in this lecture that some nations, like the Chinese, "did not develop a science because they completely lacked scientific creativity." Later, in the 1930's, Wrzosek purportedly studied the relationship between racial type and intelligence, and

concluded, e.g., that among Polish students (of Catholic faith) the most intelligent had a nordic racial type and the less intelligent the oriental racial type [56].

6 The psychologist Stefan Szuman criticized Wrzosek's definition because, he explained, Wrzosek did not take into consideration the normative character of medical knowledge [25].

7 Wrzosek's list of "eminent Polish physicians" was quite short because he was obliged to eliminate from it all those who were not religious or, even worse, were of Jewish origin. It also, at least in one case (Biegański), included a physician who according to the testimony of his wife was never a practicing Catholic ([1], p. 279).

8 In his article, 'La philosophie de la médecine - son histoire, son essence, sa dénomination et sa définition' [36], Szumowski cited among the thinkers that brought about the shift to a more spiritual understanding of medicine Hans Dreisch and Henri Bergson, but also Carlo Lombroso (the criminologist) and Alexis Carrel (a French physician who worked in the USA, and who was known not only for his pioneering work on transplantation surgery and tissue culture, but also for his militant Catholicism and his sympathies with the political movements of the far Right). For Szumowski, Carrel was a paradigmatic case of a great physician and scientist inspired by the principles of faith. Szumowski also mentioned with appreciation several German physicians (Oswald Bumke, Auguste Bier) who published popular books in the early 1940's on the spiritual dimension of medicine.

9 Philosophical reflections on the practice of medicine practically disappeared from the *Archives* in the 1930's, and were partially replaced by articles written by professional philosophers: e.g., [43].

10 In an long autobiographical article written in 1978, Bilikiewicz hardly mentioned his doctoral thesis and did not discuss its content at all [7]. In 1937, Bilikiewicz warmly praised the state of the history of medicine in Germany and the evolution of the Leipzig Institute, but without mentioning the fact that the majority of its previous staff (including Sigerist, Temkin and Ackerknecht) was now in exile [6].

11 Bilikiewicz fought for institutional recognition of research in the history of medicine, and he strongly recommended the establishment of institutes for the history of medicine (such as the Leipzig Institute) in Poland. He maintained that the establishment of such institutes was the very condition for the further development of this discipline [3, 5].

12 The tendency to return to "classic" academic subjects in the history of medicine, and to abandon reflection on the practice of medicine, was perhaps related to the institutionalization of the history of medicine in Poland in 1920, and its transformation

into an academic discipline. As Szumowski already noted in 1926, "Unfortunately, one cannot say that from the time that the philosophy of medicine became a university discipline this field evolved in a significant way in Poland" [30].

13 Although Sigerist, in 1928, warmly praised the institutionalization of the teaching of the history of medicine in Poland, Polish authors had a much less optimistic outlook; they strongly criticized the stagnation of research in the history of medicine and stressed the hopeless future for this discipline in Poland [3, 24].

BIBLIOGRAPHY

[1] Biegańska, M.: 1930, *Wladyslaw Bieganski: Life and Work,* Wydawnictwo Kasy im. Mianowskiego, Warsaw, p. 279.

[2] Bilikiewicz, T.: 1928, *The Problem of Life in Light of the Principles of Comparative Psychology [Zagadnienie życia w świetle zasad psychologji porównawczej],* S.A. Krzyzanowski, Krakow.

[3] Bilikiewicz,T.: 1928, 'On the Institute of History of Medicine in Leipzig' ['O instytucie historji medycyny w Lipsku'], *Polska Gazeta Lekarska* **47** (7).

[4] Bilikiewicz, T.: 1932, *Die Embryologie im Zeitalter des Barock und des Rococo,* Georg Thieme, Leipzig.

[5] Bilikiewicz, T.: 1932, 'Some Remarks on the Research and Teaching of the History of Medicine' ['Kilka uwag o historji medycyny jako przedmiocie badań i nauczania'], *Polska Gazeta Lekarska* **11** (3).

[6] Bilikiewicz, T.: 1937, 'Berlin and Liepzig - Two Centers of History of Medicine in Germany' ['Berlin i Lipsk - dwa ośrodki historji medycyny w Niemczech'], *Polska Gazeta Lekarska* **17**, 5-6.

[7] Bilikiewicz, T.: 1978, 'Autobiographical Memories' ['Wspomnienia autobiograficzne'], *Kwartalnik Historii Nauki i Techniki* **23**, 3-51.

[8] Dunin, T.: 1899, 'Nationality and Science' ['Narodowość w nauce'], *Krytyka Lekarska* **3**, 169-176.

[9] Filar, Z.: 1962, 'Henryk Fryderyk Hoyer', *Polish Biographical Dictionary,* vol. 10, 38-40.

[10] Higgier, H.: 1928, 'The History of Medicine in Our Country' ['Historia medycyny u nas'], *Czasopismo Lekarskie*, June 6th.

[11] Hoyer, H.: 1897, 'Fundamental Scientific Concepts in the Light of the Theory of Cognition' ['Zasadnicze pojęcia naukowe w świetle teorii poznania'], *Krytyka Lekarska* **1**, 1-24.

[12] Hoyer, H.: 1898, 'On Medical Critique' ['O krytyce lekarskiej'], *Krytyka Lekarska* **2**, 129-136.

[13] Hoyer, H.: 1899, 'On the Importance of Historical Knowledge' ['O znaczniu wiedzy historycznej'], *Krytyka Lekarska* **3**, 1-6.

[14] Kramsztyk, Z.: 1898, 'In Foreign Clothes' ['W obcej szacie'], *Krytyka Lekarska* **2**, 18-36.

[15] Kramsztyk, Z.: 1907, 'Professor Henryk Hoyer - A Posthumous Memoir' ['Professor Henryk Hoyer - Wspomnienia pośmiertelne'], *Krytyka Lekarska* **11**, 38-40.

[16] Nusbaum, H.: 1895, 'Is Medicine a Science or an Art' ['Czy medycyna nauką jest czy sztuką'], *Gazeta Lekarska* **40**, 1044-1052; **41**, 1068-1075.

[17] Nusbaum, H.: 1898, 'The Importance of Philosophical Training for Physicians' ['O ważnosci wykształcenia filozoficznego dla lekarza'], *Krytyka Lekarska* **2**, 129-136.

[18] Nusbaum, H.: 1923, 'General Ideas on the Phenomenon of Life and on the Phenomenon of Disease' ['Poglad ogólny na zjawisko życia oraz na zjawisko choroby'], *Lekarz Wojskowy*, 392-424.

[19] Nusbaum, H.: 1925, 'Medicine as Art' ['Medycyna jako sztuka'], *Archiwum Historji i Filozofji Medycyny* **4**, 250-277.

[20] Nusbaum, H.: 1926, 'Medicine as Applied Science' ['Medycyna jako nauka stosowana'], *Archiwum Historji i Filozofji Medycyny* **5**, 74-93.

[21] Nusbaum, H.: 1926, *Philosophy of Medicine [Filozofia Medycyny]*, Wydawnictwo Archiwum Historji i Filozofji Medycyny, Poznan.

[22] Ostrowska, T.: 1977, 'Professor Adam Wrzosek and his Scientific Achievements' ['Professor Adam Wrzosek i jego dorobek naukowy'], *Wiadomości Lekarskie* **30** (1), 63-66.

[23] Ostrowska, T.: 1978, 'Henryk Nusbaum', *Polish Biographical Dictionary,* vol. 23, pp. 412-417.

[24] Sigerist, H.: 1927, 'Die Geschichte der Medizin im akademischen Unterricht', *Kyklos* p. 155.

[25] Szuman, S.: 1924, 'The Definition of the Concepts "Disease" and "Health"' ['Definicja pojęć "choroba" i "zdrowie"'), *Archiwum Historji i Filozofji Medycyny* **1**, 78-109.

[26] Szumowski, W.: 1911, 'The Biological Point of View and Historical Point of View' ['Pogląd przyrodniczy a pogląd historyczny'], *Gazeta Lekarska.*

[27] Szumowski, W.: 1917, 'A Few Words on the Polish Philosophico-Medical School' ['Parę słów o polskiej szkole filozoficzno-medycznej'], *Polski Miesiecznik Lekarski,* Kijow, 5-6.

[28] Szumowski, W.: 1919, 'On a Subject of Medical Studies Called History and Philosophy of Medicine' ['O przedmiocie studiow lekarskich zwanym historią i filozofią medycyny'], *Gazeta Lekarska* **11**.

[29] Szumowski, W.: 1920, 'The Philosophy of Medicine as a University Discipline' ['Filozofia medycyny jako przedmiot uniwersytecki'], *Przeglad Filozoficzny* **23**.

[30] Szumowski, W.: 1926, 'The Nearest Goals of Polish History of Medicine' ['Najbliższe w Polsce zadania historji medycyny'], *Archiwum Historji i Filozofji Medycyny* **4**, 53-59; 70-73.

[31] Szumowski, W.: 1927, 'Medicine as a Science and as an Art' ['Medycyna jako nauka i jako sztuka'], *Archiwum Historji i Filozofji Medycyny* **7**, 193-209.

[32] Szumowski, W.: 1930-1935, *The History of Medicine from a Philosophical Point of View [Historia medycyny filozoficznie ujeta],* Krakow.

[33] Szumowski, W.: 1933, 'L'école polonaise médico-philosophique', in *La Pologne au VIIème Congrès International des Sciences Historiques,* Societé Polonaise d'Histoire, Warsaw, pp. 3-9.

[34] Szumowski, W.: 1936, 'Medicine After Two and a Half Thousand Years' ['Bilans medycyny za półtrzecia tysiąca lat'], in W. Szumowski, *On Castrates and Castration, and Other Essays [O Enuchach i kastracji i inne pomniejsze artykuły],* Gebethner & Wolf, Krakow, pp. 114-126.

[35] Szumowski, W.: 1948, *Philosophy of Medicine [Filozofia medycyny],* Gebetner & Wolf, Krakow.

[36] Szumowski, W.: 1949, 'La philosophie de la médecine - son histoire, son essence, sa dénomination et sa définition, *Archives Internationales d'Histoire des Sciences* **2**, 1097-1139 (presented at the 5th International Congress of the History of Science, Lausanne, October, 1947).

[37] Trzebiński, S.: 1925, 'Rationality and Rationalism in Medicine' ['Racjonalność i racjonalizm w medycynie'], *Archiwum Historji i Filozofji Medycyny* **2**, 91-103.

[38] Trzebiński, S.: 1925, 'Superstition in Medicine' ['Przesąd w medycynie'], *Archiwum Historji i Filozofji Medycyny* **4**, 94-105.

[39] Trzebiński, S.: 1927, 'The Absurd in Medicine' ['Absurdalność w medycynie'], *Archiwum Historji i Filozofji Medycyny* **7**, 72-89.

[40] Trzebiński, S.: 1927, 'Observation, Experiment and Statistics in Bieganski's Medical Logic' ['Obserwacja, eksperyment i statystyka w Logice medycyny Władysława Biegańskiego'], *Archiwum Historji i Filozofji Medycyny* **6**, 173-194.

[41] Trzebiński, S.: 1929, 'Chałubiński's Views on Life-Process and on Disease' ['Zapatrywania Chałubińskiego na procesy życiowe i chorobę]', *Archiwum Historji i Filozofji Medycyny* **9**, 202-217.

[42] Trzebiński, S.: 1930, 'The Epidemiological Concepts of Thomas Sydenham and Chalubiński's Work "Malaria"' ['Tomasza Sydenhama koncepcje epidemiologiczne a Chałubińskiego "Zimnica"'], *Archiwum Historji i Filozofji Medycyny* **10**, 20-33.

[43] Wize, K.: 1931, 'The Scope of the Philosophy of Medicine' ['Zasięg filozofji medycyny'], *Archiwum Historji i Filozofji Medycyny* **11**, 82-107.

[44] Wrzosek, A.: 1900, 'Scepticism and Criticism in Medicine' ['Sceptycyzm i krytyka w medycynie'], *Krytyka Lekarska* **3**, 223-226.

[45] Wrzosek, A.: 1900, 'On the Direction of Modern Medicine' ['O kierunkach w współczesnej medycynie'], *Krytyka Lekarska* **4**, 313-317.

[46] Wrzosek, A.: 1900, 'On Healers and on the Means to Combat Them' ['O zanchorstwie i o środkach walki z niemi'], *Krytyka Lekarska* **4**, 105-110.

[47] Wrzosek, A.: 1900, 'Specialization' ['Specializacya'], *Krytyka Lekarska* **4**, 10-12.

[48] Wrzosek, A.: 1901, 'On Philosopher-Physicians and on Physician-Philosophers' ['O filozofach-medykach i medykach-filozofach'], *Krytyka Lekarska* **5**, 121-124.

[49] Wrzosek, A.: 1910, *Jedrzej Śniadecki,* Krakow.

[50] Wrzosek, A.: 1911, *Introductory Lecture on the History of Medicine [Wykład Wstępny o Historji Medycyny],* Drukarnia Piller-Neumann, Lwow.

[51] Wrzosek, A.: 1918, *On the Necessity of Philosophical Education for Physicians [O potrzebie studiow filozoficznych dla lekarzy],* Drukarnia Uniwersytetu Jagiellonskiego, Krakow.

[52] Wrzosek, A.: 1919, *On the Reform of Medical Schools [Myśl o reformie wydziałow lekarskich],* Drukarnia Jozefa Winiewicza, Poznan, pp. 101-118.

[53] Wrzosek, A.: 1924, 'Goals and Tasks of the Polish Archives of History and Philosophy of Medicine' ['Zadania i zamierzenia polskiego Archiwum Historji i Filozofji Medycyny'], *Archiwum Historji i Filozofji Medycyny* **1**, 1-13.

[54] Wrzosek, A.: 1924, 'The Concept of Disease' ['Określenie pojęcia choroby'], *Archiwum Historji i Filozofji Medycyny* **1**, 72-77.

[55] Wrzosek, A.: 1927, 'The Faith of our Eminent Physicians in the 19th-Century' ['Religijność naszych znakomitych lekarzy w XIX wieku'], *Archiwum Historji i Filozofji Medycyny* **7**, 159-167.

[56] Wrzosek, A.: 1931, 'On the Relationship of Certain Anthropological Parameters and Racial Types with Intellectual Capacities' ['O stosunku niektórych pomiarów antropologicznych i typów rasowych do sprawności umysłowej'], *Przęglad Antropologiczny* **5**, 1-15.

[57] Wrzosek, A.: 1934, 'Henryk Hoyer, on the 100th Anniversary of his Birth' ['Henryk Hoyer - z powodu setnej rocznicy urodzin'], *Archiwum Historji i Filozofji Medycyny* **14**, 195-216.

[58] Wrzosek, A.: 1935, 'The Life and Work of Stanisław Trzebiński' ['Życie i działanosc Stanisława Trzebińskiego'], *Archiwum Historji i Filozofji Medycyny* **15**, 96-116.

[59] Wrzosek, A.: 1939-1947, 'Nine Years Later: the New Series of Archives of the History and Philosophy of Medicine' ['Po dziewieciu latach: wznowienie wydania Archiwum Historii i Filozofii Medycyny'], *Archiwum Historii i Filozofii Medycyny* **18**, 314-316.

[60] Ziemski, S.: 1977, 'The Philosophy of Wladyslaw Bieganski' ['Władysław Biegański jako filozof'], in B. Skarga, *Polish Philosophical and Social Thought, [Polska Myśl Filozoficzna i Społeczna],* Ksiażka i Wiedza, Warsaw, pp. 499-531.

CHAPTER VI.A

TEXTS OF ADAM WRZOSEK AND STANISLAW TRZEBINSKI

ADAM WRZOSEK

'TRENDS IN CONTEMPORARY MEDICINE'

(*Krytyka Lekarska*, 1900)

The empiricism, which impregnated 19th-century medicine and modified its direction, had already existed in ancient medicine. In 3 BC there existed in Alexandria a medical school known as the empirical school, which stressed the importance of the observation of phenomena and of direct experience.

Later on, however, the tradition of the Alexandria school was lost, and a long period of dogmatism ensued, which resulted in a standstill in the development of science. Only in the 19th-century did medicine return to an empirical trend. That was the beginning of the most brilliant developments in medicine. The major contribution to this development was made by the Germans. The new trend was initiated by Johannes Muller. Although he was not absolutely free of the metaphysical influences of the preceding centuries, and although a vitalistic hypothesis played a substantial role in the evolution of his concepts, he was quite conscious of the extreme importance of experimentation in the natural sciences and thus in medicine as well. He viewed the theories of his time as hypotheses which required confirmation by facts. Cutting his ties with dogmatism, he turned to experiment, which was a foundation on which it was possible to build a lasting edifice of science, most importantly in biology. He succeeded in transmitting his ideas to many of his gifted students, including Schwann, Henle, Reichert, Remak, E. du Bois-Reymond, Virchow, Helmholtz, Brucke, Ludwig, Traube, A. v. Graefe, Max Schultze, etc. Continuing the work of their teacher, they have raised medicine to its present level. Other nations also contributed to scholarly efforts, but there is little doubt that the Germans played the most important role in 19th-century medicine. The new trend, however, was not definitively accepted without opposition. Even in Helmholtz's time there were physicians who laboriously avoided particular methods of physical examination of the patient. They believed that they could do without percussion and auscultation as well as the checking of temperature, and they viewed the ophthalmoscope as an entirely superfluous device for a physician with a good eye [1]. However, the new trend has gradually won over minds, and today the experimental method governs medicine. The only difference is that at first this method was at the service of a pleiade of famous students of J. Muller, men with philosophical background for whom empiricism was not the alpha and omega, the goal of research but only an instrument for the construction of true scientific theories. Today, however, the experimental method may become an experimental mania in the hands of minor scientists and be practiced for the sake of experiment in itself, or even, as we shall see later, become an instrument of torture.

There exist cases of misuse of empiricism by compromising or dishonest scholars. Not long ago, we were shaken by the experiments of professor Neisser, who injected syphilitic toxin into healthy men. And we had not quite recovered from the impression made by Neisser's experiments when we learned about the experiments of Dr. Stubell, described by himself in *Archiv für klinische Medizin,* vol. 62. He performed his experiments on patients suffering from diabetes insipidus. Here are several passages from his paper, in more or less exact translation:

> In the very first few days I found out that an accurate investigation could not be carried out on the first patient, Hertel, if he were not kept in isolation.
>
> He was moved to a small room in the attic which has two windows provided with solid iron bars and a door closed with a good lock, the key being always in my pocket. But if I had thought these to be a sufficient protection from cheating, I was wrong. Two or three times, when I had obtained inconsistent results, the patient admitted, after an energetic questioning, that he had kept the water vessel outside the window during a rainstorm and obtained almost half a litre of rain water from the gutter. Once I found that the patient had drunk water supplied to him for washing, since then I have prevented him from washing during the investigation. Another time, the patient could not withstand his thirst, and had drunk 1400 cc of his own urine. On the last day of the experiment, when very little drink was provided for him, he pulled the iron bars from the window and got out on the roof, from which, after pulling iron bars from another window, he got into a charwoman's room and there was caught just in time before he could reach the water tap.

> With my second case I used the isolation room again but had the bars strengthened threefold.

And the dispassionate report of the author continued:

> As a result of the experiment, the patient was substantially weakened. He spent the night from June 30th to July 1st in awful pain. At seven o'clock in the morning, when he had to climb two storeys up for a weight check and blood test, his strength failed him completely; his face was as if dried up, his eyes and cheeks were hollow, the pulse was almost imperceptible, the whole body aching, the joints looked ossified.

This is how Dr. Stubell, the first medical assistant in a Jena Clinic, described how he had tortured his patients.

Does this require any comment?

What has medicine become today? In ancient times it was mainly an art; in later centuries, and in particular in the 19th-century, it was a science. And now, as we approach the 20th-century, it is both art and science, and unfortunately very often a trade; for all these undiscriminating, routinized physicians are actually craftsmen, as are some professors of medicine cloistered within their own, sometimes very narrow specialties.

The trend towards excessively intense specialization, that can be observed in many medical disciplines, has a negative effect on medicine. Today a scientist is often a producer of scientific papers, while laboratories have become the factories where he manufactures his products. To become a scientist is not an extremely difficult task, with all these magnificent laboratories and libraries at hand. The only thing one needs is material resources which relieve the scientist from the need to work hard during his entire life to make a living. But someone with his living secured, someone who can for years study, e.g., a nerve, say the trifid nerve, can publish the fruit of his work in a few dozen papers. He is likely to become an expert on this one tiny detail, without having ideas on any other problem, and he will see no objection to calling himself a scientist. Even among the professors one can find such specialized, or should we rather say narrow-minded scientists. Don't we sometimes hear of an authority on heart diseases or someone enjoying a high reputation in the matter of biliary stones, etc.? This has an impact on the students, who naturally try to follow their teachers.

However, a professor should be first of all a good teacher, which means that he should have a general knowledge of medicine. How ridiculous are those professors who criticize all the other disciplines in order to praise their own fields. I remember a professor of internal medicine ridiculing pharmacy, and a surgeon who used to scoff at internists, etc. This can only mean that they were not able to conceive of anything but their own trade.

Besides specialization, contemporary medicine is characterized by the central place of empiricism. An experimental mania has developed. These vehement experimentors do not always use truly scientific methods, nor are then fully conscious of what they are after. Empiricism, however, can be of value only if it is accompanied by deep thought; only when the analysis is followed by synthesis. Otherwise, experiments will be subject to chaos. Unfortunately, it is easier to carry out experiments and to reveal them to the world in hundreds of specialized magazines than to experiment for the sake of some clearly defined object, that is, to experiment with a well-defined idea in mind. If the work of these passionate experimentors were performed exclusively on animals, it would not be so much of a problem. Unfortunately, physicians seem to forget the great motto of Hippocrates: *primum non nocere*. Experiments are undertaken not only on healthy men, as in the case of professor Neisser, but also on patients, as has also been mentioned. We learned about Dr. Stubell because he was so shameless as to publish his barbarous experiments in a medical journal. But how many such experimenters escape our attention, in particular, when patients have suffered too much from their experiments. Should this continue, we could apply to the physicians Lessing's verse:

"Ein Arzt, der keinen todt gemacht;
• • • • • • • • • • • • • • • • • • • •
Wo trifft man die? Vielleicht im Mond,
Wo jedes Hirngespinste wohnt."

One should also note the haste with which medical scholars carry out and publish their research. They put the results in print in the form of undigested matter which lacks comprehensive analysis. Sound works, written with talent and based on scientific methods, are few and far between. The era of the Claude Bernards, Pasteurs and Virchows is coming to an end with the period of Kochs,[1] Schenks, Neissers, etc. Textbooks such as those of J. Muller, or the physiology textbooks of Brucke, are not easy to find today, and the textbooks which are issued nowadays are very often undiscriminating, dull compilations.

There is still one very positive trend in contemporary medicine: prophylaxis. Much good can be expected from this direction. In general, physicians are more and more interested in questions of hygiene and disease prevention. Many valuable discoveries and inventions have been made in this field which, however, often cannot be applied for social and economics reasons. Rational medical aid, modernized hospitals, sanitary policies, the attention paid to occupational hygiene in schools and in factories - these are the areas which in our country still leave much to be desired.

The task of a physician should not be limited to healing. In the future it will be more important to prevent diseases, but the physician's most important task will be education of the people. Physicians will take care of the health of children and students, will prevent them from working overly hard, and, in addition, will give lectures on hygiene which perhaps will become as important for the education of children as catechism and religious instruction is today.

It is not enough to develop science. One should change society as well. What good may come from a device designed to protect the industrial workers if the factories refuse to introduce them? What good may come from teaching the rules of hygiene if the people are not able, or not willing, to obey them?

"Oh, precious good health,
what a savory wealth
you are, will find none,
until you have gone." (J. Kochanowski)

We can expect that a higher level of civilization will provide solutions that can be used in different social conditions, and which today are often out of reach for a physician possessing only theoretical knowledge.

The medicine of the future will be a social medicine that aims at protecting the health of the largest social groups.

NOTE

1 I have no intention whatsoever of refusing Dr. Koch the high credit due him as a scientist. Nonetheless, I should like to point out the untimeliness of his publication which presented tuberculin as a means to cure tuberculosis. Hundreds of consumptives were attracted to Berlin, where tuberculin injections speeded up their death by the dozen.

BIBLIOGRAPHY

[1] Helmholtz, H.: 1878, *Das Denken in der Medizin*, Berlin.

STANISLAW TRZEBINSKI
'RATIONALITY AND 'RATIONALISM' IN MEDICINE'

A lecture delivered at the academic year's inauguration ceremony at the Stefan Batory University, Vilnaus, 11 Oct, 1924
(Archiwum Historji i Filozofi Medycyny, 1925)

Please let me begin with a personal, albeit remote, reminiscence. It was, as far as I can remember, in 1885. I was a student of medicine in Heidelberg and an intern at the internal diseases clinic directed by Professor V. Dusch, an old and experienced doctor, and also a quite reasonable man. We visited the patients downtown, and later we reported to the professor the results of treatment, which we were entrusted to carry out under supervision of clinic assistants. It once happened that one of my fellow students, who had just administered Fowler's solution to the TB patient that was in his charge, stated with pride that in this way he had introduced a rational treatment of tuberculosis. The professor, generally known as a soft-spoken, good-natured man, reacted to this statement in an unexpected way. He interrupted the report and shouted that he would never endure such statements, since, as he explained, up to then a rational treatment of pulmonary tuberculosis, and particularly a treatment performed with drugs, was absolutely out of the question. As for me, I was completely taken aback. I knew, of course, that treatment of a given patient, even when applied by the best physicians, would not always necessarily yield positive results; on the other hand, my colleagues and I were convinced that a treatment should and could be rationally justified. Automatically I started to doubt whether only tuberculosis could not be rationally dealt with, and I wondered whether this rule didn't apply to other diseases, too.

I have remembered the Heidelberg incident many times in my life and in particular it came back to my mind when, a dozen years later, I was reading some of the philosophico-medical works of Edmund Biernacki, a philosopher and research worker, a man of high merit who unfortunately met untimely death.

If I take the liberty to ask you on this solemn day for the patience to listen to my reflections on this bygone event, it is neither an account of an old custom, which allows even doctors to poke fun at medicine, nor an account of a new one allowing one to poke fun at rationality. On the contrary, having much respect for the one and the other, I am still convinced that a discussion on rationality in medicine would make good sense. Of course, it has been done before - perhaps better and more thoroughly than I can do it now. But it so happens that things from the past should sometimes be recollected; so let this be my excuse.

To begin with, let us agree on the meaning of the terms we are going to use. Otherwise, we might misunderstand each other. We shall speak of "consistency" when a process of reasoning is conducted in a logical way, whereas no judgments are expressed on the veracity or falsehood of the premises or conclusions. If we should take the well-known syllogism: "All men are mortal - Caïus is a man - Caïus is mortal" and change it as follows: "All men are immortal - Caïus is a man - Caïus is immortal", the syllogism would remain valid and consistent, although the statement expressed in it is not true.

When speaking of "rationality" we expect something more, namely, that the statement is true as well as that the reasoning is correct. However, its veracity can be relative, or predicated on its time and place. In the periods of theurgic and demonologic medicine, the majority of diseases were believed to result from individual and direct intervention of divinity or demons, and were treated with prayers or magical procedures. This therapy, which is not considered rational today, had in those times all the attributes of rationality.

The idea of "rationalism", by contrast, belongs to a different domain. It has nothing to do with logic, but describes a certain mental trend. Besides, in different circumstances, this term can be used with different meanings. It will mean something different for a theologian than for a philosopher. We are going to use it as a colloquialism related to a certain outlook on life, or practical wisdom which makes us believe that intelligence is a sufficient means to solve all life problems that can be considered as solvable, without the need to resort to irrational powers of mind such as imagination, intuition, inspiration, etc.

What then is rationality in medicine? In relation to medicine I am speaking first of all of the skill of diagnosing diseases - it should be immediately stressed that this is something other than describing disease - and of the art of their treatment. In other words, it means to use the rules of rationality in order to find therapeutic indications and to carry them out.

All the rational doctors of different ages have had this very goal before their eyes; and Tytus Chałubiński, a famous professor at Warszawska Szkoła Główna, has collected the methods leading to this goal into a system, which he describes in a very concise, clear, and perspicacious way in his book *A Method of Finding Therapeutic Indications.*

But what is disease? It is functional disorder, whereas function is a reaction to stimuli, which implies that we should take into consideration two factors: the stimulus and the organism. Factors that influence functional disorders have been called by Chalubinski disease moments, while according to him the art of finding suitable therapeutic indications consists of a skillful discovery of these disease moments.

Let us begin with pathogenic stimuli. These can be primary or secondary, or even of some other order. Obviously, the most rational indication would be to treat the primary stimulus. Is this at all possible? And under what conditions? Of course, sometimes but not always it is possible. This task will be accomplished in the most precise way, not when we treat diseases, but when we prevent them. A good example of such a procedure would be the deliberate management of childbirth, or surgery in a way that keeps the probability of infection as low as possible.

It is in general more difficult to treat the primary stimulus of an established disease. Even if by removing a foreign body primary stimulus, say a splinter, which induces pathological symptoms, we should perform our action with utmost care, for we can never be sure that removal of the disease-inducing stimulus has been complete. For example, if the splinter has been contaminated with germs, these could already have been spreading over the patient's body, and thus, in spite of the elimination of the splinter, the disease will not disappear.

Similar difficulties could arise when we try to remove or to kill parasites, whether of animal or vegetal origin, for something else can still happen. We might be dealing not with the stimulus, which might not still exist, but with the effects of this stimulus manifesting themselves in modifications of the organism which are so important that they play the role of a new secondary stimulus, while the primary one becomes totally out of reach. This is the case in intoxication, the common cold, mechanical or thermal injuries, and so on.

Sometimes the stimuli may be unattainable for different reasons. In many cases of infection the microorganisms become unattainable although we know they are present, because the means of destruction of these organisms would, if applied, kill the patient first.

Finally, very often the primary stimulus cannot be reached because we do not know which stimulus was the primary one. For example, in disturbances in a metabolic process we only have the secondary stimuli to deal with. The therapeutic method might then consist in attempts at neutralization of, e.g., bacteriotoxins produced by the Loeffler bacilli, by treating them with antitoxic serum, or in attempts to cure wounds, chilblains, burns, etc. In these cases too, we execute rational indications, but our rationality is not of the first but rather of the second degree. Sometimes, however, we are not able to afford even this, since the secondary pathological moments might also become unattainable. Antitoxic treatment which is possible in certain cases might not be so in others. In such cases one should look for moments which are even less directly connected with the primary pathological stimuli, and whose links with the latter are indirect, unattainable, or even unknown. The treatment of such remote pathological moments can, however, be sometimes considered as a rational indication, should it be found that these moments were harmful and were not produced by the organism as an act of self-defence. We try to relieve a cough with narcotic drugs, and to reduce fever or pain. However, in each case of treating the symptoms which are substantially remote from the primary stimulus, it is worthwhile to contemplate whether the patient will gain or lose if we fight them. In other words, rationality in such cases is relative, contrary to the absolute indication of removing the primary stimulus.

Let us pass to the second category of indications, i.e., to the problem of how to modify the patient's organism in such a way that it reacts properly to normal stimuli and resists pathological stimuli. This can be achieved in two ways. In the first place we can try to enhance the general endurance of the patient by favoring the best possible conditions for extensive development of the body. In other words, we either try to prevent disease through better immunity, through hygienic habits, or to improve overall resistance through, e.g., if possible, intensive nutrition. Such conduct will certainly conform to the rule of rationality.

In the second place we can try to incline the patient's organism to fight certain specific pathological stimuli effectively by developing appropriate functions, specially designed for this purpose. The latter method is called immunization, while the condition obtained is called the active immunity of the body.

The preventive methods, such as protective inoculation with vaccines of typhoid, cholera, diphtheria, etc., are of great importance from the practical point of view, whereas in spite of the efforts made by many doctors for many years, the respective curative methods are still to be worked out.

But we are not out of the woods yet. We should deal now with the problem of *hormones,* which is closely related to the above. Originally, this word denoted substances produced by different organs in order to stimulate their functions. Nowadays, we use this term also when we are thinking of substances which reduce the activity of certain systems of the organism. In this wider meaning, hormones are identical with internal secretions, which in turn belong to another, more extensive realm, i.e., the problem of the *constitution.*

This problem has bothered people for ages, and the result has been the evolution of the old ideas of eucrasia, dyscrasia, and temperament. Philosophical/cosmogonic ideas, and the analogy with the elements that were thought to constitute the world, led to the idea that the human body, being also the world "*en miniature*" - microcosmos in opposition to macrocosmos - was constituted of equivalents to the macrocosmic elements. Thus, health and illness, or correct and incorrect reactions to stimuli, were the consequence of correct or incorrect relations between the elements.

Temperaments were viewed as nothing else but types of constitution. The main idea was quite correct, but the facts required to make practical use of this idea were inadequate. And indeed, the therapeutic progress obtained over several thousand years by the art of diagnosing and treating diseases has, in general, been rather limited. In fact, the extensive notion of temperament as the type of patient reaction to stimuli has gradually become less extensive, and its present meaning covers the psyche only rather than the organism as a whole.

In the Virchow period, the period of pathological anatomy, the main goal of medical thought was to find in which organs or cells the illness was located. It was only natural that at that time the concept of constitution, viewed as a system of connections between the various parts of the body, giving certain common features to these parts, was abandoned by the majority of scientists. However, the concept of constitution has been resurrected following the development of the scientific disciplines dealing with internal secretion, bacteriology and endocrinology, and following the accumulation of new facts concerning the functions of the sympathetic nervous system. This concept has returned to the old goal, namely, the classification of reaction types. Points of view that would allow such classification are now in sight. Thus, instead of the prevalence of "blood" or "phlegm" or "bile" or "melancholy", we will have certain serologic properties, which can already be demonstrated in relatively simple reactions and which can bring practical benefits - e.g., in blood transfusion. We will have neurological classification - vagatonia and sympathicony - and we will also have the typology of metabolism and internal secretion, for example, hyperthyroidism and hypothyroidism, enhanced or attenuated metabolisms of proteins, of carbohydrates, etc.

The idea of constitution will certainly disintegrate into notions of function of individual organs or systems which already can be tested, or which would be possible to test with adequate methods.

In this way we will probably be able to extend substantially the range of therapeutic indications aiming at an effective control of the body's reaction to stimuli. Whether this control will have perfectly rational grounds in the above mentioned sense, or whether it will exert influence on the primary pathological moments, or not, this is another question.

Considering what we have come to know so far, we can presume that Theophilus Bordeu was right when, as early as the 17th-century, he insisted that all organs interact among themselves. But if this is true, then the function of each tissue or even of each individual cell is the result of so many factors that locating of the primary act and the primary cause in such chaos certainly surpasses human powers. For if we know today how to recognize the state of vagotonia of a parasympathetic system, and even if we can exert influence on certain symptoms of this disorder, we still do not have reliable information on its original cause, and even if we found what this cause consists in, say, the hyperfunction of some internal secretion gland, then it would logically follow to ask why the function of this gland has become enhanced. And then, if we insisted on searching further, and proved that this hyperfunction results from the hyperfunction of another organ of internal secretion, a new question would still arise as to what this new hyperfunction was caused by.

So we are forced to the conclusion that in such circumstances the hope of discovering the Archimedian "*poi sto*" that could be used to dream up truly rational therapeutic indications, is at best a long way off; as it is now, we may at most aim at purely symptomatic indications or at a very relative rationality. And although hypothyroidism can be effectively treated nowadays, the rationality of this treatment should not by any means be considered as better than that of treating a dry, oppressive cough with codeine, or of using morphine injections for the temporary relief of pain.

If we now remind ourselves that forty years ago we knew much less about Koch's bacillus and its effect on the human body than we know at present, we will probably not be so astonished at the indignation of the old Heidelberg professor when he was told of "rational treatment of lung consumption with Fowler's solution".

Empirical medicine has usually been put in opposition to rational medicine. *Habent sua fata vocabula.* There were times when the empirical schools were in opposition to all the other schools. But even in those times no dogmatist was a pure dogmatist, no taxonomist a pure taxonomist, and no empiricist a pure empiricist. The former has always used experience, whether his own or someone else's, in a more or less extensive way; the second one has also always tried to explain his experience or to build up a theory to match his practical abilities. For precise classification is possible only within the realm of ideas, but in real life there are only phenomena which pass gradually and imperceptibly from one to another. However, when speaking of empiricists on one hand, and of physicians who think that their methods of treatment are rational, on the other hand, we usually refer to differences in professional training.

A physician who is well-acquainted with the up-to-date requirements of medicine and who applies these requirements in his practice, is viewed as rational; but the one who does not know these requirements and applies a treatment which is in agreement with old-fashioned methods, is rather viewed as an empiricist. No wonder that the latter category includes first of all persons who are in the medical profession without having a diploma, such as barber-surgeons, quack doctors, etc. It so happens that these "empiricists", devoid of medical education, are more popular among certain social classes than are certified physicians. This is not the place to enumerate all the reasons for this phenomenon, and I shall limit myself to calling your attention to a single one, closely related to our present discussion: the tendency of the "learned" doctors to mistake rationality for rationalism. I have to admit that they are pushed in this direction by the educational system, in which, all the way from the preparatory courses through the high schools and universities and until the moment they receive their diploma, too little attention is paid to irrational factors. These factors, however, play too big a role in our life to be neglected.

There is no doubt that mathematics is the most rational and intellectual realm of the human mind, and one can hardly imagine anything closer to rationalism than mathematics. But let me remind you that one of the most outstanding scholars of the recent past, H. Poincaré, has classified mathematicians into two groups: those who are gifted with imagination or irrational mental power, and those lacking it, and he has claimed that the results of the work of these two categories are different from each other.

But let us return to medicine, viewing it this time not as an art but a science. It is not possible to understand this science without anatomy; but let us hear the advice of a distinguished professor from the Medical Faculty at Cracow University, the late Ludwik Bierkowski, on the method for studying this discipline. He suggested, first, to imagine that all the parts of the human body, i.e., bones, muscles, vessels, nerves and internal organs, were thrown into a disorderly heap, and second to select in one's mind the bones, and gradually build up a skeleton, simultaneously connecting its constituents with ligaments. Then one should connect the vessels and nerves, and fix the muscles into their proper places. Finally, internal organs can be located in the prepared cavities of the structure, and the whole body can be covered with skin in order to arrive at a complete human body. This method has also been suggested by Wrzosek and Sokołowski in their book, *Medical Propedeutics.* All those authors are certainly right; it is a very good exercise, provided that the individual who tries to apply it has an imagination equal to that of a painter or sculptor. Otherwise, if pure intellect instead of blood circulates in his veins, he will not be able to use Bierkowski's suggestion effectively, nor will he be able to learn the anatomy he needs for his professional life.

This is even more true in practical medicine. It is known that very often the most outstanding diagnostitians are able to give a correct diagnosis as a result of reasoning which is so unconscious and abridged that it can be viewed as an example of direct recognition or of insight. In addition, the whole area of relationships between the physician and the patient, which is to a great extent responsible for the effects of treatment, is often governed by irrational factors and by emotions.

Therefore, besides the accurate examination of the patient, the first task of a doctor who intends to behave in a truly rational way should be a genuine effort to establish psychological contact with the patient. The basis of such contact is the patient's faith that the physician is not only capable of but also willing with all his heart to help him. If such confidence does not exist, the treatment cannot proceed without obstruction - whether conscious or not - either from the patient himself or from his family, and this obstruction will certainly hinder the effects of the therapy.

In the cases in which a doctor finds his rationality defeated by the rough empiricism of a barber-surgeon or a quack doctor, one should give serious thought to the question whether it has not happened because the behavior of the unlearned took into consideration a whole range of very important aspects of the disease neglected by the physician, and thus, from a certain point of view, has proven to be more rational than the physician's behavior.

One of the books which every practicing physician should not only possess but also read frequently, is *Aphorisms about Medical Ethics,* by the late Wladislaw Biegański, who died prematurely with great loss to Polish science. Here is one of his aphorisms:

> "Modify the system of medical education. Not only knowledge should be praised, but also emotion. Build up a feeling of sympathy in the students, inculcate a sense of duty in them, teach them that a patient is not a more or less interesting pathological case, but our fellow, an unfortunate human being. And do not teach with words; be the example!"

STANISLAW TRZEBINSKI
'ABSURDITY IN MEDICINE'

(*Archiwum Historji i Filozofi Medycyny,* 1927)

It was a practical principle in the Republic of Babin - a 16th-century Polish association of humorists - that offices were attributed to those who under normal circumstances would be perfectly unable to perform their function. According to this principle of absurdity, the appointment of the Honorable Walerian Trepka as Doctor of the Republic was announced in the Republic's documents on the 4th of May, 1604, since he had advised a feverish patient to drink a few gallons of malmsey wine, claiming that he had once cured his father by this method.

It is obvious, then, that at that time in Poland the administration of wine to feverish patients was considered as absurd. What is "absurd"? This notion, originating from "*absurdo*" - from a deaf person - which refers to comments expressed by a deaf person who is unable to understand what others say, and who enters into conversation in a nonsensical way - has a restricted and a broad meaning.

In the first place, the term refers to a method commonly used in dialectics which consists of revealing in the opponent's arguments enunciations that either would be contradictory to those that he has used before, or that he would have to agree with although he has not put them in words, i.e., reduction to absurdity. Here the absurd is a logical slip. This might be called the absurd proper.

In the second place, we are dealing with a very common colloquialism in which an absurdity is an opinion contradictory to what is generally recognized as true, without giving any consideration to the question whether the opponent has made a logical slip, or if his error was an error of information and resulted from an inadequate orientation or lack of experience. In such a case, a logical slip might not even exist, and the error would not qualify as an absurdity proper. In this case we may call it an apparent absurdity and it may be regarded merely as a paradox. If, however, someone has expressed an opinion which contradicts a conclusion logically drawn from an assumption which has been always, everywhere, and by everybody considered as true, it would be justified to impute to him an absurdity proper in the restricted meaning of this term, i.e., a logical slip.

Thus, if Copernicus had not preceded his statement about the Earth rotating around the Sun (rather than the other way round) with evidence that our common experience, which seemed to be an objective reflection of the daily migration of the Sun and stars, consisted in delusion, his contemporaries would have been justified in accusing him of an absurdity proper in the sense of making a logical slip.

Colloquially, we usually tend to define the absurd as a sin against "common sense", taking of course for granted that our own way of reasoning is the most sensible.

But let us return to the Republic of Babin. Had they good enough reasons to appoint the Honorable Trepka as their Doctor? The Republic's members were not naturalists or physicians, but surely they were educated and well-off men who had frequent contacts with doctors and thus were more or less familiar with the recognized views of pathology of their time. According to these views, wine enhanced fever. It could not be otherwise, because everybody knew that drinking wine brings warmth to the human body. It was also known that in such circumstances, a person has the impression of being made warm. Since, for them, feeling warm and having an impression of warming up were also the most essential symptoms of a fever, it is no wonder they considered wine as a fever-enhancing device and decided that its administration to feverish patients was an absurdity, or even an "absurdity proper". Today, however, nobody would see it as absurd, although recently the administration of alcoholic beverages to feverish patients has been rejected by many doctors. We know that although alcohol, which is digested quickly, generates at first substantial quantities of heat in the human body, it also dilates the skin vessels. These circumstances produce a sensation of intense warmth, but at the same time heat loss increases, and consequently the body's temperature in fact decreases.

If one tried to formulate the past and the present views in the form of a syllogism, it might look as follows:

1- In the past:

Fever is an increase of the body's temperature.
Wine increases the body's temperature.
Wine increases fever.

2- Today:

Fever is an increase of the body's temperature.
It is not the case that wine increases the body's temperature.
It is not the case that wine increases fever.

Therefore, if we do not think that the views of the Honorable Trepka are absurd, it is not because we handle the notions in a different way than did the citizens of the Republic of Babin, but because in a syllogism derived from an identical primary premise we have replaced the second affirmative premise with a negative premise.

But the same conclusion could be reached another way -- the one used by the Honorable Trepka to arrive at his statement. He did not involve himself in theoretical discussion over the nature of the fever but derived by way of analogy his final conclusion from the starting point of his own experience - albeit insufficient because based on a single case: I cured my feverish father with malmsey, so I certainly will cure another feverish patient with it, too.

Perhaps the Honorable Trepka did not know that he might have found some defenders of his therapeutic views even among the outstanding doctors of his time. One of the most famous among them, van Helmont - who lived at the turn of the 16th and 17th-centuries - also recommended wine for the treatment of fever. What were his reasons? After all, alcohol was still considered as enhancing the body's temperature. Van Helmont did not fight these opinions. He just neglected them, taking an entirely different point of view. Van Helmont was a vitalist. He believed in metaphysical vital elements *[archei]*, which dwelt in various organs and controlled their vital functions. The deviations of these functional elements from their normal status resulted in the illness of their carrier. Fever resulted from impairment of the main Archeus *[Arch. influus]* - and wine was thought to stimulate it; therefore, wine should reduce fever. From van Helmont's point of view such reasoning was correct, and since he enjoyed the reputation of an excellent doctor, his premises were not considered as absurd even by those who disagreed with them.

But let us take another, more familiar example. It is a well-known fact that as late as the first half of the last century blood-letting was recognized as a remedy for fever and sometimes was applied five, six, or even seven times over a period of a few weeks. Daudetowski, "le terrible docteur Gringatoire", described in "Tartarin of Tarascon" as the one who "extorquerait du sang même à une rave", belonged to this school of medicine. The basis for such behavior was the notion of the so-called "plethora" established by Erasistratos (4th-3rd-centuries BC). Although this notion was modified here and there over the centuries, it was never rejected and it usually enjoyed high esteem. Naturally, under such circumstances it would have been absurd to refrain from blood-letting in the case of, e.g., pneumonia, which in addition required the application of emetics, and thus those who shared such an opinion, based on the notion of plethora, were reasoning in a perfectly correct way.

Only during the last decade of the first half of the 19th-century did our compatriot Jozef Dietl, later a professor at Cracow University, prove through a large-scale experiment in a Vienna hospital that the above-mentioned methods did not improve the outcome of pneumonia, but at best would not make it worse. In Dietl's experiment the fatality rate among patients treated by traditional methods was even slightly higher than that of the patients who had received a quite neutral treatment. Thus, in this case a therapeutic dogma derived from the premises established by Erasistratos was refuted, not because its groundlessness was proven but because the fallaciousness of the therapeutic conclusion was proven experimentally. No wonder that from that time on blood-letting has been recognized as an absurdity, first in pneumonia, later in some other fevers, and finally in all fevers. As a result, blood-letting was abandoned even in cases where it brought genuine relief to the patient.

Later, the concept of anemia was recognized as one of the main causes of disease symptoms, and it replaced the dethroned plethora. In at least nine out of ten cases, not only a physician who addressed himself to the family, but also the professor discussing the therapy of a patient with his assistants, would terminate his description of the planned treatment with the words: "adequate nutrition" or "strengthening diet". Translated into common language, this means as much meat and alcohol in one or another form as possible. The patients did not always improve as a result of this treatment; from a rational point of view, if anyone who recognized the disease as "anemia" acted in a different way, he would rightly be guilt of absurd behavior. Luckily, at that time blood tests were developed, and they demonstrated that in the majority of diseases no anemia was to be found. On the other hand, in the majority of the cases of true, ascertained anemia, neither meat nor wine were of any help.

Thus, if syllogisms were to be utilized again to sum up the above, they would read more or less as follows:

1- Before:

a. A majority of patients suffer from anemia.
A strengthening diet removes anemia.
A majority of patients require a strengthening diet.

b. A strengthening diet is composed of meat and wine.
A majority of patients require a strengthening diet.
A majority of patients require meat and wine.

2- Today:

a. A majority of patients do not suffer from anemia.
A strengthening diet is a remedy for anemia.
It is not certain whether a majority of patients need a strengthening diet.

b. Some patients suffer from true anemia.
A strengthening diet is of no use in true anemia.
Prescribing a strengthening diet in true anemia is useless.

c. Sometimes a strengthening diet is necessary.
Meat and wine do not necessarily make a good strengthening diet for everyone.
Meat and wine should not always be given a patient who requires a strengthening diet.

Again, a dogma has fallen through and so has the absurdity of those who did not adhere to it.

Instead, another dogma has immediately evolved, according to which strong drinks are absolutely noxious, and in the opinion of many doctors they should not be recommended to patients under any circumstances or in any case, since they can always be replaced by other, less harmful remedies. However, this time I have not yet heard of anyone imputing absurd behavior to doctors who continue to recommend strong drinks. We do, however, frequently meet this reproach in other circumstances.

It is commonly known from immemorial times that a tendency has existed to draw a parallel between the universe, or macrocosmos, and the human being, or microcosmos. And, since the world was viewed as constructed from a certain number of elements, say four, to which there adhered certain basic attributes - dryness for earth, cold for air, dampness for water, heat for fire - these attributes were transferred to the human body, and their reflection was found in the four humors, or basic juices: black bile, blood, mucus and yellow bile. The correct mixture of the juices resulted in health - *eucrasia* - while incorrect mixture was responsible for illness - *discrasia*. If the proportions of the humors were not too far from normal, their mixture revealed itself also in respective temperments. All pathological processes might be approximately explained by the interaction between the individual constitution and different external effects. As a consequence, the therapeutic means were classified in agreement with their proportion of the four elements, in order to create a reasonable system of treatment.

When Galen's or Celsus's or the Arab author's writings revealed that, e.g., a given patient suffered from pathological excess of cold and dry elements and from a deficiency of the other elements, a medicine would be prescribed which contained more of the deficient elements and lacked, or at least contained as little as possible of the elements, excess in the patient's body. The method of finding such combinations, which followed the entirely rational rule of supplying the human body with the elements that were deficient in it and avoiding the elements that were in excess, was called allopathy.

Allopathy originated from quite correct reasoning. Although many of its aspects might be criticized from the present point of view, its absurdity was certainly not to be understood in the restricted sense of the term, but only in the fact that it was a pure speculation resting on premises that lacked experimental support, and thus could not be viewed as a scientific hypothesis, let alone a sufficiently justified fact. But in those days medicine had very few scientific facts and hypotheses, and made use of what was available. The trouble began when the number of therapeutic substances, which were mostly of animal or vegetable origin, increased, and at the same time deductive speculations allowed the establishment of more and more new and sophisticated combinations of elements. At the end of the Middle Ages, there were a few dozen different "simplicia" that could be combined in order to introduce, say, heat and dryness elements in such quantities as were seemingly needed. This, in turn, resulted in such a muddle that nothing seemed to be reliable enough.

Such was the state of the art found in the 16th-century by Paracelsus, who immediately declared the previous medicine to be absurd, and openly burned the works of Galen and Avicenna. Paracelsus rejected galenic therapy and the rule of juice mixing, and recommended acting directly on the metaphysical, vital element - - the Archeus - - through therapeutic means called by him "the arcana" because of their mysterious nature.

However, referring to cosmic parallels Paracelsus had established a close relation between macrocosmos and microcosmos, which, after all, was quite similar to the one made by the founders of the humoral school. He also believed that there must exist a treatment for every illness (i.e., for each deviation of function of Archeus), the properties of which could be determined through the revelation of certain inner powers in the form of external symbolic features. These features were his famous "signatures". For example, the yellow sap of celandine was a sure sign of its jaundice-curing properties. Furthermore, since the Archeus had manifested itself by the life process, and the life process was parallel to the process of burning - a presentiment of the scientific fact established by Lavoisier two and a half centuries later, and perhaps the soundest foundation of Paracelsus' fame - and since the products of burning were salt, sulphur, and mercury - these of course should be understood in their metaphysical-alchemical rather than present chemical meaning - spagiric or alchemic substances taken in their majority from the mineral world were recommended as having a positive effect in case of illness.

Of course, this system was worked out from its basic premises in a way that was no less logical than the system of humoral pathology and therapy. The premises of the latter were in fact neither better nor worse than those of the former, although in practice Paracelsus' teachings eliminated one substantial source of harm from the medicine of those times: extremely complicated medications.

Some three hundred years later, the influence of another doctor on the medical art was no less positive, although he was accused of being as absurd as the old Galenian therapy had been by Paracelsus. I refer here to Hahneman and homeopathy. Too much space and time would be required to explain this theory; the appropriate thing will thus be to point to two basic premises of the Hahneman medical conception. These were: first, treatment should be performed in the opposite way from that used by allopathic doctors, who assumed the material nature of the illness and tried to complement the elements that were deficient and to remove those that were in excess in the patient's constitution according to the rule of "contraria contrariis". According to the rule of "similia similibus", the treatment should use medications producing symptoms similar to those produced by the illness itself; and these medications should be used in small quantities.

The first premise, which is based on conclusions drawn through analogy, does not seem absurd nowadays, since such analogy is quite commonly used in preparational preventive or curative vaccines. In Hahneman's day this principle was first used in variolation; then it was followed by Jenner's vaccination. However, the founder of the science of homeopathy preferred to refer to his own rather questionable perceptions in another domain. Besides, the principle "similia similibus" was an old concept which could be traced to medical views in different periods, and which was not automatically characterized as absurd, although it was quite frequently accused of being so.

The second premise was less sound. At first the doses applied by Hahneman were indeed small, but remained in some proportion to those generally in use. Gradually, however, he started to recommend dilutions which were expressed in figures of an order of 10^{-60}, asserting that the smaller the dose the more powerful would be the effect. Admittedly, today we observe more analogies than did Hahneman's colleagues, and we know that certain substances may have an effect on the organism even when applied in minute doses, but it is hard to understand the effects of the 17th or even the 10th "power" of Hahneman's dilution. Thus, one should not blame the scoffers who used to say that according to the homeopathic principles, the most effective medicine would be water from Geneva Lake, in which had been dissolved a drop of suitable medication. Besides, Hahneman's argumentation on this point was very poor. Sometimes he referred to the extraordinary, almost miraculous significance of extremely accurate mixing employed in preparation of homeopathic drugs; at other times he would turn to the mystic, dynamic effect of these drugs on the vital powers, and then again he would point out that this effect must surely be possible because it occurs (without quoting any convincing argument). Besides, this premise contradicted another, frequently expressed statement by Hahneman, namely, that efforts should be made to use a medication able to induce an artificial illness, slightly stronger but as little as possible, than the one it is supposed to fight, since in the case of low dilutions, one should rather avoid the most powerful ones. Here we have a logical slip, or absurdity proper.

It is intriguing why this science, in which only one premise was based on an analogy, and moreover on an analogy which at that time had a very weak foundation, while other premises had almost no basis at all, and were in clear-cut contradiction with a former statement by the very same author, has survived for more than a hundred years and still is developing and attracting not only those ladies taking a fancy to such treatments, but also a number of educated and sensible doctors.

This phenomenon may seem even more bizarre if one takes into consideration the fact that other medical doctrines, even those based on seemingly better grounds - one of which will be discussed below - fell into total oblivion after a couple of dozen years.

Let us try to find an explanation for this phenomenon. First, homeopathy is a totally harmless curative method, with the exception of the case of patients who, either because they chose their own medication or because they were treated by a non-physician, delayed an indispensable surgery. Second, the treatment is very simple. A person who practices homeopathic treatment does not need to have a medical education and does not need to bother about diagnosis. All that is needed is to select the most self-evident symptom, to refer oneself to the respective table, and to decide on a treatment. The drugs are applied one at a time. This has been an important trump card, especially in the past, at times when the official doctors took a fancy, even more than nowadays, to mixing various drugs. Finally, the adherents of homeopathy among the physicians do not keep strictly to Hahneman's text. They draw a general line of behavior from the homeopathic rules, but in practice they adjust to the circumstances; they follow the general progress of medical science and use the current diagnostic methods. So the shortcomings of theory are compensated by practice.

Another medical system, which was devised a little before homeopathy, but which now has for a long time been relegated to the history of medicine, took an entirely different route. I am referring here to Brown's system.

The basic principle of this system was an affirmation that all vital functions were the reactions of an organism to external stimuli, and that without these stimuli and reactions life would be completely unimaginable.

The capacity of the organism to react to stimuli was called excitability by Brown, and the condition induced by this action was called excitation. Moderate excitation meant health, while excessive or insufficient excitation corresponded to illness.

All diseases, at least when they were not clearly local, could be classified according to the previous assumptions into stenic, i.e., those originating from excessive excitation, and astenic, i.e., those originating from insufficient excitation. Suitable classification, made by the authors allowed a doctor who made a diagnosis to know immediately whether the illness was stenic or astenic. The treatments also either enhanced or attenuated the excitation. One was informed by the author whether a drug belonged to one or another group. Obviously astenic diseases required exciting treatments, while stenic required calming methods.

As we can see, the basic premise was absolutely true, and further development of Brown's reasoning also seemed correct, since if we once agree to the perception of life as reactions to stimuli, we have to agree to the understanding of pathological processes and methods of healing from the same point of view. Shortcomings of Brown's reasoning became apparent, however, when practical conclusions had to be drawn from correct but general biological concepts, because practical conclusions had to be grounded in clearly determined statements. It very quickly turned out that when doctors tried to treat patients through the application of Brownian rules, the result was in general clearly negative. How disappointing this experience was we can gather from the position later taken by Jozef Frank, once one of the most enthusiastic advocates of the system. Let me present a passage from the doctoral thesis of

Collignon - Vilnaus, 1808 - condemning Brownism, an example of Frank's opinion about this system years later. The passage, written under the inspiration of Frank, evoked a protest from Jan Śniadecki, the rector of the University at the time, and eventually Frank had to agree to more placid wording. Even in this mild form we read: "What then should be thought of the crowds of followers of Brown, that plague of humanity", etc. How might this passage have read before the rector's censorship?

On the other hand, the concepts which were the starting point of Brown's principles can be found in Sniadecki's book, *The Theory of Organized Beings,* in which the author not only devoted a full chapter to them, but also clearly admits that he has actually been stimulated by them to write the book.

The system of François Broussais, a well-known French doctor who took Brown's premises but modified them and constructed his own system, met a similar fate. His concept of the limitation of excessive local excitation or irritation of internal organs by so-called reductive methods - such as blood-letting - although at first received very enthusiastically by Vilna - e.g., by Rymkiewicz - was quickly abandoned, when it was found that the mortality in the wards in which Broussais had applied his methods had increased instead of diminished. It therefore became a discouraging example of careless jumping to conclusions by the author. Thus, in both cases described above, medical systems were discarded not because of inadequacy of theory, but because they did not withstand practical tests. These systems shared, after all, the fate of many other different systems having a better or worse theoretical background, which did not survive for the same reasons, such as, for example, the systems of iatrophysicians and iatrochemists in the 17th-century. Once these systems were discredited for the reasons explained above, they were readily viewed as absurd, without their rational structure being taken into consideration. Here we are dealing with the second *sit venia verbo* mode of deciding on absurdity in medicine; not from the premise but from the conclusion.

An opposite example would be a case in which the main claims were at first viewed as absurd, and furthermore were initially devoid of any theoretical grounds, being based solely on the personal experience of the author. However, later on they lost their absurdity and won high social esteem, although their theoretical aspects might still be deficient. This was precisely what happened to

Jenner's protective vaccination. Jenner worked out a method solely by observation of the facts, and without seeking its theoretical grounds; he was therefore blamed of absurdity by his adversaries. Their arguments included allegations about the danger resulting from induction, through vaccination, of animal matter into the human body. Efforts were made to provide convincing illustration of this danger in pamphlets describing how lymph-vaccinated children had mooed like calves, or through pictures showing these children with the heads of beasts. Nevertheless, within a very short time the method spread over Europe; it waited for an explanation for almost a hundred years until the elaboration of immunity theories were able to explain the positive results of such a practice. Even after a hundred years, the acceptance of these principles was not always easy. In his memoirs, printed recently in *Polska Gazeta Lekarska [Polish Medical Journal]*, professor Bujwid recalls that after the publication of Pasteur's report about curing rabies with vaccination, a famous Viennese surgeon Billeroth printed in *Neue Freie Presse* an article explaining that Pasteur's experiments were in contradiction with the most evident medical logic, because it is impossible to cure an illness recently induced by a dog bite using a vaccination with the very same germ. Further, he referred to Frisch's experiments and his allegations that Pasteur's method was lethal, and finally he recommended closing the Ullman Pasteur Clinic in Vienna, which soon was done. A new clinic was opened only when the emperor Franz Joseph came to the Lwow Exhibition, took a close look at the model of the Pasteur Institute in Cracow, and listened to the detailed and exhaustive explanations on this subject.

And here is another example. Speaking about Paracelsus, I mentioned the so-called "signatures", i.e., some mostly external features of certain drugs matched with similar external properties of certain diseases, against which they were said to be effective due to a mysterious, metaphysical relation. This concept was in fact much older than Paracelsus' writing. A great number of very similar "signatures", accompanied with detailed explanations, are found in ancient Chinese medical documents; while the "arcana" described by Paracelsus are colloquially present nowadays as "sympathetic" drugs which are believed to be more medically effective, the more repugnant they happen to be. As long as these mysterious relations between macrocosmos and microcosmos were mentally acceptable, the "sympathetic" drugs had as good or even better logical grounds as any other drugs. But later, when official medicine abandoned the notions of "sympathy" and "arcana", they were rejected by physicians as absurd. The non-medical world, however, continued to use them. I should like to present here a passage from a book - *Mór w Polsce [Plague in Poland]* by professor Giedroyc - in order to show what methods were sometimes used; the note on page 129 contains the text of a report of an 18th-century municipal physician from Danzig entitled "Einiger Medicorum Schreyben", that reads: "I have been informed by one Great Poland Reformed theologian that it was a practice in Warsaw, when nothing could cure a disease, to cut out bubos from the dead, dry them up, grind them down and give them to the patients, who indeed immediately improved. Having seen this, poor people were so encouraged that they ate pus from ripe bubos with spoons. Two or three patients saved their lives by sucking pus from their own bubos, in front of the theologian's eyes. This is true, believe me. The theologian's name is Tobiany."

It is absolutely certain that any representative of the official medicine of that time would call such a practice an abominable absurdity - and would be entirely entitled to do so from his point of view. However, knowing now what we know, and presuming what we can guess, we would not light-heartedly give so explicit a qualification to this practice. Our patients are not given pus to eat, but in some cases it has been injected subdermally into some patients, while everybody, even without being a doctor, knows about therapeutic injections of serum. Here, the periodicity is quite evident. First dogma, next absurdity, and then non-absurdity again.

Almost all the cases discussed so far have belonged to the category of apparent absurdities. The absurdity proper, although not so frequent, has also occurred in the history of medicine. I have already mentioned Jozef Dietl, who revolutionized the treatment of pneumonia and then of other fevers. He and Hamernik, both students of Skoda, a famous Viennese clinicist, went even further in their criticism of the contemporary state of therapeutic skill than did their teacher, also known for his rather dogmatic opinion on this problem. They claimed that in the face of the impotence of therapy the doctor should be required not to heal patients but to observe them, leaving the cure to the forces of nature. This statement, which represents so-called "medical nihilism", can be considered as an absurdity proper because it is not only based on a paradox, but also on a logical slip. This logical slip could have been avoided by the authors only if they had said at the same time that doctors are totally unnecessary and no one should become a doctor, because a person who makes observations is not a doctor performing the goal-oriented, expedient art of healing, but a scientist investigating pathological phenomena. Such a person would be completely useless for patients and would have no right to call himself a doctor.

To summarize, we can draw the following conclusion. Instances of absurdity have occurred quite often in the history of medicine. They have usually been made at times when opinions on essential problems were modified, either as a result of the development of theoretical reasoning originating from a different starting point, or because personal experience seemed to prove that the opponent's curative methods were harmful.

Immense difficulties which obstruct the way to medical knowledge on the one hand, and emotionally-loaded moments of medical practice on the other, seem to constitute the major reasons for bringing forward charges of absurdity. These charges probably occurred more frequently in the history of medicine than in the history of the purely reconstructive sciences. These charges, however, refer in most cases not to absurdity proper, but to apparent absurdity; not to the logical slip, but to a paradox related to the prevailing views at a given time and place. The absurdity proper, or logical slip, occurs much less frequently and is usually the result of an oversight, which in turn is due to a unilateral way of viewing a given problem at a given moment.

CHAPTER VII

LUDWIK FLECK
FROM PHILOSOPHY OF MEDICINE TO A CONSTRUCTIVIST AND RELATIVIST EPISTEMOLOGY

Ludwik Fleck (1896-1961) is today widely recognized as a pioneer of the sociologically-oriented constructivist approach to history and philosophy of science. Fleck was not, however, trained as an historian, a philosopher, or a sociologist; he was a physician who specialized in microbiology and immunology. Fleck studied medicine at Lwow University, and immediately upon receiving a medical degree (in 1920) he became assistant to a well-known typhus specialist, Rudolf Weigl, and worked in the latter's laboratory in Przmysl, near Lwow. In 1921, Weigl received the chair of bacteriology at Lwow University, and Fleck followed him to Lwow. However, because of personal conflicts with faculty members, or perhaps due to the antisemitism that prevailed at Lwow during this period, or both, he was obliged to leave the university in 1923. From 1923 to 1928, Fleck worked as a bacteriologist in General City Hospital in Lwow. He first joined the Department of Internal Medicine, then became head of the bacteriological and serological Laboratory of the Department of Skin and Venereal Diseases. In 1923, he also created a private laboratory of bacteriology and serology in which he performed routine analyses and conducted personal research. In the years 1928-35, Fleck headed the bacteriology laboratory of the Lwow Sick Fund [Lwowska Kasa Chorych]; but he lost this job in 1935, and between 1935 and 1939 he worked exclusively in his private laboratory. Although from 1923-39 Fleck worked in a routine analyses laboratory, he continued to consider himself a scientist. He published numerous scientific articles in professional journals in Poland and abroad, most of them based on his laboratory practice. During the same period, he formulated a highly original epistemologial position [23, 28].

The formulation of theory of knowledge by a physician and bacteriologist working at the laboratory bench is not a common event. My thesis is that Fleck's epistemological thought did not emerge from a vacuum. It was influenced by seminal ideas from the Polish School of Philosophy of Medicine. In a way Fleck can be viewed as the epigone of the Polish School. Moreover, even if Fleck never formally acknowledged specific ties to the Polish School of Philosophy of Medicine (nor, to be exact, with any other school of thought since he considered himself an original thinker), he probably remained more faithful (albeit in his own way) to the spirit of this School than other thinkers of his generation who explicitly presented themselves as disciples of the Polish School.

Fleck, to be sure, did not view himself as a philosopher of medicine. Although the majority of the examples in his works are taken from his professional specialty - medical microbiology and immunology - these examples underscore, according to him, the development of scientific knowledge in general and not simply the development of medical or biological knowledge. He took, however, his professional experience as bacteriologist and as immunologist as the starting point of his epistemological reflections.[1] But in addition, his reflections on his own professional experience were, at least in the first stages of the evolution of his thought, influenced by the ideas developed by the Polish School of Philosophy of Medicine.

Fleck's first epistemological study, 'Some Specific Features of the Medical Way of Thinking', was presented in a lecture to the Society of Amateurs in the History of Medicine in Lwow (April 1926). [This lecture was later published in the *Archives*] [5]. In this work, Fleck argued in favor of a holistic approach to pathological phenomena. He explained that such phenomena cannot be understood from a simple, reductionist point of view. More specifically, he objected to the widely accepted belief that infectious diseases had a single etiological cause, namely, pathogenic microorganisms. For him, an infectious disease was a highly complex entity, the result of multifactorial and multidimensional interactions between the microorganism and its host. Fleck affirmed that such a complex event was by definition impossible to describe in a simple, clear cut manner. Diseases, he argued, should thus be observed from several different points of view - chemical, bacteriological, psychological, etc., because neither cellular nor humoral theory, nor the functional understanding of diseases alone, nor their 'psychogenic' conditioning, by themselves would ever exhaust the entire wealth of morbid phenomena. Physicians, however, often refuse to admit to this complexity of pathological phenomena. Sometimes they attempt to find simple, global (and false) "logical" explanations for morbid phenomena. In the short term they may believe that they are successful: "it is nowhere easier to get such pseudological explanations than in medicine, because the more complex the set of phenomena, the easier is it to get a law verifiable in the short term, and the more difficult it is to reach an embracing idea."

Another approach used by physicians in order to simplify the complex field of pathology is the classification of pathological manifestations into distinct units called "diseases". But, Fleck argued, in nature there is no such thing as "diseases"; only individual pathological phenomena exist. The so-called "diseases" are in fact constructed by physicians: they are "ideal fictitious pictures, known as morbid units, around which both the individual and the variable morbid phenomena are grouped, without, however, ever corresponding completely to them." These pictures are created in order to permit the development of the practical science of healing; they are undoubtedly useful, on condition that one not confuse these fictitious representations with real pathological states.

Later, Fleck generalized this view of the social construction of diseases to science as a whole, and argued that *all scientific knowledge is socially constructed.* One can follow the evolution of this generalization in the second article, published by Fleck in 1929: 'On the Crisis of Reality' [6]. In this article, Fleck affirmed that not only diseases, but also their causal agents, bacteria, are at least partially constructed concepts. Fleck probably developed this view under the influence of the unorthodox ideas of his teacher, Rudolf Weigl. Weigl belonged to the small minority of microbiologists who, as late as the 1920's, believed that bacteria do not form fixed species and are highly variable. The bacteriologists who supported this view thought that the same bacterium can have a different morphology and physiology within different environments. Some of them also thought that morphologically and physiologically different bacteria can represent different stages of a (putative) complex life cycle of the very same organism. These bacteriologists believed that the generally accepted classification of bacteria, based on observations made in a test tube, was a laboratory artifact [20].

This view of bacteria as ever-changing proteiform organisms was in accord with Fleck's holistic approach to pathological phenomena and his notion of disease as a complex, dynamic interaction between the parasite and the host. He assumed that classifications of bacteria depended on the conditions in which they were observed. But, he argued, if such was the case, these classifications had no objective validity whatsoever; rather, every classification depended on its purpose and the process by which it was made. In 'On the Crisis of "Reality"', Fleck explained that it is well known that the same bacterium can be classified in different ways, e.g., by fundamental scientists and by epidemiologists. For example, basic scientists involved in a biochemical investigation prefer to form a very stringent definition of the bacterium they study, because they are much less worried over the exclusion from their sample of some relevant bacteria than over the inclusion of some irrelevant ones. For epidemiologists the opposite is true. They wish above all to avoid false-negative results, and to mistake a dangerous bacterium for a benign one. They thus prefer to provide a less stringent definition of the same pathogenic bacterium. This discrepancy in classification had usually been interpreted as an epidemiologist's modification of the true definition given by scientists to his own, practical goals. This view was rejected by Fleck, who affirmed that both classifications - that of the basic scientist and that of the practice-oriented epidemiologist - were equally valid and equally "true". Applied science, he claimed, is no less "scientific" than basic science, for both aim at discovering the truth. But the truths they discover are divergent and not interchangeable, each depending on the specific goal of the investigation that generated it and on the *thought-style* of the professional community that produced the given truth. Here, in all likelihood, lies the origin of one of the principal epistemological concepts developed by Fleck: the idea that truth is relative to a thought-style.

In his magnum opus - *Genesis and Development of a Scientific Fact* - published in 1935 [9], and in articles published at the same period, Fleck further developed the idea that truths elaborated by different thought collectives were incommensurable. In this book Fleck explicated the history of syphilis, and showed that the view of this disease radically changed over time. And, Fleck added, one should not view these consecutive modifications merely as the result of "progress of knowledge". Each view of syphilis was *true* in its time, and each contained some interesting cognitive elements that were lost in the next view. Moreover, according to Fleck, different and incommensurable views of a given disease can exist not only diachronically in different cultures, but also synchronically within the same culture. In his article, 'Some Specific Features of the Medical Way of Thinking', Fleck had already explained that it was impossible to develop a single, global point of view capable of explaining all pathological phenomena; it is impossible, that is, for a single theory to account for the wealth of pathological phenomena. "The results in the incommensurability of ideas which develop from the varying ways of grasping morbid phenomena, and which give rise to the fact that a uniform understanding of morbidity is impossible" [5]. Fleck later argued that this was true not only with respect to pathological manifestations but for every natural phenomenon. Natural phenomena are observed by different thought-collectives, each arriving at a specific point of view. This point of view is in agreement with the specific thought-style of a given thought collective, and is incommensurable with the points of view of other thought-collectives. But, as is the case with different points of view on pathological phenomena, these incommensurable points of view can all be equally true.

Fleck claimed that all observations depend on the *a priori* ideas of the observer. For example, old anatomical drawings reflect nonexistent details that conformed to the anatomical ideas of the period in which the book was written ([9], pp. 135-145). Fleck also analyzed the way medical specialists "see" pathological phenomena. His conclusion was that the ability of a physician to perceive certain phenomena depends on the training he has received. For example, a bacteriologist trained to observe microscopic preparations is unable to see pathological deformations of the skin the way a dermatologist does, and a surgeon frequently does not know how to view a bacteriological preparation. Moreover, the capacity to recognize some specific phenomena is necessarily accompanied by the loss of ability to perceive other phenomena: "One might believe that the hypothetical research worker of Poincaré, having an infinite amount of time at his disposal, would simply be a specialist of all trades, all sciences, thus being able to perceive specific forms in all fields. However, this is psychological nonsense, since we know that the formation of powers of perceiving certain forms is accompanied by the disappearance of the faculties of perceiving some others" [7]. Fleck generalized this conclusion to apply to the way scientists observe natural phenomena. For him, each thought-style "made possible the perception of many forms as well as the establishment of many applicable facts. But it also rendered the recognition of other forms and other facts impossible" ([9], p. 93).

An analysis of Fleck's first epistemological articles reveals that his principal ideas - the non-existence of "objective" observations, the dependency of the latter on the previous training and preconceived ideas of the observer, the dependency of scientific truth on the thought-style of the scientific community that generates it, and the incommensurability of truths generated by distinct thought-collectives - emerged from his reflections on his own medical practice. But if one takes a closer look at the principal ideas that led him to his epistemological conclusions, one can see that many of these ideas are closely related to the concepts developed by members of the Polish School of Philosophy of Medicine. Thus one of the starting points of Fleck's epistemological reflections is the idea that diseases do not exist "out there" in nature, but are *constructed* by physicians. This point was already made, albeit usually in a less radical form, in the writings of the Polish School. Recall that the principal thinkers of this school, starting with its founder Chalubinski, developed a holistic approach to medicine and affirmed that there was no such thing as "disease" in general, only specific pathological states in individual patients. Consequently, they viewed the classification of diseases as an artificial construction, perhaps indispensable for the transmission of knowledge (Bieganski), but which should only be used with caution, in order to avoid confusion between the fictitious and schematic textbook representation of diseases and actual pathological states observed in patients (Biernacki, Kramsztyk). Some Polish philosophers of medicine (Kramsztyk, Bieganski, Trzebinski) also stressed the modifications in medical classifications and in medical perception in different historical periods, and noted that in each period physicians were convinced that their theories represented the absolute truth. But when a new approach was taken, the same theories were declared inadequate, or even viewed as absurd.

Fleck also maintained that absolutely objective observations do not exist. Even anatomical drawings reflected the knowledge of a given period and were not "objective descriptions" of what the physician had really seen during a dissection ([9], pp. 141-142). Observations, he affirmed, are dependent on the thought-style of the observer, and each thought-collective has its own thought-style that generated different observations. He stressed the influence of the observer's training (his example concerned his medical sub-discipline) over the nature of the observations he makes. But the idea that observations depend on the *a priori* ideas of the observer had already been developed by other philosophers of medicine (Bieganski, Kramsztyk, Trzebinski). Kramsztyk in particular explained that the observer noticed above all those phenomena which were consistent with his previous knowledge and overlooked the others. Moreover, he might even observe non-existing phenomena if they conformed to his previous views [16, 17]. He also affirmed that the drawings in medical textbooks, supposedly objective descriptions of pathological states, in fact represented the collective point of view of a given period. For this reason, the utility of such drawings is limited to the period during which they had been produced [18].[2]

Another important idea developed by Fleck - the possibility of *the simultaneous coexistence of equivalent and incommensurable truths* - has its roots in his affirmation that the truth of an investigation depends on its final goal. Thus a clinical scientist (in his example, an epidemiologist) and a fundamental scientist (a biochemist) can arrive at divergent and incommensurable but nevertheless equally valid truths. This conclusion, crucial to the evolution of his *constructivist* and *relativist epistemology*, can also be viewed as a sophisticated answer to the question central to the thought of Polish philosophers of medicine: 'Is Medicine an Art or a Science?'

Polish philosophers of medicine attempted to legitimate their medical practice and to demonstrate that it was as valid as any biomedical scientific research glorified by the experts in scientific medicine. Each of them did it in his own way. Biernacki affirmed that clinical medicine (i.e., for him, the art of healing) of his time was not a science, but that it could become one in the future, by incorporating new approaches originating in the biomedical sciences. Kramsztyk explained that medicine was an art (i.e., for him, a technology), but added that this did not mean that it was in any way inferior to basic science. For him, science and practical knowledge represented two different but equally important human activities. Bieganski developed an entire philosophical system aimed (among other things) at showing that every science (and, in fact, every cognition) was goal-oriented. Therefore, for him, medicine was not fundamentally different from the most abstract or "pure" natural sciences. Finally, Fleck explained that clinical research and fundamental research represent two distinct and usually mutually exclusive thought styles: "It happens more frequently that a physician simultaneously pursues studies of a disease from a clinico-medical or bacteriological point of view along with the history of civilization than from a clinico-medical or bacteriological one together with a purely chemical one" ([9], p. 111). Clinical and basic research often arrive at different and incommensurable truths but, Fleck added, incommensurability is a general characteristic of human cognition, and the truth of the physician is no less valid than the truth of the basic scientist.

It is interesting to note that Bieganski and Kramsztyk tried to legitimate clinical medicine (the art of healing) and argued that it was no less valid than clinical science (e.g., histology, bacteriology), while Biernacki praised the advent of medical immunology as a discipline able to reconcile therapy and science. In contrast, a generation later Fleck attempted to legitimate the clinical disciplines (bacteriology, clinical immunology, epidemiology, etc.) and argued that they were not inferior to fundamental disciplines (e.g., biochemistry). This difference illustrates the rapid evolution of scientific medicine and the growing impact of the basic sciences, if not on medical practice than at least on the self-image of physicians.

How familiar was Fleck with the works of the other members of the Polish School? It is impossible to answer this question with absolute certainty. However, although Fleck neither made explicit his relationship with the Polish school of philosophy of medicine, nor quoted from their works in his writings, the evidence strongly points to the possibility that he was (directly or indirectly) acquainted with at least a few of their works. Recall that Fleck studied medicine at the University of Lwow, where the history of medicine was taught by Szumowski, a zealous propagator of the ideas of the Polish School. Later, Fleck was a founding member of the Lwow Society of Amateurs of Medicine, and participated in all its meetings in 1925 and 1926.[3] Indeed, his first epistemological study was presented at a Society meeting, (February 1926), and was published in the *Archives* a year later. In the years 1924-1926, the *Archives* (which, one can reasonably assume, were read by Fleck) published numerous articles discussing the ideas of the Polish School and referring to previous studies by Chalubinski, Bieganski, and Kramsztyk. In addition, Fleck's 1926 lecture was perfectly in accord with works published in the *Archives* in 1924 and 1925. Thus in 1924, Wrzosek published in the *Archives* an article discussing the variability of the concept of disease in different historical periods. Trzebinski argued, in 1925, that truth in medicine was relative and depended on the historical period and on the environment in which it was produced. In another article (published in the same year), he argued that observations could be dependent on ideology; for example, during the Middle-Ages theology strongly influenced scientific observations. And Nusbaum in 1925 evoked the view that pathological processes were unique events, and that there was no such thing as "disease" in general [21, 26, 27].

Fleck's influence by the Polish school of philosophy of medicine was in all probability limited to the first stages of the formation of his quite original epistemology. After 1927, Fleck no longer maintained a close relationship with the *Archives*. Although he wrote several other epistemological articles in Polish, he published them either in general philosophical journals, or in medical journals, but not in *Archives*.[4] Fleck was in all likelihood rejected by the professional Polish historians of medicine of his time; his numerous publications on this subject notwithstanding, he was never given the opportunity to teach history or philosophy of medicine. In an article published in 1939, Bilikiewicz denied Fleck the title of historian of medicine. He argued that although Fleck's reflections were based on historical material, his ideas belonged to medicine and biology, not to the history of science, which was a humanistic discipline [2]. And the report in the *Archives* (in the Section on History and Philosophy of Medicine of the 15th National Medical Congress that convened in Lwow, in July, 1937, and written by Bilikiewicz) summarized all the presentations, with the exception of the one presented by Fleck [1]. This lack of recognition of Fleck by Polish historians and philosophers of medicine can perhaps be explained on the basis of personal conflicts, or perhaps by the growing nationalist-Catholic orientation of *Archives,* or by both.

The principal concepts of the Polish school of philosophy of medicine were in all probability but one influence among the intellectual forces that contributed to the genesis and development of Fleck's original epistemology. Others included the thought of the Lwow school of philosophy (Twardowski, Adjukiewicz, Chwistek) [15, 24], Gestalt psychology, with which Fleck probably became familiar during his stay in Vienna, in 1928 [23],[5] sociological and anthropological studies (Durkheim, Lévy-Bruhl, Jerusalem) and the theory of complementarity promulgated by Niels Bohr and known to Fleck through Kurt Reizler's popularized version, published in 1928 [22].[6] The thought of the Polish School had, in all probability, a decisive impact on Fleck's ideas in the very first stages of the evolution of his epistemology. The views of thinkers like Chalubinski, Bieganski, Biernacki, and Kramsztyk not only attracted Fleck's attention to problems that later became central to his constructivist and relativist epistemology, but - and this is perhaps the most important point - they contributed to the shaping of Fleck's broad approach to the study of medical knowledge, and later to the study of scientific knowledge in general.

In the 19th-century, many philosophically-oriented physicians, fascinated by the development of scientific medicine, tried to define what medicine (or rather medical science) should be. In contrast, the starting point for the critical reflections of the Polish philosophers of medicine was the actual status of medical practice in their times. They were less interested in knowing how a theoretically-oriented physician should behave than in understanding how physicians actually behaved at the bedside. Their central interest in clinical practice was perhaps the most important lesson that they were able to pass on to Fleck. Fleck went further in his own investigations of the ways medicine and science work than any other Polish philosopher of medicine. Unlike his predecessors, he did not content himself with analyses of specific phenomena, but systematized his reflections on this question, and proposed a method, which he called "comparative epistemology", of studying the social, institutional, cultural, and linguistic influences on the formation of scientific knowledge. But the fundamental attitude of Fleck was not very different from the attitude developed by the Polish School. The strong interest of this school in clinical practice may have thus played a role in Fleck's turn toward a sociologically-oriented epistemology.

NOTES

1 In previous works [10, 19, 20] I have argued that the roots of Fleck's original epistemology can be found in his professional practice. This claim needs, however, to be completed.

2 Fleck's student and collaborator, Prof. W. Kunicki-Goldfinger, recalled that Fleck had strongly recommended that he read Kramsztyk's works. W. Kunicki-Goldfinger, personal communication, Warsaw, May 28th, 1988.

3 This association was affiliated with the Polish Association of Amateurs of Medicine which edited the *Archiwum* [11, 12, 13, 14] and was active in the years 1925-1926.

4 A complete bibliography of Fleck's works, prepared by T. Schnelle, can be found in ([3], pp. 445-457).

5 Gestalt psychology allowed Fleck to radicalize the view (present in the writings of the Polish School) that observations were dependent on the *a priori* ideas of the observer. While Kramsztyk and Bieganski believed this was usually the case (in particular in medicine), but that it was possible to eliminate, or at least limit the impact of one's *a priori* ideas on one's observations, Fleck affirmed that *a priori* - free observations could not exist - in modern terms, that every observation was theory-laden.

6 Fleck's article 'On the Crisis of Reality' [6], published in the same journal, was an answer to this article. See [4]. The complementarity theory led to Fleck's claim that there was no such thing as a single "true" vision of a given phenomenon.

BIBLIOGRAPHY

[1] Bilikiewicz, T.: 1937, 'Report from the Meeting of Polish Historians and Philosophers of Medicine in Lwow, 4-7 July 1937' ['Sprawozdanie z zjazdu polskich historykow i filozofow medycyny w Lwowie, 4-7 lipca, 1937'], *Archiwum Historji i Filozofji Medycyny* **16**, 244-252.

[2] Bilikiewicz, T.: 1939, 'Some Remarks on the Article by Ludwik Fleck "Science and Environment"' ['Uwagi nad artykulem Ludwika Flecka "Nauka i środowisko"'], *Przegląd Współczesny* **9-19**, p. 15.

[3] Cohen, R.S., and Schnelle , T. (eds.), 1986, *Cognition and Fact, Material on Ludwik Fleck*, Reidel, Dordrecht.

[4] Elkana, Y.: 1986, 'Is There a Distinction Between External and Internal Sociology of Science,' in R.S. Cohen and T. Schnelle (eds.), *Cognition and Fact, Material on Ludwik Fleck*, Reidel, Dordrecht, pp. 309-316.

[5] Fleck, L.: 1986 (1927), 'Some Specific Features of the Medical Way of Thinking' ['O niektorych swoistych cechach myślenia lekarskiego'], *Archiwum Historji i Filozofji Medycyny* **6**, 55-64. English translation in R.S. Cohen and T. Schnelle (eds.), *Cognition and Fact, Material on Ludwik Fleck*, Reidel, Dordrecht, pp. 39-46.

[6] Fleck, L.: 1986 (1929), 'Zur Krise der "Wirklichkeit"', *Naturwissenschaften* **18**, 425-430. English translation in R.S. Cohen and T. Schnelle (eds.), *Cognition and Fact, Material on Ludwik Fleck*, Reidel, Dordrecht, pp. 47-57.

[7] Fleck, L.: 1986 (1935), 'Scientific Observation and Perception in General' ['O obserwacji naukowej i postrzeganiu w ogóle'], *Przeglad Filozoficzny* **38**, 56-76. English translation in R.S. Cohen and T. Schnelle (eds.), *Cognition and Fact, Material on Ludwik Fleck*, Reidel, Dordrecht, pp. 59-78.

[8] Fleck, L.: 1937, 'Some Specific Features of the Serological Way of Thinking' ['O swoistych cechach myślenia serologicznego'], W. Nowicki and D. Szymkiewicz (eds.), *Report of the XVth Meeting of Polish Physicians*, Lwow, 4-7 July 1937.

[9] Fleck, L.: 1976 (1935), *Genesis and Development of a Scientific Fact*, University of Chicago Press, Chicago, transl. by F. Bradley and T.J. Trenn (first edition, L. Fleck,*Entstehung und Entwicklung einer wissenschaftlichen Tatsache*, B. Schwabe, Basel).

[10] Freudenthal, G. and Löwy, I.: 1988, 'Ludwik Fleck's Roles in Society: A Case Study Using Joseph Ben-David's Paradigm for a Sociology of Knowledge', *Social Studies of Science* **18**, 625-652.

[11] Fritz, J.: 1925, 'Report of the Foundation Meeting of the Lwow Society of Amateurs of History of Medicine' (24 October, 1925) ['Sprawozdanie z posiedzenia organizacyjnego towarzystwa miłosnikow historji medycyny w Lwowie z dnia 24 pazdziernika 1925'], *Archiwum Historji i Filozofji Medycyny* **4**, 154-156.

[12] Fritz, J.: 1925, 'Notes from the 2nd Meeting of the Lwow Society of Amateurs of History of Medicine' ['Notatki z II posiedzenia towarzystwa miłosnikow historji medycyny w Lwowie'], *Archiwum Historji i Filozofji Medycyny* **4**, 331-332.

[13] Fritz, J.: 1926, 'Notes from the 3rd Meeting of the Lwow Society of Amateurs of History of Medicine', *Archiwum Historji i Filozofji Medycyny* **5**, 299.

[14] Fritz, J.: 1926, 'Notes from the 4th Meeting of the Lwow Society of Amateurs of History of Medicine', *Archiwum Historji i Filozofji Medycyny* **5**, 149.

[15] Giedymin, J.: 1986, 'Polish Philosophy in the Inter-war Period and Ludwik Fleck's Theory of Thought-styles and Thought-collectives', in R.S. Cohen and T. Schnelle (eds.), *Cognition and Fact, Material on Ludwik Fleck*, Reidel, Dordrecht, pp. 179-216.

[16] Kramsztyk, Z.: 1898, 'Clinical Fact' ['Fakt kliniczny'], *Krytyka Lekarska* **2**, 29-32.

[17] Kramsztyk, Z.: 1906, 'Difficulties in the Observation of Chronic Diseases' ['Trudności w badaniu chorob przewlekłych'], *Krytyka Lekarska* **10**, 149-157.

[18] Kramsztyk, Z.: 1899, 'On Clinical Descriptions' ['O kazuistyce klinicznej'], in Z. Kramsztyk, *Critical Notes on Medical Subjects,* E. Wende, Warsaw, pp. 55-58.

[19] Löwy, I.: 1986, 'The Sociology of Knowledge of a Sociologist of Knowledge: Ludwik Fleck's Professional Outlook and its Relationship to his Philosophical Works', in R. S. Cohen and T. Schnelle (eds.), *Cognition and Fact, Material on Ludwik Fleck,* Reidel, Dordrecht, pp. 421-442.

[20] Löwy, I.: 1988, 'The Scientific Roots of the Constructivist Epistemologies of Hélène Metzger and Ludwik Fleck', *Corpus* (Paris) 8/9, 219-235.

[21] Nusbaum, H.: 1925, 'Medicine as Art', *Archiwum Historji i Filozofji Medycyny* **4**, 250-277.

[22] Reizler, K.: 1928, 'Die Krise der Wirklichkeit', *Naturwissenschaften* **16**, 705-712.

[23] Schnelle, T.: 1986, 'Microbiology and Philosophy of Science, Lwow, and the German Holocaust: Stations of a Life - Ludwik Fleck 1896-1961', in R. S. Cohen and T. Schnelle (eds.), *Cognition and Fact, Materials on Ludwik Fleck,* Reidel, Dordrecht, pp. 3-36.

[24] Schnelle, T.: 1986, 'Ludwik Fleck and the Influence of the Philosophy of Lwow', in R.S. Cohen and T. Schnelle (eds.), *Cognition and Fact, Materials on Ludwik Fleck,* Reidel, Dordrecht, pp. 231-266.

[25] Trzebiński, S.: 1925, 'Rationality and Rationalism in Medicine', *Archiwum Historji i Filozofji Medycyny* **2**, 91-103.

[26] Trzebiński, S.: 1925, 'Superstition in Medicine', *Archiwum Historji i Filozofji Medycyny* **4**, 94-105.

[27] Wrzosek, A.: 1924, 'Definition of the Concept of Disease', *Archiwum Historji i Filozofji Medycyny* **1**, 72-77.

[28] Zwoźniak, W.: 1964, 'History of the Medical Department of Lwow University' ['Historia wydziału lekarskiego uniwersytetu lwowskego'], *Archiwum Historii Medycyny* **28**, 57-85.

CHAPTER VII.A

TEXTS OF LUDWIK FLECK AND TADEUSZ BILIKIEWICZ

LUDWIK FLECK
'SOME SPECIFIC FEATURES OF THE MEDICAL WAY OF THINKING'

(Archiwum Historji i Filozofji Medycyny, 1927)*

Medical science, whose range is as vast as its history is old, has led to the formation of a specific style in the grasping of its problems and of a specific way of treating medical phenomena, i.e. to a specific type of thinking. In substance such separateness of the way of thinking is nothing extraordinary. One has only to realize the difference between the way of thinking of a scientist and that of a humanist, even if the subject in question is the same: for example, how great is the difference, and how great is the impossibility of a direct juxtaposition, between psychology as science and as a branch of philosophy. Even the very subject of medical cognition differs in principle from that a scientific cognition. A scientist looks for typical, normal phenomena, while a medical man studies precisely the atypical, abnormal, morbid phenomena. And it is evident that he finds on this road a great wealth and range of individuality of these phenomena which form a great number, without distinctly delimited units, and abounding in transitional, boundary states. There exists no strict boundary between what is healthy and what is diseased, and one never finds exactly the same clinical picture again. But this extremely rich wealth of forever different variants is to be surmounted mentally, for such is the cognitive task of medicine. How does one find a law for irregular phenomena? - this is the fundamental problem of medical thinking. In what way should they be grasped and what relations should be adopted between them in order to obtain a rational understanding?

---

* English translation of this article was first published in R.S. Cohen and T. Schnelle (eds.), *Cognition and Fact, Materials on Ludwik Fleck*, Reidel, Dordrecht, pp. 39-46.

One begins to look for types among the phenomena which at first appear to be atypical. For instance, the normal, typical action of the heart has such and such characteristics. There exist individual differences as regards the duration and intensity of each component of that action and the sequential rhythm of these components. However, these differences are physiologically minute. It is only the morbid action of the heart that yields a tremendous wealth of pictures which are more and more different. It becomes unavoidable that we broaden the observations to include peripheral vessels, capillary vessels, skin, the endocrine glands, the vegetative system, the development relations, etc.

A tremendous wealth of material is produced. It is the task of medicine to find, in this primordial chaos, some laws, relationships, some types of higher order.

In principle, this goal is attained. We know, from the calculus of probability, that even an accidental case, even events lacking mutual relations, can be embraced in certain laws, and so one should not wonder that event these abnormal morbid phenomena are grouped round certain types, producing laws of a higher order, because they are more beautiful and more general than the normal phenomena which suddenly become profoundly intelligible. These types, these ideal, fictictous pictures, known as morbid units, round which both the individual and variable morbide phenomena are grouped, without, however, ever corresponding completely to them - are produced by the medical way of thinking, one the one hand by specific, far-reaching abstraction, by rejection of some observed data, and on the other hand, by the specific construction of hypotheses, i.e. by guessing of non-observed relations. In this case we use the statistical juxtaposition and comparison of many such phenomena, i.e. that which I could call simply statistical observation, which is the only method of finding the type among a number of individuals. The role of statistics in medicine is immense. It is only numerous, very numerous, observations that eliminate the individual character of the morbid element, and in such abstruse fields as pathology and sociology the individual feature is identical with an event and ought to be removed. However, the statistical observation itself does not create the fundamental concept of our knowledge, which is the concept of the clinical unit.

There come into play here many elusive - as far as logic is concerned - imponderable factors which enable one to foresee (in a way to forebode !) the course of problems which determine the development of a given field of thought and create its style peculiar to the epoch. I venture to call the factor the specific intuition. I am unable to dwell here in more detail on the problem of intuition, as this only becomes possible in the light of the history of science; however, I have to stress here that, without this concept, i.e. if we admitted that the development of science is only a matter of time, technical possibilities and accident, we would never understand science; in the first place we would be unable to graps why the developmental stages possess a specific style of thinking, why a phenomenon which is accessible to everybody had been observed at the given moment for the first time, and even almost stimutaneously by several researchers. Thus, in a certain developmental stage, there arise certain definite clinical units, and this way of their genesis explains some of their specific features. Nowhere outside medicine does one find so many qualifications, pseudo- and para, e.g. typhoid - para-typhoid, psoriasis - para-psoriasis, vaccine - para-vaccine, aneamia - pseudo-anaemia, paralysis pseudobulbaris, pseudo-croup, pseudo-neuritis optica, pseudoptosis, pseudo-sclerosis, pseudotabes; next, meningitis - meningismus, Parkinson - Parkinsonism, etc... These specific names are found in medicine, because, with the progress of medical knowledge, it became necessary to single out, in the definite idealistic clinical type, the individual sub-types, e.g. typhoid-para-typhoid which sometimes proved to be completely unrelated: tabes-pseudotabes. The further does medical knowledge progress the more such definitions, such proofs of departures from the original way of dealing with a situation, do and will arise, since the original approach is found to be too abstract, too ideal.

As regards the role played in medical thinking by intuition, even in simple diagnosis, this can be seen best from the act that we really lack almost always a pathognomic symptom which, by itself, would suffice to determine the clinical state: even the typhoid bacillus cultivated from feces does not prove that the given individual suffers from typhoid fever; the individual may be only the germ carrier. It is only the combinations of symptoms, the habitus, the entire *status praesens* of the patient that is conclusive. Why even the best diagnosticians are most frequently unable to give a specific basis for their diagnosis; they only explain that the entire appearance is characteristic of such or another disease.

As soon as the medical thinking has found a certain ideal type in an finite plurality of apparently atypical morbid phenomena, it faces a novel problem:

how to reduce them to a common determinator, to obtain, by way of analysis, certain common elements, some component bricks from which the observed phenomena could be reproduced. In this way elements of morbid anatomy and morbid physiology arise. However, combinations of the motifs obtained in this way and repeating themselves again and again (inflammation, degeneration, atrophy, hypertrophy, hypofunction, hyperfunction etc...) never do adequate justice to the entire wealth of the individual features of the disease. The specific, most characteristic features remain always outside such handling, and they prove that the elements of morbid anatomy and physiology are too general.

This is again the specific feature of medicine. Nowhere outside medicine, in any other branch of science, have its species so many specific features, i.e. non-analysable features that cannot be reduced to common elements. In this way the abstracting process that has been carried very far produces the notion of the species whose fictitiousness is considerably greater than in any other field of science, and a notion of the element (or property) with an equally specific generality. This results in a characteristic divergence between book knowledge and live observations, but not the divergence between medical art and science, because in chemistry also one witnesses a certain incommensurability between science and applied art. However, there no observation can be incompatible with theory, or even be included in it. On the other hand, one can use in medicine the celebrated saying: "In der Theorie zwar unmöglich, in der Praxis kommt es aber vor."

In practice one cannot do without such definitions as 'chill', 'rheumatic' or 'neuralgic' pain, which have nothing in common with this bookish rheumatism or neuralgia. There exist various morbid states and syndromes of subjective symptoms that up to now have failed to find a place and are likely not to find it at any time. This divergence between theory and practice is still more evident in therapy, and even more so in attempts to explain the action of drugs, where it leads to a peculiar pseudo-logic. Not long ago the administration of camphor in the case of hemoptysis was forbidden - and a reason for it was found. Today camphor is recommended, and a 'logical' motivation has been found. Every therapeutic method, including homeopathy and psychoanalysis, has a 'strict, logical, almost mathematical' motivation, mostly the more exact the shorter its life. It is nowhere easier to get such a pseudo-logical explanation than in medicine because the more complex the set of phenomena the easier it is to get a law verifiable for a short term, and the more difficult it is to reach an embracing idea. It is in medicine that one encounters a unique case: the worse the physician the 'more logical' his therapy. The point is that, in medicine, one is able to stimulate almost everything, which proves that, up to now, we have indeed failed to explain anything.

Beside the fundamental notions of species and element, medical thinking possesses also the equally specific notion of the relationship of morbid phenomena. This extremely complex field presents an epistemologically unique picture. Along with the natural sciences, medical thinking recognizes causal relations (though it is generally accepted that a physician says always 'afterwards', but almost never 'because of that'). Just as in biology, the conditioning of phenomena in medicine can be developmental, correlative, substituting, synergetic and antagonistic. A completely specific factor, which explains the morbid phenomena, one finds in medical thinking in the notion of internal disposition and of external substrate, i.e. of conditions which, as if in *potentia,* comprise the given morbid phenomenon. Besides, we have the epidemiological grasping of morbid phenomena and, by far not the last - teleology. Thus medical phenomena are mutually related by means of a tremendous number of relations, as the result of, and compensation for, their original atypical character.

However, this plurality whose elements are so multiply conditioned is irrational if we examine it as a whole and consistently from the same standpoint. We admit causal relationships, but the result is never proportional to the cause, nor is it always the same. The action of the pathogenic cause is a resultant of its intensity and disposition, i.e. to the causal relationships are added the dispositional factors which are incommensurable with the former. However, even if both of these active series are taken in account, one cannot deduce anything in medicine, since an antagonistic reaction may appear. For example, 'dermographismus albus' points, according to some, to the hyperfunction of the suprarenal gland, while, according to others (with equal logic), to the hypofunction, in view of the antagonism between skin and intestines. Schultze's law of the action of stimuli, the different action of the small and medium doses of atropine, the variable reaction of pupils in the case of anaesthetization - such are further examples of this irrationality.

Even a thorough familiarity with the anatomy and physiology of the bladder and a thorough familiarity with the tuberculor processes would not enable one to foresee the interesting phenomenon, viz. that bladder TB recedes after resection of the tuberculous kidney. Similarly, even familiarity with the physiology of speech would not enable one to deduce the fact that one can learn to speak even after the complete removal of the larynx. According to the classical theory of the Wassermann reaction one may deduce that, when working with an active serum, one would obtain more negative results, whereas in reality a contrary result is obtained. In this case medicine has its own motivation which, however, does not lie along the line of classical theory, but requires a change in mental attitude.

This is what one encounters in the case of any medical problem: it becomes ever and ever necessary to alter the angle of vision, and to retreat from a consistent mental attitude. Only in this way does the world of morbid phenomena, which is irrational in its entirety, become rational in its details. Just as, on the one hand, the far-reaching abstracting action enables medical thinking to find types among atypical phenomena, so also, on the other hand, it is only the renouncing of consequences that enables one to apply a law to irregular phenomena. This results in the incommensurability of ideas which develop from the varying ways of grasping morbid phenomena and which gives rise to the fact that a uniform understanding of morbidity is not possible. Neither cellular nor humoral theory, nor the functional understanding of diseases alone, nor their 'psychogenic' conditioning, by themselves will ever exhaust the entire wealth of morbid phenomena.

However, much as it is impossible to get in medicine an idea that would embrace the entirety, like atomism in chemistry or energetics in physics, yet one witnesses that a new methodical idea, a certain keynote for grasping medical phenomena, comes to the fore. This is a specifically temporal and dynamic grasping of morbid phenomena. The object of medical thinking - illness - is not an enduring state, but a process which changes continually, and which has its temporal genesis, its course and decline. This scientific illusion, this fiction, this individual entity created by abstraction based on statistics and intuition, the entity called the disease which is virtually irrational, elusive and undefinable univocally, becomes a substantial unit only when grasped temporally. Never a *status praesens*, but always only the *historia morbi* actually create the clinical unit. The former yields at the very most a syndrome of symptoms, such as Banti syndrome or Horner's syndrome, which modern medical thinking does distinguish carefully from the disease. This historic, temporal nature of the notion of the disease is unique. Since the disease is a change, which develops in time, of life functions which have likewise their temporal course, it is obvious that, being a *sui generis* variety of life variations, it is doubly dependent on the instant. If one may use a comparison from a distant field, the disease has a relation to normal fonctions just as acceleration has to velocity. Life as such has its temporal course. The course of the disease takes place within that course, being somewhat independent on it. A child develops in accordance with a known pattern. Simutaneously, its TB develops in its own tempr and according to its own laws. Thus this disease obtains its double, or practically quadruple genesis.

Thus in the first place the pathogenesis of a single definite case; its disposition, diathesis, constitution or habitus, its infection, original symptom, the origin of allergy, the development of pathological symptoms etc. I would call this the detailed ontogenesis of the disease. Next the general pathogenesis of the single TB case, i.e., for example the disposition factors and the progress of TB or typhoid or uric acid diathesis in childhood, puberty, climacteric etc. This I would call the general ontogenesis of the disease. Thirdly, the independent history of the disease in a certain social or geographical environment, the history of a certain epidemic or of a certain degeneration. This I would call the detailed phylogenesis of the disease. Finally, the independent history throughout the ages of the disease, its appearance in mankind and its change. This I would call the general phylogenesis of the disease. I do not know any other field of scientific thought in which the fundamental idea would allow of so many different genetic investigations. Embryology or paleontology, history or sociology - recognize only a development in one direction. In pathology, two developmental series are combined: the onto- and phylogenetical development of the living creature and the development of the disease. This historic formulation of the disease idea becomes more and more clear-cut.

I wish to point to two relevant, fully modern and fertile ideas: that of 'hygieogenesis' and of the latent infection ('inapparente Infektion' - Weil); and also to the idea of the latent disease, e.g. *lues latens*. The relevant processes can be numbered neither among the former ideas of health nor among the ideas of disease. In their light, health is a certain mutual attitude of the patho- and hygieogenetic processes, and any other attitude in any direction is a disease. Since the most different organs and glands can replace one another, and certain diseases compensate one another, producing a more advantageous state, one should, to all intents and purposes, define, or specify, health consistently, though paradoxically, as an illness which is the most profitable at a given moment. Thus a specifically dynamic grasping of the subject does arise, where, instead of constant causes, we have mutually influencing processes. The relations between these processes are different and incommensurable, depending upon the always necessary change of the viewpoint. If one adds to it the specific abstract nature of the idea of the clinical unit, we obtain a general picture of the medical way of the formulation of the problem.

Let me use a figurative comparison: medical thinking differs in principle from scientific thinking in that it uses Gauss's coordinate system, while the latter uses the Cartesian system. Medical observation is not a point but a small circle. It is placed not in the system of coordinate straight lines inclined to one another at a constant angle, but in a system of optional, mutually intersecting, curves which we do not know closely.

A certain correction is introduced into this picture by the fact that, strictly speaking, the multiplicity of medical phenomena can be only approximately rendered by means of Gauss's system since its points are not univocally determinable. To all intents and purposes, scientific thinking uses, for small ranges, the Cartesian system, and for large ranges, Gauss's system (as in the theory of relativity). On the contrary, medical thinking uses Gauss's system for small ranges, while in the entirety it does not find any consistent and rational way to grasp phenomena."

LUDWIK FLECK
'ON THE CRISIS OF 'REALITY''
(*Naturwissenschaften,* 1929)*

In exploring the sources of cognition *(Erkenntnis),* we frequently commit the mistake of regarding them as much to simple.

We forget the simple truth that what we are acquainted with *(Kenntnisse)* consists rather of what we reach by learning *(Erlerntes)* than of what we arrive at by knowing *(Erkanntes).* Yet this is a momentous fact, since in the short span from the teacher's mouth to the student's ear the content of the knowledge transmitted is always slightly distorded. Thus, in the course of decades or even centuries and millenia, divergences develop to an extent that it sometimes becomes doubtful whether anything of the original has been preserved at all.

In the circumstances the content of knowledge is by and large to be evaluated as a free creation of culture. It resembles a traditional myth.

Unfortunately, however, we characteristically regard old, habitual trends of thought as particularly self-evident, so that no proof is required or even admissible for these. They constitute the firm foundations on which further construction is allowed.

In addition to this, the physiology of our cognition *(Erkenntnisphysiologie)* has a second, important characteristic, by virtue of which every new cognitive activity depends on the previously accumulated store of knowledge, since the weight of that which is already known changes the internal and external conditions of newly acquired cognition.

In this manner three systems of factors come into being, that contribute to every process of cognition *(Erkennen),* are interrelated and interacting: the burden of Tradition, the weight of Education, and the effect of the Sequence of the acts of cognition.

---

* English translation of this article was first published in R.S. Cohen and T. Schnelle (eds.), *Cognition and Fact, Materials on Ludwik Fleck*, Reidel, Dordrecht, pp. 47-58

These are social factors, and any new epistemology must, therefore, be brought into a social and cultural-historical context, lest it seriously contradict the history of cognition and the everyday experience of teacher and student.

At no time do we resemble a blank page, nor are we in a state of a *tabula rasa* as is the screen before a film is projected on it. Cognition has no discernible beginning, certainly not at the moment of birth or even in the womb, because the capacities for feeling, and feeling as such, originate in a parallel and synchronous way through interaction. It is equally impossible to establish the phylogenetic beginnings of cognition.

In individual life there take place not just one but numerous epistemological births and embryonic developments. We undergo rebirth for each new situation and bring with us a more or less complete mechanism of birth and more or less ready-made predispositions that are decisive for our reactions and for the contents of our cognition.

Wherever and whenever we touch on something, we are always in the very midst and never at the beginning of cognition. Therefore I am at a loss to see how one could possibly construct epistemology out of sensations as building blocks.

An experienced teacher has found that only a small minority of students independently notice something new without having their attention explicitly drawn to it, and that even then only a few see it immediately as it is shown to them. They first have to learn to see it. Even the adult, facing something new for the first time - perhaps an abstract picture, a strange landscape, or looking through a microscope - 'does not know what he is supposed to see'. He is looking for similarities with something familiar, thus overlooking the new, which is incomparable, specific. He, too, must first learn to see. How many examples could be adduced here from the history of science! And yet, it is just this 'seeing' that one first has to learn, which makes for the progress of any science, the progress which thus is again and again given its social imprint.

If one wanted to solve the problem of the genesis *(Entstehung)* of cognition by the traditional method as the individual concern of a symbolic 'human being', one would have to subscribe not only to the proposition *'nihil est in intellectu, quod non fuerit in sensu'* but also to its reverse form: *'nihil est in sensu, quod non fuerit in intellectu'*. And beyond this one cannot make any progress.

Consequently I do not know why and wherefore should I make a difference between a first and a second reality, as Riezler, among others, describes them.

Just so, the social factor in the genesis of cognition must not be disregarded.

Thus every thinking individual, being a member of some society, has his own reality in which and according to which he lives. Everybody has even many, sometimes contradictory, realities: the reality of everyday life, a professional, a religious, a political, and a small scientific reality. And secretly he has a superstitiously fateful personal reality that renders the actual I exceptional.

Each cognition, each system of cognitions, each entry into the social realm has ist own corresponding reality. This is the only correct position.

Otherwise, how could I understand that, e.g., a person with a humanistic education will never completely grasp the science of the naturalist? Or even the theologian? Should I regard the others as fools, as regrettably happens so often?

They encounter the greatest difficulty not in the solutions of the problems, but in comprehending the origins and the significance of the problems as such; not the concepts but their evolution and function.

Every knowing *(Wissen)* has its own style of thought with its specific tradition and education. Out of the almost infinite multitude of possibilities, every day of knowing selects different questions, connects them according to different rules and to different purposes. Members of different scientific communities live in their own scientific and also professional reality. In their daily lives these people can get along with each other in perfect harmony, for they may have a common everyday reality. There are cultures, as e.g. the Chinese culture, which in important fields, such as medicine, arrived at quite different realities from those of us occidentals. Shall we punish them for this with pity? They had a different history, different aspirations and demands that are decisive for their cognition.

For cognition is neither passive contemplation nor acquisition of the only possible insight into something given. It is an active, live interrelationship, a reshaping and being reshaped, in short, an act of creation. Neither the 'subject' nor the 'object' receive a reality of their own; all existence is based upon interaction and its relative.

Just as everything that is socially conditioned, that is perceived, has its own life, independent of the particular individual, so it has its special characteristics, its style in time and space, and consequently its own destiny.

Even the schizophrenic from whose asocial momentary reality spring utterances as such as '1-2-3 this is pharmacy, this is boxhedge, Bucks, Rio de Janeiro' employs concepts of social origin. Yet his reality remains inaccessible to others - and probably in the next moment also to himself. It is probably of no enduring importance to anybody.

However, there exist realities with style *(stilvoll)* that are founded upon serious, continual work by large groups and great men, in the spirit of which people live (or for the sake of which they die). They develop, flourish, endure, waste away leading their own lives, such as forms of government, or social arrangements. The relative independence of the cognized from the individual is well illustrated in the fact that different individuals frequently make the same discovery or invention simultaneously but independently from one another. Cognitions are formed by human beings, but also conversely they form their human beings. It would be simply foolish to ask here what is 'cause' and what 'effect'.

Once upon a time there existed a Great Science that was related to almost every branch of knowledge in that period, that was based upon solid theoretical-philosophical foundations, and that had the greatest influence on political-economic and personal life. I believe no such universally predominant science had ever existed before or after; in all areas it explained the past, defined the present and predicted the future. This science was called astrology. Nowadays it leads only a pitiful existence in the thinking of some uneducated freaks, not comparable to its erstwhile greatness, similar in a way to that of our lizard to the dinosaurs. It was replaced by a differently constructed system of thinking by society, namely by the natural sciences. Surely there had always existed thinking typical of the natural sciences. It was to be found among the artisans, the seamen, the barber-surgeons *(Wundärtze),* the leather-workers and saddlers, the gardeners and probably also among children playing. Wherever serious or playful work was done by many, where common or opposite interests met repeatedly, this uniquely democratic way of thinking was indispensable.

I am calling the thinking which is typical of the natural sciences democratic, because it is based upon organization and control at all times, it rejects the privilege of divine origin and wants to be accessible and useful to everybody. Yet we have learned from experience that evey democracy contains its little untruths, since what is wanted is an impressive, majestic government - not just a useful and wise one. This is why there are medals, titles, flags and presidents. Therefore the natural sciences have their own natural philosophy and their *Weltanschauung*.

When talking of natural sciences, one often forgets that there exists a living scientific practise and, parallel to this, an official *Gestalt* of Science on paper.

These two worlds, however, are frequently as different from one another as are the practise of demacratic government and its official theory. Certainly this cannot be helped, but this natural disharmony gives rise to serious misunderstandings. One mistakes the natural sciences as they are for the natural sciences as they ought to be, or rather as one would like them to be. But the practice of natural sciences cannot be learned from any book, for silence is kept about its ways and means. It contains the small "divergences' wich are not taken into account, the 'exceptions' that should only confirm the rule, the 'accidental', the 'unessential', the 'unavoidables mistakes'. These are the usual figures of speech that are always available when one wants and has to preserve the regularities.

These phrases are indispensable, even though one squeezes the rich, free stream of possibilities through the narrow passage of conceptual and material instruments (based on the responsibility of the fathers).

All this gives occasion for a distinct, though slight, change in comparison with what is officially demanded. The slight changes are integrated, and in this way they become larger, for they are not chaotic but bear the imprint of tradition, of the scientific moment, and of the researcher's personal style of thinking - as everybody knows in practice but forgets in theory. For the next generation they already become facts.

Everyday practice also teaches us that even the 'simplest' (i.e., today simplest) activity, such as measuring and weighing, is an art that has to be taught and that is sometimes never learned. Even the Wassermann reaction, so clearly worked out and frequently applied, is ultimately an art whose value depends much more on the practician than on the method by which it is carried out - as was recently stated by one of the leading serologists (Eisenberg).

Not only the ways and means of the solutions are subject to the scientific style, but also the choice of problems, and at that event to a higher degree. But the course of science is immensely influenced by the sequence of the solutions, for it determines the development of technical possibilities, the education of the researchers of the future, and the formation of scientific concepts and comparisons.

Here it is needless to adduce examples, since everybody knows thousands and could mention entire series of cognitions *(Erkenntnisreihen)* bearing the stamp of the epoch and the imprint of the scientist's personality on the method and style of their solutions. If the individual was strong enough, and his qualities were not only those of a pioneer but those of a leader, then his style became universally accepted into the body of science. In this way the scientific style and recognized scientific practices became codeterminant agents in the formation of scientific reality. How much conventionality, tact, intuition is contained in this agency follow from the simple truth that there may exist too much consistency that leads to one-sidedness, and too much criticism that causes sterility. One must practice moderation; the goal of the research determines this. Even weighing and measuring are done in different ways, according to the purpose to be served. And even though the 'greatest exactitude' today appears applicable (though uneconomical) for all purposes, I still believe that quite a few laws, such as e.g. the law of Boyle and Mariotte, the law of the conservation of matter, or the laws of classical mechanics, would never have been discovered, had the inexactness of observation necessary for whether a law was not formulated at all, or whether it is 'supplemented' or 'restricted' after it has for many years influenced the formation of reality and of men.

Even more striking is the purpose-dependency of scientific truths in areas in which nowadays one arrives at divergent and not exchangeable truths, depending on the purpose of the investigation: for example in bacteriology, where there exists a botanical-genetic and a medical-epidemiological aspect. As an illustration of the epidemiological standpoint I would mention Professor Friedemann's article on the problem of scarlet fever. (*Klin Wschr* (1928), N. 48, 2280). In the author's opinion the convalescents are no longer infectious after three consecutive negative bacteriological tests.

> There have also been, though, some divergent pieces of evidence. At the Königsberg congress on scarlet fever Elkeles reported that three out of seven cases came from patients who had been discharged after three negative tests. I would venture to surmise that the explanation of this result which is at variance with our and others' experiences lies in the method applied by Elkeles. For it seems striking that Elkeles found heamolytic streptococci only in 84 % of fresh cases of scarlet fever, whilst almost all other authors were able to prove the presence of haemolytic streptococci in nearly 100% of the cases.
> Elkeles indicates that he regarded as haemolytic streptococci only those which showed an absolutely indisputable heamolytic ring on the microscopic slide. In view of the practical purpose of these examinations it seems to be more correct in doubtful cases rather to diagnose the presence of haemolytic streptococci than the absence. For if we disregard them, any error is liable to have serious consequences, and it appears to me likely that, as a result of his rigorous rejection of all doubtful cultures, Elkeles indeed disregarded scarlatina streptococci that were actually present.

Thus the overly rigorous and therefore one-sided position is useless for practical epidemiology. Here the 'unavoidable mistake' is compensated for purposefully and knowingly by another one.

Yet, in addition to this dependence on the special purpose of an investigation which has a contributing effect on the scientist (as on every other) reality, there also exists a general effect of observation and investigation as such:

> The quantum postulate means that every observation of atomic phenomena constitutes a non-negligeable interaction with the measuring instrument, and that no independent physical reality in the ordinary sense can be ascribed either to the phenomena or to the means of observation. Generally speaking, the concept of observation has an element of arbitrariness, in that it essentially depends on what objects have to be included in the system that is being observed. (Bohr, *Naturwiss* (1928), N° 15).

This statement applies to all observations of any phenomenon whatever, but the mutual relationship with the means of observation is relatively negligible in most cases. Yet, if the 'treatment' of the phenomena, with whatever instruments, goes on ever centuries, will the effect not become significant? *To observe, to cognize (erkennen) is always to test and thus litterally to change the object of investigation.*

This is the day-to-day *praxis* of science. Here the social and the historical-traditional element is predominant. In great creative moments, however, the newly emerging science is simply an artistic creation that one can only admire and never 'prove' and determine 'objectively'. For there never existed nor exists a scientific demand for fundamental content. And at the given moment a yardstick for greatness is never to be found.

I am thinking, for instance, of Vesalius' idea to dispense with a completely elaborated, a hundred percent consistent, highly respected science, and consistently to build a new one from confused, unstable, changeable, interwined masses of flesh, such as the scolars of that time would have considered it to be beneath their dignity even to touch.

For a correct appreciation of this we should call to mind the moment when for the first time we stood before a corpse. Did not the medical examiner then appear to us as a sculptor who is just modelling the intended structure of the body, carving it out of the corpse, throwing away kilogrammes of 'unimportant' matter in order to bring out thin, hardly visible threads of tissue that he declares to be of sole importance, giving it high-sounding names and thus calling it into existence for the first time? Were we not at that time much more aware of our little bit of book-learning of anatomy than of this practical art of dissection?

Today's pathologist is just an imitator of his teacher. But Vesalius had no teacher. He had to perform his modelling according to his own intuition, battling against the far more obvious knowledge of the mighty scientists of his time, against his own mystical dread of the corpse which is still discernible in the grouping of his figures, and against his deeply ingrained veneration for Galen and tradition, which sometimes obscures his judgment.[1]

Thus he did his shaping and cut off everything that became inessential for a long time to come: the fatty and connecting tissue - and old emotional contexts that fell away as 'superstition' thanks to his work. Thus he formed the structure of the body and scientific concepts.

This was a creative act, unproven by bookish syllogisms or by intellectual reasoning. The reigning science had no demand for this, since it wanted to persist in the wealth of thought of its *Anatomia imaginabilis*. Thus, for instance, Bartholomaeus Eustachius wrote in 1516 that he would rather err with Galen that accept the truth from the innovators. Johann Phil. Ingressias (around 1600) wanted "*in quibus omnibus veteres defendere interpretando, elucidare atque excusare...*". And the ancients have been defended by thousands of cunning devices. Bauhin, e.g., attributed to Galen his own discovery of the Valvula - only in order to avoid taking a stand against him.

This was no struggle about details, about 'facts'; what was at issue was the familiar reality itself, the sacred faith that needed to be defended, not to be proven. Then comes an innovator, blasphemously relying on his own powers and through simple labour constructs, controls, and develops a science in place of the ready-made immutable doctrine of the divinely inspired master, which had so many deep connections with the totality of knowledge. As against this, how poor was the anatomy of Vesalius!

This was a battle for democratic values, and Vesalius had created for it the method, the style of thinking, and thus he had laid the foundations of the democratic general reality that was free of deep mysticism, of sentimental poetry, of grand affections.

Natural science is the art of shaping a democratic reality and being directed by it - thus being reshaped by it. It is an eternal, synthetic rather than analytic, never-ending labour - eternal because it resembles that of a river that is cutting its own bed.[2]

That is the true, living natural science. One must not be oblivious to its creative-synthetic and social-historical elements.

Its official ideal image is different: it is naive and beautiful. The absolute, Riezler's third reality belongs within this concext. The first one is the life and work of the researcher, the second is his religion.

It is beautiful if an artist in the course of his work has a vision of his own creation in its unattainable perfection. But it is naive not to know that this vision is not something absolute but that it mainly depends on the subject and the moment of time. One must nor forget that there exists no fully completed science but only that is becoming. Every solution constitutes a fresh problem, just as conversely every formulation of a problem already contains part of its own solution. Several fields of natural science are lying fallow after years of intensive development, such as anatomy today, or astronomy that was so much alive in Kepler's and Tycho Brahe's times. They appear to be done with, dead.

But one day they will be revived, illuminated from a different angle, taken up again with new conceptions, desirable due to new requirements - and then they will be as fresh and "as marvellous as on the first day" *(herrlich wie am ersten Tag)*.

We approach the ideal, 'absolute' reality not even asymptotically since it changes incessantly, renews itself and moves away from us at the same pace as we are advancing. It is an imaginary ideal whose content is solely determined by negation, by longing for something else. Does it not possess as little or as much reality as the ideals of beauty or goodness? Is it not just as dependent on time, place, culture and person? A few centuries ago something else was regarded as good as well as true than what we consider good and true today. Have we arrived at the terminal point of evolution today? Surely not. Fortunately not. But even if this were the case, our ideals would be conditioned by their historic development, therefore never absolute.

The striving for to know, to gain knowledge *(Erkennen)* of the absolute is based upon a strange misunderstanding: is it not the same as if one wanted to open up a pristine jungle, without changing its condition?

It is impossible to deduce an absolute reality from the laws of natural science, whose contents cannot be derived from the mere philosophically trained intelligence of the contemporary European. After all, there also exist ethical laws, commercial customs, political abuses[3] that are not derivable from any contemporary intellect. Should I believe also here in an 'absolute existence', in a *deus ex machina* whose image is reflected in those laws and regulations? I see no principal difference, for there is no law without its exceptions, they are all conditioned by culture, therefore dependent on development, replaceable by others; they are sensible or senseless, according to the critic's viewpoint.

Of what ought the absolute reality to be independent? If one wished it to be independent of man, one ought to consider that in this event it would also be of no use to man.

If one wished it to be independent of the individual, one should construe it as socially consitioned, and therefore dependent on the collaboration and communication of many individuals, as many as possible. One should construct it democratically, taking into account that it would then become much less dependent on time, because the collective *(die Masse)* develops much more slowly, but also more consistently. This is the way of the natural sciences.

If one wanted to make it independent of so-called 'appearances', one ought to consider that all 'appearances' are nothing but the expression of the interrelations between a number of elements of cognition. The same expression, when magnified, will become what is called an 'iron law'. There exists no fundamental difference between 'appearance' and 'truth', the difference is one of development *(Entwicklungsunterscheid).*

If an object appears small from afar and large when close by, generally one should not ask how it is 'in reality'. The natural sciences deduce from this perception of appearances the laws of perspective and solve the question through comparison with a yardstick at the same distance as the object. This is, of course, not the expected solution, for now one could ask how large really is a meter-stick, as large as I see it in the distance or as it appears close by? And this, just like all insistence on the 'essences and things', like all search for the 'thing in itself' *(Dinge an sich),* would not be natural science at all, because there can be no democratic, generally applicable uninfluenced *(affektfrei)* answer to it. This question demands the miracle of faith, the experience, related to the singular person *(Als-Ich-Erleben);* but scientific thinking does not provide for this, as otherwise it would become undemocratic and useless in everyday life.

Finally, if one wanted to understand 'absolute reality' as the most comprehensive sum-total of reality *(Sammelwirklichkeit)* from which every other reality could be derived, then one would either have to renounce the law of contradiction or to admit a general principle of 'reciprocal uncertainty'. Our logic would have to be revised and the future will have to decide on that.

Therefore, I believe one should highly esteem, even love, the ideal of an absolute reality as a vision of the next working day, but it must never be applied as a yardstick for the previous day. For this we need rather an 'image of knowledge' *(Wissenanschauung)* than a *Weltanschauung.*

At the present time we are so fortunate as to witness the spectacle of the birth, the creation, of a new style of thinking. Let us give free rein to the creators, the experts!

Sooner or later much will change: the law of causality, the concepts of objectivity and subjectivity. Something else will be demanded from scientific solutions and different problems will be regarded as important. Much that has been proven will be found improven, and much of what was never proven will turn out to be superfluous.

Education for life will be different; life and art will be given a different form. A new, up to date reality will be created.

What is the use of awkward metaphysics, if tomorrow's physics will transcend all phantasy?

Let us leave a free hand to the experts and reserve room in our thinking for the future!"

NOTES

1 Cf. *M. rectus abdominis* and *Mm. scaleni* in Tables Nos 3 and 6 of his *Anatomy*.

2 Vesalius's example is very simple. Just compare with it the tortuous roads of the birth of chemistry (phlogiston!), of its development in the materialist age, and today. How much could have been quite different, if only the discoveries had been made in another sequence. Surely one could have formulated entirely different concepts, e.g., the concept of the element, could have connected them in another manner, i.e., could have built a different reality without getting into conflict with any 'fact'.

The concept of weight - so important for such a long time - was introduced by Lavoisier as something self-evident, without any 'proof' or substantiation, despite the fact that at the time Spielmann (1763) rightly rejected drawing any conclusion from a loss or increase of weight, "since up to now the cause of weight is still unknown to the physicists." Also Sage was right, according to the state of science in those times, in that he regarded Lavoisier's theory of the composition of water as untenable, because "then the inflammable air must be regarded simultaneously as the son and the father of water."

Lavoisier simply created his own concept of an element - not at all the only possible one - and his own concept of composition. Subsequently both proved generally acceptable and have remained so to this day.

3 I.e., descriptive, not normative laws of the commercial or the political reality.

LUDWIK FLECK
'SCIENCE AND SOCIAL CONTEXT'
(Przegląd Współczesny, 1939)

The problem of the dependence of science on the epoch and social context of its production is particularly actual today. The problem is not only the dependence of the conditions of scientific work on social context, but above all the dependence of the very substance of science, its problems, views and factual data. Undoubtedly, both aspects of the dependence of science on the epoch and social context are interrelated, because modified working conditions change the result of work. However, the scientists are mainly fascinated and worried by the increasingly imposing relation between scientific views and a given social context.

Historians of philosophy have long situated philosophical systems against a background of general cultural features of a given epoch and have taken into account the relations between philosophy and the natural world, contemporary art, and the politics of a given country. However, historians of science entertained a belief that certain elements of "true science" are independent of time and place. "True science", they believed, particularly the empirical sciences, which have developed and expanded in a consistent way since approximately the 16th-century and achieved triumphs in the 19th and 20th-centuries, should be somehow definitive.

Meanwhile, even among the scientists themselves there appeared increasingly numerous reflections on the influence of social context on scientific views. I may recall here the work of Schrödinger "Ist die Naturwissenschaft milieubedingt?" (1932) in which this outstanding physicist shows the relation between modern physics and certain features of modern art or certain features of our social life.

I might mention an interesting article on "The Theory of Evolution in Biological Sciences" by Jan Dembowski, in which the author concludes that, "science is not a greenhouse flower grown in isolation from the world. It is created by people that live in society, and for this reason the great social events cannot but be of significance in exerting enormous and permanent influence on human psychology and, consequently, on the psychology of the scientist too."

The relation between science and the whole of cultural life of a given epoch has been described in a particularly beautiful and detailed manner by Tadeusz Bilikiewicz in his work *On Embryology in the Period of Baroque and Rococo* [1], which reveals a "parallelism between views in this special branch of science and contemporary stages of the development of general cultural environment." This work enables one "to follow the complex and mysterious phenomenon of collective intellectual life from a truly specialized point of view". Since political absolutism began to collapse at the turn of the baroque and rococo periods, parallel to the drive towards individual freedom in social life, embryology yielded the discovery of the spermatozoon. Leeuwenhoeck, Hattsoeker, Andry, and other animalculists viewed spermatozoon as an independent *vita propria* that leads the living organisms.

The rise in importance of women - the 18th-century is called "the woman century" - led to the development of ovulists (Vallisnieri and Bourguet), and "Buffon goes as far in granting sexual equality of rights as to claim to have discovered female spermatozoa." The struggle between preformism and epigenetism, between vitalism and mechanism, finds its own political, artistic, and philosophical background. Every stage of development of science is the result of a system of forces and symptoms of the epoch.

Thus, the phenomenon of the dependence of the content of science on the time and the context, is more distinct the longer the period of the development of science that is studied, and the more extreme is the variation of social and political conditions at the present turbulent time. The phenomenon should be utilized for the benefit of human knowledge. It has to be treated in such a way as to have a heuristic and not a destructive value, to provide a starting point for useful investigations and not be a source of pointless phrases on the non-existence of "voraussetzungslose Wissenschaft", or melancholic reflections on the "uncertainty of all human knowledge."

We are at a marvelous turning point on the path of expansion of science, from which unfolds vistas of an entirely inconceivable and exotic landscape. Whereas "well-educated", that is, conservative scientists view with some anxiousness the emerging new aspects, clever politicians process as rapidly as possible the newly-acquired information into demagogic slogans. The sociological, collective nature of knowledge was first turned into a political slogan involving the social and class conditioning of science, and then the competing political trends created the spirit of nation and race to provide a mythical world-view propagated through the ages. Now, since all science depends on the environment, the process should be reversed and a suitable science should be developed to fit the artificially modified environment. There is no objective science, anyway. Consequently, left-wing or right-wing, proletarian or national, physics or chemistry, etc., should be "made" as soon as possible. We will fabricate evidence to obtain politically necessary and predicted results. We will introduce planned management of thought, free creation will be replaced by bureaucratic centres, and propaganda will replace the autonomous penetration of society.

These manipulations would be ridiculous if they were not so dangerous. An ignoramus, or rather a half-ignoramus who had heard something about breeding horses, tries to breed a winged horse. But, first of all, the following danger is imminent: a new generation is growing up of future scientific workers who believe that truth in the good, old professional sense of this word does not exist anymore.

Some become fanatics because they have been deprived of confidence in reason: the others become cynical because they see that there is no absurdity big enough not to be believed, if only it is promoted by shrewd and intensive propaganda. That is why the study of the dependence of science on environment and epoch is particularly important today.

However, the approach employed by the majority of authors is artistic and literary rather than scientific; it consists in intuitive feeling of similarities (Schrödinger compares smooth surfaces in architecture - with empty and unfilled areas in science, and Bilikiewicz compares the "struggle between understanding and feeling" in life with the struggle between mechanism and vitalism in science. Preformationism and mechanism in biology was viewed as analogous to contemporary political absolutism (*viz:* the relation between the development of individualism and the discovery of spermatozoa). Such intuitive approaches are not yet sufficiently elaborated to be studied. The approach is still too literary and indeterminate: the statements, when extracted from rhetoric formulations and considered soberly, do not convince anyone. The subject of study is shattered and disappears like a ghostly phantom in daylight.

When considered composedly, it seems inconceivable that a relation can exist, for example, between individualism and the observation of a spermatozoon. The latter is only a question of having a good look at the sperm through a microscope. What is individualism doing here?

It is true that today a look is sufficient. But discovery and first observation do not take place in a peaceful state of mind. A peculiar, anxious mood is necessary to look for something new. A directed intellectual readiness is necessary to perceive something new. This anxiousness and readiness result from the influence of the environment. To upgrade a hazy outline of a new observation to the status of a deliberate subject of research, to focus obstinately on it and to isolate, name, and describe it so that the description arouses some reflections in others, all these activities are influenced by the environment.

The environment involves the words one hears, the views which are exchanged every day with the surrounding people, the shocks and the impulses of everyday life, the education acquired in school, etc. All these create a directed intellectual readiness which is a researcher's contribution to his work. He thought in terms of an independent and free personality, he was prepared to see it everywhere, and consequently, he discovered freely movable, "unconstrained" and independent spermatozoa. It should be stressed that freedom then meant unconstrained motions. In a different mood, or in different surroundings, the researcher would attach no importance to those "mobile commas", he would not examine or describe them, and he would forget the first unclear picture formed among so many others. A directed, collective, cognitive mood leading to a common thought-style provides science with an adequately formulated subject of investigation.

That is why it seems to me that an adequate starting point for a positive study of the effect of the epoch on science can be found only in the general sociology of thought. This necessarily leads to the concept of thought-community and thought-style, which are subject to modifications over time.

It seems to me that historians overestimate the importance of individual epochs. Undoubtedly there are many common features of an entire community at a certain time, but this community is easily overestimated from an historical perspective, the more so as an epoch is defined on the basis of some representative individuals. We can far more realistically evaluate the history of intellectual life if we consider individual thought-communities and their development, interrelations, counter-actions, and co-operation in different periods.

First at all, in this way we may come to learn the origin and development of different thought-styles, e.g., the thought-styles of chemistry, anatomy, astronomy, etc. We may come closer to the problems of a given field and their solutions, and to a scientific fact and its discovery. We learn the style-dependent aura of concepts; archaic, incomprehensible sentences acquire a sense, and we find how the present meaning has been derived from an old one. In this way artistic impressions, intuitive conjectures, and subjective feelings can be changed into the relations derived from independent laws of the sociology of thought and the theory of the development of thought. We avoid being pushed into dry ideological doctrine and we attain a science of cognition rich in details and capable of cultural growth. It seems to me that it is less important to study entire concepts and theories, such as, for example, the embryological evolutionism of the 18th-century, than to analyze individual sentences of a text as one analyzes an unkown code.

We cannot render the essence of a concept from a bygone epoch using contemporary words, since the individual concepts of that epoch are incomparable with the present ones. The notion of "nucleus" in the view of the 18th century is something entirely different from "nucleus" found in the present style of embryology. It is demonstrated perfectly in the chapter of Bilikiewicz's work, describing "shake of evolutionism and mechanism", how the entire long-standing dispute suddenly at a moment of change of thought-style becomes a dispute concerning the definition and meaning of words.[1]

The style aura of concepts undergoes change and is followed by changes in the concepts themselves. So, first one should examine the atmosphere of concepts and their style-coloring, which is reflected in the linguistic use of certain words, and one should pay particular attention to their metaphorical use. Only this opens the way to the study of the thought-style of a given epoch.

The study of the relation between science and social context is timely for yet another reason: perhaps less impressive but even more important. The study of the sociology of science is becoming increasingly important. With the increase in specialization, resulting from the enormous expansion of science, problems of organization, pedagogy, interdisciplinary branches, craftmanship, popularization, general syntheses, and the like arise. These are by no means simple problems. The absolute and relative (in terms of the density of population) number of productive centers of science is increasing, and similarly the areas of consumption of science are increasing with the increase of the level of general education. The number of individual papers is increasing and new and different journals with different perspectives are appearing. The paths along which scientific thought propagates are becoming longer and more complicated. A book which would embody the essence of a field (in the form of a codex) is becoming increasingly impossible, since it would become obsolete even before having been written and published. In many fields the present state of knowledge is either an obsolete general synthesis or an unsolved dispute among authors. Only a good specialist can understand it.

The more scientific knowledge becomes generalized, the more it becomes exclusive and difficult to attain. It was once thought that science would make that mysterious and intricate complex, called Nature, somehow simple and easier to grasp for the mind. Instead, science has become a structure which as a whole is not at all simpler than Nature, but which is much harder to attain. It is easier to find one's way in the woods than in botany. It is also easier to cure a patient than to know what he really is suffering from. A discovery which has been made by a specialist has to be processed stage by stage through a second and third discovery, made by successive ever more popular popularizers, before it gets within the range of comprehension of a layman or a specialist in another field. Consequently, intermediaries are necessary for whom the imminent goals of a branch of science are far less important than an independent, merely social problem of mediation, adjustment, and equalization. The work of these intermediaries, however, affects a specialist who, beyond his own specialty, is entirely dependent on it, and draws from it his general ideas, views, and impulses. Science has become an intricate formation governed by its autonomous laws, hitherto unknown to us, and substantially irrespective of the intentions and opinions of an individual researcher.

The life of science reveals characteristic features which seem to be irrational to the researcher, mainly because he cannot ever predict the state of his own research results after they have passed through the enormous mill of the collective group. Will they be followed, left unacknowledged, or transformed one way or another? This is frequently determined not by the substance, but by the form of presentation of the results. A word which was intended to be only the name of a thing, said at a certain moment, becomes, to the astonishment of its creator, a slogan only loosely connected with that thing. Such a slogan causes unexpected reactions in the scientific community. This is no longer the problem of the general cultural context, but that of the collective nature of scientific work itself.

A researcher may feel a deep disharmony between an entirely modern practice of scientific work, and an entirely obsolete theory of scientific knowledge which sticks to the conventional individual "epistemological subject" remaining unchanged in time and space and having only two organs, i.e., the eye - a photographic camera - and the brain, a recorder of photographs. What a primitive picture! How naively it is related even to a photographic camera!

This disharmony discourages a specialist who studies general problems, and makes him indifferent to or believing in miracles. We face a strange phenomenon involving an ideological crisis and discouragement among specialists, and at the same time an increased interest in science among the general public, reflected in the growing demand for popularized scientific books. This phenomenon is not without consequences, mainly because under these conditions popularization is becoming the domain of incompetent people.

This situation can be changed only by a thorough study of science, based primarily on the sociology of knowledge. Both these problems, i.e., the problem of the dependence of science on the general context and the epoch, and, on the other hand, on the more restricted scientific collective, are thus related to each other in a formulation provided by the theory of thought-collectives and thought-styles.

NOTE

1 Cf. the attempt to analyze the pair of concepts of heat and cold, taken from the "Andrzej of Kobylin Talks on the Activity of Human Limbs", in the author's work: *Przegląd Filozoficzny,* vol. 33, issue 1. Cf. also an attempt to analyze the word 'phosphor' in the

author's work: *Entstehung und Entwicklung einer wissenschaftlichen Tatsache*, Basel, 1935, pp. 136-141.

BIBLIOGRAPHY

[1] Tadeusz Bilikiewicz, *Die Embryologie im Zeitalter des Barock und des Rokoko*, Georg Thieme, Leipzig, 1932.

TADEUSZ BILIKIEWICZ 'COMMENTS ON LUDWIK FLECK'S 'SCIENCE AND SOCIAL CONTEXT'"

(Przegląd Współczesny, 1939)

The problem of science and context may be divided into two issues, clearly presented in the valuable remarks of Ludwik Fleck. One is the dependence of scientific work on the influence of context, while the other is the dependence of the content of science on this influence.

It is clear that the content of science can be affected only through scientific work: the latter should also include psychological and epistemological conditions that affect the mind of a researcher. After the suspension of these factors, what can be called the essence of the science would remain .

It may not be an oversimplification of my task if I say that the basic difference between my views and those of Ludwik Fleck is just that I have mainly attempted to describe the effect of the environment on the content of science, whereas Fleck predominantly studies its effect on the cognitive conditions of the mind, including in those epistemological considerations the significant sociological aspects which, unfortunately, have not been sufficiently taken into account until now. Thus, the effect of the socio-cultural context on the content of science is studied by Fleck through a epistemological prism, or, to put it outright, through the prism of the sociological criticism of knowledge.

Although I consider Fleck's epistemological account to be correct and fruitful, I am of the opinion that my discussion would not be honest if I did not raise certain objections to some of his generalizations [1, 2, 3, 4, 5].

I shall probably not be too far from the truth in seeing in Fleck's views a note of transcendental idealism as expressed by some Neo-Kantians. Besides, his approach goes beyond the forms of cognition accepted by Kant, actually covering (in the psychological, not "pure" sense) the whole of the cognitive mind in relation to reality. The essence of Fleck's epistemology is that the process of cognition is not a passive, indiscriminate reproduction of a subject, but rather a form of creation of this reality, following the example of creations of culture. In this way the gap between Culture and Nature disappears, for both are generated by the cognitive mind. In this connection, it should be noted that Fleck attaches less importance to the inborn, individual, and psychogenetically understandable structure of the subject. He is attracted rather by the sociological conditioning of cognitive creation that always takes place according to the spirit and the direction of a certain "thought-style", prevailing and operating in a given scientific environment, the latter being called a "thought-collective."

One might entirely agree with that point of view if the problem were only to determine the influence of the environment, thought-collective, or style on the outcome of the cognitive process. But this is not the case. When reading Fleck's works, one comes to the conclusion that the discovery of that influence entails metaphysical consequences. One has the impression that because the image of reality is different, depending on the collective group to which a given researcher belongs and on the thought-style he is subject to, the reality studied is inherently subject to change and becomes objectively different. The reason is that reality *[Ding an sich]* does not exist absolutely or independently of a researcher. The latter creates this absolute reality through a cognitive act and makes it such as it should be according to his thought-style. These are the metaphysical consequences of Fleck's epistemological assumptions.

Before I assume an attitude toward the above issue of the philosophy of science, I should like to criticize the fundamental term in Fleck's reasoning, i.e., the term 'style'. This term is undoubtedly not applied in its usual meaning. Usually we interpret 'style' as a form that is a manifestation of some kind of creativity. In Fleck's formulation the term means a set of sociological conditions on which depend not only the form but the substance of creation, too. Certain of Fleck's statements lead to the conviction that his "style" has the same meaning as attitude, point of view, cognitive disposition, preparedness of the mind - all of which are generated and made possible thanks to acquired knowledge and new-born notions. Only given such content and such scope to the term "style" can we understand why the theory of cognition is, for Fleck, simply knowledge about thought-styles and about their historical and sociological development.

The consequence of such an approach is an inability to establish a general criterion of truth from the standpoint of classical theories of cognition and acceptance of truth merely as a current stage of modifications in thought-style. Thus, unquestionable cognitive relativism follows, which attributes equal importance to various images of reality arising from different thought-styles, even if these images are inconsistent.

I offer such a detailed description of the metaphysical basis of Fleck's ideas to stress that if it had not been for them, it would have been easy to work out and co-ordinate various views on the problem of the relationship between science and environment. In my book *Die Embryologie im Zeitalter des Barock und des Rokoko* (Leipzig, Thieme, 1932), I tried to demonstrate, using the particular example of embryological views in the 17th and 18th-centuries, and expanding the ideas of Joël and Wölflin, the influence of socio-cultural influences on the content of these views. Though I was using different terminology, I constantly called attention to the influence of certain sociological factors on science. I did it on the basis of historical material, which provided me with a better perspective and consequently with more liberty of action.

Now, I wonder how my historical observations and conclusions would appear if they were put in the light of Fleck's ideas. I have come to the conclusion that they would lose their value. This would happen because of the metaphysical aspect of Fleck's ideas. Let us take an example. I have found that the "primary harmony" of Leibniz is nothing else than the translation of theories of embryological preformationism into a philosophical language. This has not been noticed (and it could not have been noticed) hitherto by any historian of philosophy. What was needed was to compare such a special branch of science like embryology with philosophy. I have explained this phenomenon - like many similar ones - by the creation in the researchers' mind of some patterns that predetermine the form, style, and direction of a cognitive process, in a certain environment and in a given epoch. These patterns have their effects, take shape, and then vanish to give way to others. To some extent the above patterns have been and still are well-known. They have manifested themselves incessantly since time immemorial in the form of vogue, style, customs, and trends. It is not easy to explain from which source that invisible compulsion originates that forces various individuals to create according to such ideal patterns. Anyway, this compulsion implicates a number of psychological and sociological factors that are still to be investigated.

I have tried to explain at least one of these factors, i.e., the individual, and psychological one, which happens to be extremely simple. It is a set of characterological features of a man, his likes and dislikes, aesthetic and pragmatical inclinations, etc., that make activities of an individual in different fields somewhat similar to one another. This phenomenon is the basis of our practical predictions as to how any individual will behave under new conditions: these predictions - if we know human nature - in general prove correct. The same phenomenon is a keynote for biographical, criminological, and historical research, to say nothing of pedagogy, politics, etc. We can guess the author of an isolated page torn from a novel if he is known to us from somewhere else. We can recognize the character of a given individual from his nails and heels. Not only an expert but also an ordinary music lover may recognize bars beyond a doubt after a few times if the music is classical or romantic, Spanish or Russian, or if it is Tchaikowsky, Berlioz, or Debussy. It is impossible to describe precisely the "style" of those various compositions. But the style is infallibly grasped in a twinkling of an eye.

The styles mix together, harmonize, or are out of accord. The personal style of an author, if he is a true product of the environment, is in harmony with the style and the spirit of the social context or epoch. Sometimes, due to various reasons which would need separate study, it is out of step. The overlapping of styles can be complex. Chopin has his personal style which is in harmony with the overlapping Polish style which in turn is in harmony with the Slavonic style, and the latter is an element of what can be defined as the European style, as contrasted with, e.g., the Negro style.

However, my intention was to prove that the output in other domains - where aesthetics seemingly does not reach - is also shaped according to style. This style is by no means something of a different kind. It is the same style known in art. Scientific output, in its theoretical and hypothetical part as well as in its range of interest, or, using Fleck's style and language, in the preparedness of the mind, can also be submitted to those invisible, evasive, and indescribable patterns and styles of the epoch and socio-cultural context. In my paper, I tried to depict with some consistency a style, its fluctuations and changes, its prevalence in the form of a vogue, of habit and intellectual inertness, i.e., a style of nothing other than scientific theories, hypotheses, and exploratory studies in embryology, etc.

My task was confined to this aspect only. I have not tried to surpass the limits of experiment. I have assumed that there exists an objective state of affairs which is recognizable to a high degree. The purpose of science is to study this objective state of affairs, its reproducibility in a researcher's mind, and the transmission of the content of study, or the truth (as we see it) to other researchers. Going deeply into conditions that determine the accuracy of cognition, effects that blur the objectiveness of investigation and warp the direction of interest, that "dispose" affectively or intellectually the researcher's mind: investigating the effects that make the researcher look through these or those glasses - psychological or sociological - such investigation, though approaching cognitive relativism, agnosticism and scepticism, in fact sharpens the accuracy and firmness of our knowledge of reality through a criticism of the way we perceive reality.

Our studies, aiming toward the determination of the influence of socio-cultural context on science, virtually tend to remove this influence on the cognitive process. When we study these phenomena as historical material and we see the aberrations of past researchers who adapted to the prevailing style, it serves at the same time to remind us of our cognitive efforts. In order to guarantee accuracy, the cognitive process has to be liberated from all sorts of influence, or at least it has to tend to the ideal of total lack of bias. That is why I have so sharply attacked Fleck's tendency to blur the borderline between "Nature" and "Culture". Sociological criticism, together with any other criticism of cognitive powers, tends precisely to make the results of cognition different from the products of a culture. Admittedly, a historian has to accept the fact that the influence of the social context on science used to be heuristically very positive, but an epistemologist, who should take into consideration both the errors caused by the dependence of science on environment and the inability to apply the criterion of truth (even the most unquestionable) to the products of culture, should also attempt to eliminate the impact of influences of this kind on knowledge.

If, in spite of all those noble and stern reservations, researchers keep borrowing inspiration from methods pertaining to culture, this is not because it is epistemologically permitted, but because the human mind is limited and inefficient in its cognitive ability. The stern rules of scientific epistemology apply without exception to facts and conditions of objects that are simple, clear, and thoroughly explored. If the observation, description, numerical representation, condensation, and other obvious activities are sufficient, one has no need of hypotheses and theoretical and philosophical generalizations. As we move away from objects of cognition that are "easy", simple, and do not raise any doubts, toward "difficult", inaccessible, complicated, and incomprehensible ones, we are unable to make use of "reproductive" methods and we have to resort more and more to "creative" ones. As we advance toward the unknown, our mind is exposed more and more to the action of numerous sociological factors. It is a necessary evil, not a scientific necessity.

The obsolete, naïve, and abandoned epistemological situation - on the one hand passive, independent, and unbiased cognitive subject, on the other scientific reality, independent of a researcher - is not entirely wrong. It only requires some rectifications, reservations, and exceptions. I would reproach Fleck for making his epistemological generalizations on a too narrow basis, involving natural and medical experiments only. This reproach is not disproved by the fact that in his main work of 1935 he widely took into account, e.g., the history of the Wassermann reaction, since Fleck's reflections, although based on historical material, are not of a humanistic but of a philosophical and scientific character. If Fleck had, in his epistemology, used the historians' experience, he would certainly have found that the above-mentioned naïve epistemological situation happened quite frequently. A historical fact is usually independent of the researcher and does not undergo changes (as any scientific fact does) which depend on experimenting and studying: under usual conditions the historian is, or at least should be a purely receptive and reproductive subject, passive in his cognitive activity. If a historical fact somehow affects the researcher's cognitive powers themselves, such influence is of a different kind than in the natural science: e.g., an historian can experience a peculiar emotional state in relation to the object of study, and this disturbs his objectivity. In natural science this happens more rarely, though it is not impossible: e.g., the discoverer of a phenomenon becomes narrow-minded and is apt to adapt all his observations to this phenomenon. However, one can easily notice that it is rather the object that should be "blamed". In history, the object of study is most often a written source. The elementary cognitive activity is to read, and of course to understand. This elementary activity is as simple and easy as, in the natural sciences, statements or descriptions of a plain fact, e.g., a human hand has five fingers. Earlier epistemologists, creating the ideal "naïve" epistemological situation, restricted themselves to such examples. Such a situation really happens, probably more often in history than in the natural sciences, and - what is more important - it constitutes an ideal which we try to approach, advancing along a strenuous path, amidst more and more criticism of the cognitive conditions.

Both in the natural sciences and in the humanities theories and hypotheses become necessary when we move away from simple and easy facts and venture into yet unknown and complicated ones. In history, when the facts themselves are not sufficient, or if we are short of them, we resort to synthesis. The meaning of historical study is not given by a faithful reproduction of reality or close approach to the historical truth. Historical study is useful if it provides us with a cognitive orientation [6]. Individual facts are important insofar as they make this cognitive orientation possible. If they are unable to fulfill this task, we consider them unimportant and we neglect them. In natural sciences the same holds: an individual fact may not have significance because it does not arouse our interest, we do not feel a thought-preparedness towards it, it does not fit our thought-style. When a new style is employed, these neglected facts will become significant. From a retrospective historical standpoint, neglecting facts at an earlier stage is considered a mistake that has been corrected, thanks to the heuristic sense of a new style. However, one should not forget that there are countless facts which we neglect, for they are similar to one another. What is important is the essence, the summary, the law of Nature, or the synthesis.

Styles turn out to be necessary only when applied to synthesis and hypotheses, either natural or humanistic ones. Only theories, implications, formulations, constructions, and hypotheses can vary and sometimes may seem to be equally important in relation to the presumed state of affairs despite all inconsistencies between them. Equal importance given to contradictory theories that have developed from various thought-styles is always nothing but an expression of the inefficiency of the human cognitive powers. It is impossible for researchers dealing with such facts as "the normal human hand has five fingers" or "the neck joins the head and trunk", to invent two or more non-overlapping or contradictory theories even if they were members of the most distant "thought-collectives", the matter not being that of style but of inspection. [Of course, I am ignoring the opinions of the insane.] On the other hand, in the 17th-century, contradictory theories had to be constructed on the main embryological problems, since from the present point of view the investigators of that time were ignorant and tended to replace their lack of acquaintance with facts by the products of their shrewdness or imagination. In this respect, they resembled the creators of culture. And only as such were they subject to the style, if not of the epoch then (according to Fleck) of certain thought-collectives. Nevertheless, there remains an insurmountable gulf between culture and Nature, because within the scope of culture (e.g., art) one can create with impunity, without any control by reality. In art one is allowed to make mistakes (senseless qualification in art, neglect of aesthetic norms), whereas similar creation in natural science is only a trick, a way of coming to the truth, an attempt to attain cognitive knowledge, and nothing else. Anything beyond epistemological purpose does not belong to science but at most to culture. If such a product of scientific generalization does not sustain criticism, it becomes a mistake and acquires only "historical" meaning, i.e., it loses its significance altogether. The loss of significance in Nature is not an obstacle for such a product to have value in culture. For creation in science is not a creation, but only reproduction. If it appears that the supposed reproductiveness has been only free creation, the whole intellectual effort is ruined, though it must be admitted that before it happens, the mistake may - intentionally or unintentionally - have provided significant knowledge of the reality.

In order to exhaust the subject of the difference between Culture and Nature, we have to decide definitively if considering the problem of the influence of the environment on science is to be made using an historical or a scientific method.

To be even more precise, we may ask whether these considerations belong to history or to natural science. I must stress that both the studies of Jöel and Wölflin and mine were historical studies. If such studies give an impression of being artistic, literary, intuitive, and subjective rather than scientific ones, such an impression springs from the fact that we are moving in the realm of facts that are extremely difficult to determine. Grasping the relations between individual products of the same style is by no means as simple as it would seem if one were to follow certain of Fleck's remarks.

For example, it is impossible to introduce the accuracy that would be required by "different laws of the sociology of thought and development of thought" into historical studies of this kind. The similarity of different products of the same style is the result of extremely complicated psychological processes. Admittedly, we can assume that these processes occur in accordance with certain exact laws in which the sociological element plays a considerable part, but it is a long way from this assumption to the discovery of causes and mechanisms of interaction between people. The historian necessarily depends on a subjective opinion concerning the fact that the socio-cultural context affects the work of individual scientists. When examining historical material, the historian sees the finished results of this influence in the form of similarities, analogies, or the community of the content. The intuitiveness and subjectivity of the observations made in this way are not in disagreement with the historical method, though, to be sure, in terms of accuracy such methods are inferior to the scientific method. At the same time, one should not forget that historical studies, though not as accurate as studies of Nature, are nonetheless scientific studies. The charge made against historical studies - - lack of a scientific approach - - can originate only in an environment where a scientific style of thought predominates.

On the other hand, making use of historical material in order to detect certain natural, psychological, or sociological laws is a quite different problem. I have already said that, though Fleck frequently employs historical material, not for a moment does he cease to be a natural scientist. For the purpose and meaning of his studies is not historical, but naturalistic, sociological, and psychological. That is why I said in the beginning that if my studies on embryology during the 17th and 18th-centuries were translated into Fleck's language and ideas, they would lose their value. For they would cease then to be historical studies and would only be a basis for naturalistic and sociological considerations. For Fleck, the similarity between preformationism and the "primary harmony" of Leibniz must be unimportant, or at least useless for his ideas. This similarity, though very important for a historian, would seem to Fleck to be an expression of the artistic, subjective, and thus unscientific opinion of a researcher.

On the other hand, a historian is not very pleased even with the heuristic value of "thought-collective" studies on historical material, if their only purpose is to reveal or prove certain sociological and psychological laws, and not to provide any humanistic advantage. If some of Fleck's suggestions were actually realized, historical studies would cease to be what they were and would become a mine of examples, sources, and proofs for a sociologically saturated epistemology. The historian would be busy investigating thought-collectives in their historical development and their influence through the ages, and would neglect the proper task of a humanist that, in spite of all improvements, should always be a guide in historical studies. This task consists in acquiring cognitive knowledge with regard to past events and not in drawing up general laws of the sociology of thought.

Having stressed this discrepancy between the humanistic interest and the naturalistic one in the problem we are considering, it is clear that in order to be fruitful these studies must to be based on the immediate co-operation between the two camps. However, there is a great difference between co-operation and the absence of a borderline (distinction) between aims and tasks in fields so different as the humanities and the natural sciences.

BIBLIOGRAPHY

[1] Fleck, L.: 1929, 'Zur Krise der "Wirklichkeit"', *Naturwissenschaften* **17**, 423 ff.

[2] Fleck, L.: 1935, 'On Scientific Observation and Perception in General' ['O observacji naukowej i postrzeganiu w ogole'], *Przeglad Filozoficzny*, R. XXXVIII., p. 58 ff.

[3] Fleck, L.: 1935, 'Zur Frage der Grundlagen der medizinischen Erkenntnis', *Klinische Wochenschrift* **14**, p. 1255.

[4] Fleck, L.: 1936, 'The Problem of Epistemology' ['Zagadnienie teorii poznawania'], *Przeglad Filozoficzny* **39**, issue 1.

[5] Fleck, L.: 1935, *Entstehung und Entwicklung einer wissenschaftlichen Tatsache: Einführung in die Lehre von Denkstil und Denkkollektiv*, Benno Schwabe & Co, Basel.

[6] Bilikiewicz, T.: 'Reflections on the "Meaning" of History' ['Z rozważan nad "sensem" historii'], *Przeglad Wspołczesny* **189**, pp. 114-126.

LUDWIK FLECK
'REJOINDER TO THE COMMENTS OF TADEUSZ BILIKIEWICZ'

(*Przeglad Wspolczesny*, 1939)

I have read through the profound reflections contained in the comments of my colleague, Dr. Tadeusz Bilikiewicz, with great interest. What can be more enjoyable for a scientist than the discussion of things that are his concern? What can be more fruitful for science than the exposition of a problem in the light of different standpoints? The more so when the problem dealt with is of such great consequence for theory and for practice as the relation of science to environment. Perhaps other specialists, humanists, naturalists, and philosophers would also like to take part in this discussion. This way, the aim of my work will be attained.

However, I wish to rectify a few mistakes. There is nothing so distant from my mind as metaphysics. I am simply unable to understand the statement that "a thing by itself" exists absolutely, irrespective of the witness, nor the contrary statement that "a thing by itself" does not exist absolutely. I use the word "reality" only for grammatical reasons, as a grammatical object indispensable in a sentence dealing with the act of cognition. I am all for avoiding any ontological statements about reality, since I find to be fruitless statements about the existence of numbers, whether or not a number exists, irrespective of mathematicians. Hence, I have nothing in common with idealism, neither in its old nor in its new form. But I certainly know from both my and other people's experience that nothing could be examined without being acted upon, hence without being subjected to change. After all, even the most simple visual examination would require at least lighting and, as it has been very clearly stressed by physicists since the Heisinger proposition, light itself will modify the object.

It cannot be claimed that reality will "become objectively different" according to the thought-style, exactly as it cannot be claimed that numbers have become objectively different from those which existed before the development of the calculus of infinite numbers. In these cases any ontologic conclusions are precisely unnecessary metaphysics, and this is true also as concerns "reality by itself", which seems to be the aim of the cognition which interests Bilikiewicz.

No cognitive relativism results from the theory of thought-styles. "The truth" as a current stage of transformation of thought-style is always only one: it is completely determined within a style. The variety of images of reality is only an effect of the variety of objects of cognition. I am not claiming that "one and the same statement" can be both true and untrue for "A" and "B", respectively. Provided "A" and "B" participate in the same thought-style, the statement will be for both either true or untrue. But when they have different thought-styles, there will be no such thing as "the same statement", for one of them will either interpret the other's statement in a different way or will be unable to understand it at all.

In the 18th-century, it was claimed that a man with an empty stomach is heavier than a man after a meal. There was even a theoretical explanation to this, viz: "the spirit" becomes more fiery after a meal and it lifts the flesh up, deducting in a way a proportion of its weight.[1]

Today we are very confident that the contrary is true: a man becomes heavier after a meal. Here, at first sight we have two contradictory statements. An historian will quote the earlier statement as curious, a naturalist will laugh patronizingly. But both these standpoints are futile. The analysis of thought-styles leads to a different approach: not, of course, because the truth was relative, not because the man in the 18th-century was "really heavier" with an empty stomach than after a meal. But by making it obvious: 1) that "heavy" had a different meaning at that time than it has now, and 2) that this obsolete meaning of the word corresponds to a concept which does not exist today and thus cannot be represented by any word used at present. Such analysis shows how several discrete present concepts have developed from the primitive one, and what relics of this primitive concept have been retained in today's expression.

Our physical concept of heaviness represented by weight was not generally used at that time. "Heavy" in its old meaning was a property that could be described today only with a number of qualifications. Steady as, for example, a rock; but also sluggish, like viscous liquid. Physically and psychically inert or tardy, as a slothful man or animal; but also reluctant to be persuaded into something, like a stubborn man. Also unhandy, like an awkward parcel or a corpse, and, then again, difficult to apprehend, like a foreign language or sophisticated idea. And overwhelming, like a stroke of bad luck. Which of the present words could communicate the essence of this obsolete concept and still cover its complete range? Different heterogeneous - as we take them today - phenomena are harnessed here into one conceptual form in which there is the nucleus of our concept of physical weight, but also a number of qualifications which at present are metaphorical. These qualifications are now the source of many colloquial phrases ("a heavy mind", "a man heavy in business") and the nucleus of numerous psychological and physical observations. The word "heavy" can by no means be translated into present-day language because by using any of the present words we would totally damage the line of thought: the analogy to a corpse, the relation with the rising of the flame ("reduction of weight by fire" - the visible symbol of life, mobility, lightness). This word can only be explained or described, but not translated because the author had not only a different language but also, first of all, a different thought-style. In order to become isolated and to develop into the present concept of weight, this primitive idea had to pass through a series of stages, *viz:* Galileo's, Newton's, Lavoisier's, Einstein's - a cycle that is certainly not yet complete. The physicists have developed and are still modifying their own style where its expressions have specific meaning. Is it "good" or "the best possible"? The style of the physicist is specialized to fit certain problems, and thus it is more precise but also more one-sided. The old, clumsy expression contained somehow the nuclei of the observations, comparisons, and combinations originating from the now-separated realms: physics, physiology, psychology. The physicist of today has a sharper view on physical phenomena, but he has become more or less blind with respect to psychological or physiological phenomena. This is where a substantial proportion of his sharp sight lies.

Bilikiewicz believes that it is impossible to create even two theories that would not entirely coincide on a fact like: "the normal human hand has five fingers", even if the authors belonged to extremely distant thought-collectives. For in his perspective it is not a question of style but of observation.

But it is not so. Our concept of "a normal hand" is highly related to style. For a doctor, the normal organ is a certain theoretical fiction, totally inaccessible to observation in the same way as the "perfect gas" is for physicists. For a layman, "normal" means "ordinary", or "the most common", hence its status is concluded from comparison with a great number of hands, but it is not directly observed. Even if we leave aside this difficulty, in order to be able to state that there are five fingers there must exist the general idea of "a finger" and the ability to count, or the idea of a number, which also is a matter of style. Thus, the whole sentence, which in our style represents an undoubted fact, can in other styles be nonsense. Such styles do exist. They are naturally very distant from our present European style, but not so distant that we would not be able to make any comment on them. Many primitive tribes have no common word for "finger", but separate names for thumb, forefinger, and every other one. They do not use the general concept of "a finger", just as they do not use the general idea of "a horse", but some dozen qualifications for different horses: a grey, bay, piebald, young, old horse, etc. In their style our sentence remains inexpressible. Many primitive communities (in Australia or South America) have separate words for numbers 1, 2, 3 only, but in any other case they say: a lot, many, multitude. Thus, the statement, "a hand has five fingers", cannot be translated into their language. They say: "a hand has many fingers", and here we have a theory that is different from ours.

Other primitive collectives (Papuans) use the word "hand" to name the number five, or "two hands" to name ten. These could not say "a hand has five fingers" without tautology, while the question: "how many fingers are on one hand?" is practically nonsense, the same as if we would ask: "what is the cost of one dollar?" or "how long does one hour last?" Aborigines from the Andaman Islands say "all" for five, and thus the sentence would be expressed as: "a hand has all fingers". And here again we have a theory which, though not contradictory, does not overlap with ours. These are not only language differences, because the words which are commensurable with number five, that is, "many", "hand", "all" have an entirely different range than the word 'five'. This is quite a different thought-style. However primitive in comparison with ours, it is still not unreal or impractical, for these men live in and cope with their world.

Thought-styles change, develop, or die away as a result of the circulation of thoughts within the limits of the respective groups and the action of specific sociological forces. I am not so sure that conscious imitation was the main reason for the community of style within the limits of a group. Usually a member of a group would simply not be able to think in a way different from his community; he could neither see nor use ideas or images, nor seek relations which were different from those used by the people he was living among. He adopts a hostile attitude toward every alien thing. His style will change very slowly, following the changes that occur in the group. The exceptionally creative individuals, or rather creative moments, are a separate problem.

Bilikiewicz believes that research on the effects of the environment on science should aim at the elimination of these effects from the process of cognition. But how can that be done?

The primary locus of every cognitive process is actually provided by the very environment, its history, and its current status. From where would the cognitive processes sprout, if there were no present situations, new ideas, or new problems which are always generated in the course of collective life? I assume that Bilikiewicz considers the environment as laymen, or the general human community do, but not the community of specialists. However, I would not agree even to the loss of stimulation provided by laymen. Do we have to return to inaccessible towers, to a language foreign to normal life, to the mysterious air of the isolated scholar of the Middle Ages? What would science be without the concepts of standard, buffer, level, threshold, reserve, depot, center, control, and so many others, borrowed from modern life, from the street, the home, and from travel? Could a recluse, having at his disposal only old books, invent any one of these? I am doubtful if Schrödinger would ever give up those elements of modern physics related to modern art and life style. There is a difference between a blind craze and a fecund and indispensable impact of social context. An isolated scholar unaffected by environmental powers, devoid of the effects of its evolution, would remain blind and thoughtless. To get rid of the effect of social context would at least mean to seriously delay the process of cognition.

Now, the relation of science to arts: here Bilikiewicz believes - surely in accordance with public opinion - that the essential difference is only in that art is there the capacity to create freely, while in science one can only reproduce by giving descriptions of something that exists independently, i.e., using sentences to transcribe something, or to be more precise, representing things by means of a system of signs.

But this is only an apparent difference or, rather, this difference is only quantitative. The artist translates his experience into certain conventional material by certain conventional methods. His individual freedom is, in fact, limited; by exceeding these limits, the work of art becomes non-existent. The scientist also translates his experience, but his methods and materials are closer to a specific (scientific) tradition. The signs (i.e., concepts, words, sentences) and the ways in which he uses these signs, are more strictly defined and are more subject to the influence of the collective; they are of more social and traditional character than those used by the artist. If we call the number of interrelations between the members of a collective "social density", then the difference between a collective of men of science and a collective of men of art will be simply the difference of their densities: the collective of science is much more dense than the collective of art. The obstacles hindering the scientist in his free creation, the so-called "hard core of reality" with which he is confronted in his work, result from this very density.

It is quite natural that a given individual scientific creation is never free because it has always to refer to other branches of science, because it requires upbringing and education, and reference to the history of science. There is no need of metaphysical considerations to distinguish myth from scientific notion: for practical purposes, the results of the cognitive process are adequately determined by the collective character of the scientific thought-style and by the continuity of its historical development.

But this is not the most important point to be gained from the theory of thought-styles. After all, it is not too consequential whether we apprehend the process of development of the sciences as the development of thought-styles or an approximation to "objective reality". We know from observation that specialists are more and more outgrowing the idea of "the thing by itself", because when they penetrate ever deeper into objects they find themselves more distant from the "things" and closer to the "methods"; the deeper in the woods, the fewer the trees, but the more the woodcutters. Nevertheless, the concept of "the thing by itself" as the unattainable ideal does not interfere with specialized cognitive work. But other gains that result from the theory of thought-styles are far more important. First, this theory allows the development of a comparative theory of cognition; second, it makes possible research on the history of the development of thought. These are perfectly sufficient reasons to make this theory worthy of interest.

I am all for what Bilikiewicz suggested, that such research will require close co-operation of specialists from many branches.

NOTE

1 It was claimed by analogy that the weight of the body would increase at the moment of death as it was giving up the ghost.

TADEUSZ BILIKIEWICZ
'REPLY TO THE REJOINDER BY LUDWIK FLECK'
(*Przeglad Wspolczesny*, 1939)

In order that our dispute will not become vacuous, which is something that can happen if points of view are too far apart, I will present quite briefly a few reservations of mine.

I would hesitate to present - as Fleck does - the theory of thought-styles as a terminological matter. For example, I see nothing surprising in the statement of the scientists of the 18th-century that "a man on an empty stomach is *heavier* than after a meal." If some historians have not understood this statement, the reason is, I think, not because they differed in their thought-style from their ancestors, but because they did not sufficiently master historical terminology. The statement mentioned above has been worded as if in a foreign language.

First, it should be translated into today's language and then it could be expressed as something like a man on an empty stomach is "languid" or "sluggish". This terminological issue is even more visible in Fleck's reasoning concerning the following statement: the normal human hand has five fingers. If a Papuan or a child lacks terms and notions that enable him to understand the content of this statement, it does not mean that both parties belong to different thought-styles. Everybody will agree about such a simple statement, which is clearly verifiable, irrespective of the thought-style to which they belong - that is, provided they speak the same language, in both a linguistic and a terminological sense. In order to agree about the statement, "the normal human hand has five fingers", one should, of course, establish the notion of norm, be able to count to 5, know elementary anatomical terms, etc. In my opinion, all this has nothing to do with a thought-style. Or, at least the denotation of thought-style should not be expanded so far. Otherwise this useful notion will become a commonplace.

I have not been convinced either by Fleck's reasoning on the influence of socio-cultural context on cognitive processes. Everybody will obviously agree that such an influence exists, that it plays a formidable role, that often thanks to it cognition is possible. The environment of both laymen and specialists contribute to this influence. It is profitable and impregnating, and sometimes it even provides the only existing heuristic basis. We agree on this point. However, if we evaluate objectively the results of a given study - if such a thing is at all possible - we have to distinctly separate what proved to be scientifically true from what proved to be false. Studies on thought-style are particularly useful for understanding the sources of mistakes in science. It appears that a given mistake was made because a researcher followed passively the thought-style of his context. In the above sense I have stated that studies on the influence of context on science should tend to render the cognitive process independent of it, i.e., maintaining active criticism where blindness may be threatening.

One could say that the influence of environment on science is permitted as long as it exerts an advantageous heuristic effect. But woe unto him who is led astray.

Mutatis mutandis, the same could be said about the consequence of a quasi-artistic creation for science. If the result is an idea, theory, or heuristically fruitful hypothesis, the creation in question can be "tolerated". In case of error, this creation becomes condemnable and will be judged on a level with fantasy or fable.

How to manage to evaluate these phenomena without an "objective reality" - I really do not know. Luckily, even the most fierce enemy of metaphysics, just like the most committed idealist or solipsist, in practice always holds to such an ontologically uncertain "hard reality", both in his own life and in his scientific research.

CHAPTER VIII

CONCLUSIONS
PHILOSOPHIZING AT THE BEDSIDE

Polish philosophers of medicine were aware of the specificity and the unique character of their school of thought. In the introduction to the second edition of *The Logic of Medicine*, Biegański reviewed other studies dealing with the philosophy of medicine: *Medizinische Logik* by Oesterlen and *Introduction à l'étude de la médecine expérimentale* by Claude Bernard. He noted that:

> both works deal only with medical knowledge, i.e., with the theoretical and scientific aspect of medicine. The practical tasks of medical cognition, namely the theory of diagnosis and of therapeutic indications, were not discussed in these works. Thus both Oesterlen's *Logic of Medicine* and Bernard's *Introduction to the Study of Experimental Medicine* are incomplete studies which do not deal with the full scope of medical knowledge. The practical aspects of medical knowledge were taken into consideration only by a series of studies in the Polish language.

Later he added that:

> the doctrine of present scientific medicine omits the practical tasks of medicine and concentrates solely on its scientific tasks.... From this limited point of view, the logic of medicine is only the logic of medical science, a critique of scientific knowledge, and is presented as such in the works of Oesterlen, Bernard, and even in the contemporary writings of Magnus. In contrast, the Polish authors, from the time of Chałubiński, have always stressed the importance of the clinical tasks of medicine.

Biegański reviewed the studies of Chałubiński, Biernacki, and Kramsztyk and remarked that the Polish journal *Medical Critique*, whose purpose was to critically analyze medical knowledge and the clinical practice of physicians, is "as far as I know, the only journal in the world which has such aims" ([2], p. VI, VII, 29).[1]

In the introduction to *Principles of Medical Understanding,* Biernacki made similiar claims. He, too, reviewed other (mostly German) works dealing with the logic of medicine, and affirmed that while practically all the energy of the German investigators was directed toward specific problems of medical science, Polish reflections on medicine during the last twenty years had been attracted to more general questions of medicine. The existence of a special journal - *Medical Critique* - attested, according to Biernacki, to the particular inclination of Polish theoreticians of medicine to "deduce general principles from well-known facts", and to make explicit "the ideas hidden in the everyday activity and behavior of a physician" ([3], p. VIII; X). When Szumowski defined the tasks of Polish philosophy of medicine, he stressed that in Poland the philosophy of medicine was "a child of the clinics" [22], while later Bilikiewicz described the principal goal of the Polish school of philosophy of medicine as the "aspiration to formalize the elements of cognition that guide the thought of a physician at the bedside" [4]. This evaluation of the unique character of Polish philosophico-medical thought was shared by foreign observers, too. Commenting on the German edition of the *Logic of Medicine,* the Viennese physician and historian of medicine, Max Neuburger, wrote to Biegański: "I always admired the fact that Polish physicians were the only ones to elaborate a brilliant theory of medical cognition."[2] And a review of the German edition of Biegański's book which appeared in the *Netherlands Medical Weekly* stressed the fact that Polish authors were the only ones who had elaborated a critical approach to medicine.[3]

The attempts by Polish philosophers of medicine to put forth their philosophical reflections on the specificity of medical knowledge and of medical practice did not, however, attract followers in other countries. The diffusion of their ideas was thus essentially limited to a Polish reading public. Moreover, even in Poland their works were gradually forgotten. Later, some isolated studies[4] discussed the practice of medicine from a philosophical standpoint. But only the growing interest in bioethics in the 1970's brought about a revival of interest in studies that deal with the philosophical principles underlying clinical medical practice. Such studies were at first included in reflections on bioethics, and studies on medico-philosophical topics were published mainly in journals or books devoted to issues in "medical ethics". But, gradually, philosophical reflections on medicine separated themselves at least partially from bioethics. The result was the publication of books dealing with this specific subject - Edmund Murphy's *The Logic of Medicine* (1976), Lester King's *Medical Thinking* (1982) [10, 17], accompanied by the birth of journals dealing with specific philosophical isues in medicine: *The Journal of Medicine and Philosophy* (founded in 1976) and *Theoretical Medicine* (founded in 1979). In 1986, Tristram Engelhardt could claim that: "a discipline has arisen with its own set of issues, its own journals and its own set of specialists. There is now a philosophy of medicine" [8].

The birth of the philosophy of medicine was closely related to the development of reflections on the ethical problems of medical practice. The burgeoning interest in bioethics stemmed from critical modifications in medical practice in the second half of the 20th-century. Some of the central clinical problems that preoccupied physicians in the past, such as childhood diseases and epidemic infectious diseases, nearly disappeared from industrialized countries - at least prior to the AIDS epidemic. Western medicine became a highly technical and very expensive enterprise. The introduction of certain new medical technologies, such as resuscitation methods, organ transplants, pre-natal diagnoses, and, more recently, *in vitro* fertilization and embryo transfer, became largely debated issues, and the attention centered on these techniques stimulated a great deal of interest in medical ethics.[5]

However, while the technical conditions of the practice of medicine were deeply modified, the more fundamental problems of medicine as a science and as a profession have varied but little in the last fifty years. The central question debated by the Polish philosophers of medicine - "medicine - a science or an art"? - has not lost its relevance today and is present in current debates among philosophers of medicine. In the first issue of *The Journal of Medicine and Philosophy,* the editor-in-chief, Edmund Pellegrino, maintained that a specific philosophy *of* medicine was necessary because medicine could not be reduced to medical science. Medicine is above all a praxis, a goal-oriented activity aiming at the cure of illness and promotion of health, and the philosophy of medicine should specifically study this aspect, i.e., medicine *qua* medicine. Ten years later, Pellegrino argued that while medicine drew heavily on a large range of disciplines, from the physical, biological, and social sciences to the humanities, it did differ essentially from these sciences, because "it draws on them from a particular perspective. As an activity, the specific practical aim of medicine is the curing, containment and prevention of illness, and the cultivation of health.... Medical science, as a science, is not synonymous with medicine, nor is it sufficient to define medicine as an activity" [18, 19].

The specificity of medical knowledge, and of the activity of physicians, was discussed by several thinkers. For example, Stephen Toulmin explained that "biomedical science had to achieve an effective union with the particularity of clinical treatment, most particularly with the individuality of the relationship between the patient and his personal physician." According to Nancy Maull, medicine is a practical science that is different from other sciences because "medicine employs a normative distinction between normal and abnormal; it has a practical therapeutical goal, and medical problem-solving involves a deliberate, cure-directed manipulation." For Ronald Munson, medicine is a specific enterprise that "possesses features that make it an inappropriate subject for reduction," while for Ineke Widderschoven-Heerding, medicine is a form of practical understanding in which "the way medical activities are performed is not external to the outcome of these activities" [14, 16, 23, 25].

Contemporary philosophers of medicine are also interested in other subjects that were debated by Polish philosophers of medicine, such as the problem of reductionism versus holism in medicine; the validity of biological models developed in laboratory animals for human therapy [9, 11, 20]; the problem of causality in medicine [1, 24]; and the problem of definition and classifications of diseases [6, 15]. This is not to claim that the writings of Polish philosophers of medicine fully anticipated the level of present reflection within this discipline. Philosophers of medicine today deal with problems that were at best barely touched upon by their predecessors: for example, the cultural aspects of medicine, the limits of ethical and legal responsibility of a physician, the definition of medical fallibility, and the problem of interpretation in medicine. It also deals with problems closely related to recent developments in medicine and society, such as the degree of responsibility a society owes toward its sick members, the informed consent of patients, and the myriad of problems related to the application of advanced medical technology. Philosophers of medicine have been able to benefit from collaboration with other humanists and social scientists: psychologists, sociologists, anthropologists, and philosophers of sciences. It is not surprising, then, that today studies by philosophers of medicine have a wider scope and are much more complex and more elaborated than can be found in the early works of the three phases of the Polish School.

This does not mean, however, that interest in the work of Polish philosophers of medicine should be limited to an evaluation of their role as "precursors" of the present-day philosophers of medicine, or to a reading of their works solely as historical documents - first-hand testimonies to the reception of the 19th-century "scientific revolution" in medicine by practicing physicians, or interesting discussions of specific problems in the philosophy of medicine. Surely there is one important lesson that contemporary philosophers of medicine may still learn from the study of the writings of the Polish School: the continuous need for critical reflection on clinical medical practice. The approach to medicine of the Polish School was based on two complementary aspects: 1) concrete (and, if one may use an anachronistic term, "sociologically-oriented") analysis of medical practice founded on detailed descriptions of the actual behavior of physicians in specific clinical situations, and 2) historical analysis of the evolution of medical concepts and of the modification of clinical medical practices in different historical periods, in particular during the 19th-century. Both approaches have contributed to the elaboration of a serious critique of medicine that was at the very center of the philosophico-medical writings of the Polish School.

In contrast to Polish philosophy of medicine, contemporary philosophy of medicine, with all its important achievements notwithstanding, frequently lacks this critical dimension. The principal factor here is perhaps a sociological one: the majority of today's specialists who call themselves "philosophers of medicine" are either professional philosophers or physicians turned "academic" philosophers. There are few philosophers at the bedside, i.e., clinicians, who combine their daily practice of medicine with a systematic philosophical reflection on it. Too often theoretical reflection on medical practice is delegated to specialists: specialists in "clinical judgment" (e.g., statisticians, decision theorists, computer scientists) on the one hand and specialists in "bioethics" (attorneys, theologians, and philosophers) on the other. Perhaps this was an unavoidable evolution: conceding philosophical debate on medical issues to specialists in this field simply reflects the general trend of specialization in medicine, and the compartmentalization of medical competences. But the fact that medicine has become such a complex enterprise in the late 20th-century does not necessarily imply that physicians should exempt themselves from reflection on the philosophical and ethical dimensions of their activity. One may argue that just the opposite is the case. The more complex problems that each physician must confront in his on her practice, the less he or she can afford the luxury of delegating the task of critical reflection on his or her practice to specialists in other disciplines. Perhaps the philosophy of medicine today is still awaiting its "Medical Critique", and its generation of philosophers at the bedside.[6]

NOTES

1 Biegański refers to a book by H. Magnus, *Kritik der medizinischen Erkenntnis* [13].

2 Letter of M. Neuberger to W. Biegański, November 7th, 1908. Quoted by T. Bilikiewicz [5].

3 *Nederlands Tijdschrift voor Geneeskunde* of August 20th, 1910, quoted by E. Stocki and W.J. Jasinski [21].

4 One of the most important among those was G. Canguilhem's study, *Le Normal et le pathologique* [7], first published in 1947.

5 Introducing a new international journal, *Bioethics,* its editors explained that interest in ethical aspects of medical practice is the direct result of the recent evolution of medicine: "we are living in a very different world from the 1950's, and we cannot go back to the less complicated past. That is why we cannot avoid bioethics" [12].

6 Philosophers at the bedside, if they are not physicians, should at least be philosophers who continue to accompany physicians to the bedside (a point I owe to Stuart Spicker).

BIBLIOGRAPHY

[1] Agasi, J.: 1976, 'Causality and Medicine', *The Journal of Medicine and Philosophy* 1, 301-317.

[2] Biegański, W.: 1908, *The Logic of Medicine*, E. Wende, Warsaw, pp. VI; 29; VIII.

[3] Biernacki, E.: 1902, *Principles of Medical Understanding*, Wende i Ska, Warsaw, pp. VIII; IX.

[4] Bilikiewicz, T.: 1957, 'Biegański as a Philosopher' ['Bieganski jako filozof'], *Archiwum Historii Medycyny* **20**, 347-355.

[5] Bilikiewicz, T.: 1929, 'The Relationships Between Władysław Biegański and Prof. Max Neuberger, as Reflected by Their Correspondance' ['Stosunek Władysława Biegańskiego do prof. Maksa Neuburgera w świetle zachowanych listów'], *Archiwum Historji i Filozofji Medycyny* **9**, 181-193.

[6] Brown, W.M.: 1985, 'On Defining "Disease"', *The Journal of Medicine and Philosophy* **10**, 311-328.

[7] Canguilhem, G.: 1966 (1947), *Le Normal et le pathologique*, Presses Universitaires de France, Paris.

[8] Engelhard, H.T. Jr: 1986, 'From Philosophy and Medicine to Philosophy of Medicine', *Journal of Medicine and Philosophy* **11**, 3-8.

[9] Kaplan, A.L.: 1986, 'Exemplary Reasoning? A Comment on Theory Structure in Biomedicine', *The Journal of Medicine and Philosophy* **11**, 93-105.

[10] King, L.: 1982, *Medical Thinking*, Princeton University Press, Princeton.

[11] Krummardt, R.: 1986, 'The "Borderzone Zone" Controversy: A Study of Theory Structure in Biomedicine', *Theoretical Medicine* **7**, 243-258.

[12] Kuhse, H., and Singer, P.: 1987, 'Bioethics: What? and Why?', *Bioethics* **1**, iii-v.

[13] Magnus, H.: 1904, *Kritik der medizinischen Erkenntnis*, Max Müller, Breslau.

[14] Maull, N.: 1981, 'The Practical Science of Medicine', *The Journal of Medicine and Philosophy* **6**, 165-182.

[15] Mersky, H.: 1986, 'Variable Meaning for the Definition of Disease', *The Journal of Medicine and Philosophy* **11**, 215-232.

[16] Munson, R.: 1981, 'Why Medicine Cannot be a Science', *The Journal of Medicine and Philosophy* **6**, 183-208.

[17] Murphy, E.A.: 1976, *The Logic of Medicine*, Johns Hopkins University Press, Baltimore.

[18] Pellegrino, E.: 1976, 'Philosophy of Medicine: Problematic and Potential', *The Journal of Medicine and Philosophy* **1**, 5-31.

[19] Pellegrino, E.: 1986, 'Philosophy of Medicine: Towards a Definition', *The Journal of Medicine and Philosophy* **11**, 9-16.

[20] Schaffner, K.F.: 1986, 'Exemplar Reasoning about Biological Models and Diseases: A Relation Between the Philosophy of Medicine and Philosophy of Science', *The Journal of Medicine and Philosophy* **11**, 63-80.

[21] Stocki, E., and Jasinski, W.J.: 1971, '*The Logic of Medicine* in Polish and Foreign Reviews', in G. Swiderski and M Stanski (eds.), *Wladyslaw Bieganski: a Physician and a Philosopher [Wladyslaw Bieganski, lekarz i filozof]*, Poznanskie Towarzystwo Przyjaciol Nauk, Poznan, p. 237.

[22] Szumowski, W.: 1926, 'The Nearest Goals of the Philosophy of Medicine', *Archiwum Historji i Filozofji Medycyny* **4**, 70-73.

[23] Toulmin, S.: 1976, 'On the Nature of the Physician's Understanding', *The Journal of Medicine and Philosophy* **1**, 32-50.

[24] Whitbeck, C.: 1977, 'Causation in Medicine: the Disease Entity Model', *Philosophy of Science* **44**, 619-637.

[25] Widderschoven-Heerding, I.: 1987, 'Medicine as a Form of Practical Understanding', *Theoretical Medicine* **8**, 179-185.

Ackerknecht, E. 10, 22, 87, 178n
Adjukiewicz, K. 165, 223
Æsculapius 65
Agasi, J. 282
Andral, G. 20n, 21n
Andry, N. 250
Aristotle 102
Arthus, N. 174
Avicenna 206

Bacon, R. 102
Baranowski, I. 13, 21n, 22, 72
Bauhin, G. 244
Bechtel, W. 22
Ben-David, J. 10, 10n
Bergson, H. 178n
Berlioz, H. 260
Bernard, C. 79, 80, 85n, 87, 120, 189, 227
Biegańska, M. 69, 87, 179
Biegański, W. 2, 5, 6, 7, 8, 19, 22, 22n, 69-120, 121, 133, 133n, 164, 166, 172, 173, 174, 178n, 199, 220, 221, 222, 224, 225n, 277, 278, 282n, 283
Bier, A. 178n
Bierkowski, L. 198, 199
Biernacki, E. 2, 3, 5, 6, 7, 8, 19, 20n, 22, 37-68, 81, 86n, 88, 119, 124, 163, 164, 166, 172, 174, 175, 191, 220, 221, 222, 224, 278
Bigelow, J. 88
Bilikiewicz, T. 10, 88, 175, 176, 178n, 179, 223, 225, 250, 251, 252, 257-266, 267, 268, 270, 273, 274-275, 278, 283
Billroth, T. 10, 211
Boerhave, H. 64
Bohr, N. 223, 243
du Bois-Reymond, E. 185
Bordeu, T. 196
Borkowski, E. 88
Bourget, E. 250
Brahe, T. 245
Broussais, F. 19, 21n, 82, 169, 210
Brown, J. 20n, 21n, 82, 209
Brown, W. 283
Brücke, E. 185
Büchner, E. 172
Bujwid, O. 211
Bumke, O. 178n
Burrows, W. 47

Cabanis, P. 86n, 88
Canguilhem, G. 22, 282n, 283
Cantani, G. 46n
Carrel, A. 178n
Celsus, 205
Charcot, J. 37
Chałubiński, T. 2, 5, 8, 13-36, 43, 46n, 71, 73, 74, 75, 81, 82, 85n, 92, 99, 120, 121, 163, 166, 172, 174, 182, 193, 220, 222, 224, 227
Chwistek, L. 69, 223
Chopin, F. 260
Codell-Carter, C. 23, 47
Cohen, R. 225
Comte, A. 102
Copernicus, N. 201

Debussy, C. 250
Dembrowski, J. 249
Descartes, R. 174
Dietl, J. 3, 18, 23, 173, 203, 212
Doll, R. 47
Dreisch, H. 178n
Dubos, J. 47
Dubos, R. 47
Duclaux, E. 47
Dunin, T. 179
Durkheim, E. 223
Dusch, V. 191
Dworzaczek, F. 13, 14, 20n, 165

Einstein, A. 269
Elkana, Y. 225
Elkeles 242, 243
Engelhardt, T. 10, 279, 283
Erasitratos 203
Erbe 37
Eustachius, B. 244

Fessel, W. 23
Filar, Z. 179
Finkel, F. 174
Fleck, L. 5, 6, 9n, 10, 69, 215-276
Flexner, A. 10
Forbes, J. 89
Frank, J. 209
Freudenthal, G. 226
Friedemann 242
Fritz, J. 10, 226

Galen 205, 244
Galileo, G. 102, 269
Galland, H. 23
Gawrychowski, S. 88
Geison, G. 10

Giedroyc, F. 177n, 212
Giedymin, J. 226
von Graefe, F. 185
Greisinger, W. 64
Gyöffry, I. 23

Haeckel, E. 172
Haeser, H. 101
Hahneman, S. 207, 208
Hamernik 212
Hattsoeker, L. 250
Haller, A. 74
Hayem, G. 37
Heisinger, W. 267
Helmholtz, H. 172, 185
van Helmont, J. 202
Henle, J. 185
Higgier, H. 180
Hippocrates 188
Hoyer, H. 3, 133n, 163, 164, 166, 172, 177n, 180, 182

Jamous, H. 10
Jarosinski, T. 23
Jasinski, W. 89, 282n, 284
Jenner, E. 211
Jerusalem, W. 223
Joël 259
de Jussieu, A. 20n

Kant, I. 257
Kaplan, A. 283
Kepler, J. 245
King, L. 23, 279, 283
Koch, Richard 1
Koch, Robert 171, 189, 190
Kohler, R. 47
Kochanowski, J. 189
Kotarbiński, T. 165
Krakowiecka, L. 89, 135
Kramsztyk, Z. 2, 6, 7, 8, 11, 19, 21n, 23, 39, 70, 80, 85n, 86n, 89, 118, 121-162, 163, 164, 166, 173, 177n, 180, 220, 221, 222, 224, 225n, 227
Krummardt, R. 283
Krupiński 174
Kühne, H. 37
Kuhse, H. 283
Kunicki-Goldfinger, W. 225n
Kus, M. 11, 135

Laënnec, T. 116
Lavoisier, A. 206, 248n, 269
Leewenhoeck, A. 250
Leibnitz, W. 259, 265
Lenoir, T. 85n, 89
Lessing, G. 188
Lévy-Bruhl, L. 223
Lindenmann, J. 89
Locke, J. 69
Lombroso, C. 178n
Lotze, A. 172
Löwy, I. 11, 226, 227
Louis, P. 20n, 69, 80
Ludwig, K. 185
Łukasiewicz, J. 69, 165

Magendie, F. 20n
Magnus, H. 118, 120, 277, 282n, 283
de Malebranche, N. 174
Maulitz, R. 26
Maull, N. 23, 280, 283
McMenemey, W. 23
McTavish, J. 23
Mersky, H. 283
Metchnikoff, E. 46n
Mickiewicz, A. 132
Mill, J.S. 70
Muller, J. 185, 186
Munson, R. 23, 280, 283
Murphy, E. 279, 283

Neisser, A. 171, 186, 189
Neuburger, M. 278, 282n
Newton, I. 102, 269
Niemyer, O. 20n
Nusbaum, H. 3, 118, 119, 166, 168, 175, 177n, 180, 223, 227

Ochorowicz, J. 48
Oesterlen, F. 72, 80, 85n, 89, 120, 277
Ostrowska, T. 11, 135, 180

Paracelsus 206, 207, 211
Paré, A. 101
Parisé, R. 85n
Pasteur, L. 189, 211
Pellegrino, E. 10, 279, 283
Phil, J. 244
Poincaré, H. 198, 219
Pollitzer, R. 48

Reichert, T. 185
Reigel, O. 37
Reinchardt, B. 177n

Reizel, K. 223
Reizler, K. 227
Remak, R. 3, 185
Rokitanski, K. 18
Rosenberg, C. 11
Rymkiewicz, K. 210
Rytal, A. 48

Sage, O. 248n
Salomon-Bayet, C. 48
Scadding, J.G. 24
Schaffner, K. 284
Schiller, J. 24
Schleiden, M. 21n
Schnelle, T. 225, 225n, 227
Schönlein, J. 64n
Schrödinger, E. 249
Schultze, M. 185
Schwann, T. 185
Shaw, G.B. 47n, 48
Sichel 121
Sigerist, H. 9n, 175, 176, 178n, 179n, 180
Singer, P. 283
Skarga, B. 183
Skarzynski, B. 89
Skoda 18, 116, 212
Skulimowski, M. 11, 24, 135
Słowiecki, M. 11, 135
Śniadecki, J. 13, 20n, 21n, 170, 172, 182, 210
Sokołowski 198
de Solla-Price, D. 48
Spicker, S. 282n
Spielmann, J. 248n
Stański, M. 88, 89
Staum, M. 89
Stocki, E. 89, 282n, 284
Strauss, H. 102
Struve, A. 69
Stubell 186
Sudhoff, K. 176
Swaroop, J. 48
Świderski, G. 88
Sydenham, T. 101, 182
Sylvius 101
Szokalski, W. 121, 122, 135
Szuman, S. 178n, 180
Szumowski, W. 1, 2, 11, 24, 46n, 47, 48, 173-175, 176, 179n, 181, 222, 278, 284

Tarski, A. 69
Tartakiewicz, W. 85n
Taylor, R. 12
Tchaikowsky, P. 260
Temkin, O. 12, 176, 178n
Thomas, L. 48
Tolstoy, L. 103
Toulmin, S. 280, 284
Traube, L. 185
Trepka, W. 200, 201, 203
Trousseau, A. 65, 66, 92
Trzebinski, S. 168-170, 172, 181, 182, 183, 191-214, 221, 223, 227
Twardowski, K. 85n, 165, 223

Vesalius, A. 244, 245, 248n
Virchow, R. 4, 21n, 172, 185, 189, 196
Vallisinieri 250

Warner, J. 12
Weigl, R. 215, 217
Weil, E. 235
Weisz, G. 12
Whitbeck, C. 48, 284
Widderschoven-Heerding, I. 24, 280, 284
Wize, K. 182
Wölfin, H. 259
Wrzosek, A. 2, 12, 20n, 24, 48, 165, 170-173, 174, 177n, 180, 182, 183, 185-190, 198, 222, 227
Wunderlich, K. 22n
Wundt, W. 172

Ziemski, S. 90, 183
Zworniak, W. 48, 227

The Philosophy and Medicine Book Series

Editors

H. Tristram Engelhardt, Jr. and Stuart F. Spicker

1. **Evaluation and Explanation in the Biomedical Sciences**
   1975, vi + 240 pp. ISBN 90–277–0553–4
2. **Philosophical Dimensions of the Neuro-Medical Sciences**
   1976, vi + 274 pp. ISBN 90–277–0672–7
3. **Philosophical Medical Ethics: Its Nature and Significance**
   1977, vi + 252 pp. ISBN 90–277–0772–3
4. **Mental Health: Philosophical Perspectives**
   1978, xxii + 302 pp. ISBN 90–277–0828–2
5. **Mental Illness: Law and Public Policy**
   1980, xvii + 254 pp. ISBN 90–277–1057–0
6. **Clinical Judgment: A Critical Appraisal**
   1979, xxvi + 278 pp. ISBN 90–277–0952–1
7. **Organism, Medicine, and Metaphysics**
   Essays in Honor of Hans Jonas on his 75th Birthday, May 10, 1978
   1978, xxvii + 330 pp. ISBN 90–277–0823–1
8. **Justice and Health Care**
   1981, xiv + 238 pp. ISBN 90–277–1207–7(HB) 90–277–1251–4(PB)
9. **The Law-Medicine Relation: A Philosophical Exploration**
   1981, xxx +•292 pp. ISBN 90–277–1217–4
10. **New Knowledge in the Biomedical Sciences**
    1982, xviii + 244 pp. ISBN 90–277–1319–7
11. **Beneficence and Health Care**
    1982, xvi + 264 pp. ISBN 90–277–1377–4
12. **Responsibility in Health Care**
    1982, xxiii + 285 pp. ISBN 90–277–1417–7
13. **Abortion and the Status of the Fetus**
    1983, xxxii + 349 pp. ISBN 90–277–1493–2
14. **The Clinical Encounter**
    1983, xvi + 309 pp. ISBN 90–277–1593–9
15. **Ethics and Mental Retardation**
    1984, xvi + 254 pp. ISBN 90–277–1630–7
16. **Health, Disease, and Causal Explanations in Medicine**
    1984, xxx + 250 pp. ISBN 90–277–1660–9
17. **Virtue and Medicine**
    Explorations in the Character of Medicine
    1985, xx + 363 pp. ISBN 90–277–1808–3
18. **Medical Ethics in Antiquity**
    Philosophical Perspectives on Abortion and Euthanasia
    1985, xxvi + 242 pp. ISBN 90–277–1825–3(HB) 90–277–1915–2(PB)
19. **Ethics and Critical Care Medicine**
    1985, xxii + 236 pp. ISBN 90–277–1820–2

20. **Theology and Bioethics**
Exploring the Foundations and Frontiers
1985, xxiv + 314 pp. ISBN 90–277–1857–1
21. **The Price of Health**
1986, xxx + 280 pp. ISBN 90–277–2285–4
22. **Sexuality and Medicine**
Volume I: Conceptual Roots
1987, xxxii + 271 pp. ISBN 90–277–2290–0(HB) 90–277–2386–9(PB)
23. **Sexuality and Medicine**
Volume II: Ethical Viewpoints in Transition
1987, xxxii + 279 pp. ISBN 1–55608–013–1(HB) 1–55608–016–6(PB)
24. **Euthanasia and the Newborn**
Conflicts Regarding Saving Lives
1987, xxx + 313 pp. ISBN 90–277–3299–4(HB) 1–55608–039–5(PB)
25. **Ethical Dimensions of Geriatric Care Value Conflicts for the 21st Century**
1987, xxxiv + 298 pp. ISBN 1–55608–027–1
26. **On the Nature of Health**
An Action-Theoretic Approach
1987, xviii + 204 pp. ISBN 1–55608–032–8
27. **The Contraceptive Ethos**
Reproductive Rights and Responsibilities
1987, xxiv + 254 pp. ISBN 1–55608–035–2
28. **The Use of Human Beings in Research**
With Special Reference to Clinical Trials
1988, xxii + 292 pp. ISBN 1–55608–043–3
29. **The Physician as Captain of the Ship**
A Critical Appraisal
1988, xvi + 254 pp. ISBN 1–55608–044–1
30. **Health Care Systems**
Moral Conflicts in European and American Public Policy
1988, xx + 368 pp. ISBN 1–55608–045X
31. **Death: Beyond Whole-Brain Criteria**
1988, x + 276 pp. ISBN 1–55608–053–0
32. **Moral Theory and Moral Judgments in Medical Ethics**
1988, vi + 232 pp. ISBN 1–55608–060–3
33. **Children and Health Care: Moral and Social Issues**
1989, xxiv + 349 pp. ISBN 1–55608–078–6
34. **Catholic Perspectives on Medical Morals**
Foundational Issues
1989, vi + 308 pp. ISBN 1–55608–083–2
35. **Suicicide and Euthanasia**
Historical and Contemporary Themes
1989, vi + 286 pp. ISBN 0–7923–0106–4
36. **The Growth of Medical Knowledge**
1990, viii + 194 pp. ISBN 0–7923–0736–4
37. **The Polish School of Philosophy of Medicine**
From Tytus Chalubinski (1820–1889)
To Ludwik Fleck (1896–1961)
1990, viii + 287 pp. ISBN 0–7923–0958–8

Batch number: 10761599

Printed by Printforce, the Netherlands

Boliang Yi

Integrated Planning and Control for Collision Avoidance Systems

Schriftenreihe
Institut für Mess- und Regelungstechnik,
Karlsruher Institut für Technologie (KIT)

Band 039

Eine Übersicht aller bisher in dieser Schriftenreihe erschienenen Bände finden Sie am Ende des Buchs.

Integrated Planning and Control for Collision Avoidance Systems

by
Boliang Yi

KIT Scientific Publishing

Dissertation, Karlsruher Institut für Technologie
KIT-Fakultät für Maschinenbau

Tag der mündlichen Prüfung: 5. Oktober 2017
Referenten: Prof. Dr.-Ing. Christoph Stiller
Prof. Dr.-Ing. Ulrich Konigorski

Impressum

KIT Scientific Publishing

Karlsruher Institut für Technologie (KIT)
KIT Scientific Publishing
Straße am Forum 2
D-76131 Karlsruhe

KIT Scientific Publishing is a registered trademark
of Karlsruhe Institute of Technology.
Reprint using the book cover is not allowed.

www.ksp.kit.edu

Print on Demand 2018 – Gedruckt auf FSC-zertifiziertem Papier

ISSN 1613-4214
ISBN 978-3-7315-0785-7
DOI 10.5445/KSP/1000081452

Integrated Planning and Control for Collision Avoidance Systems

Zur Erlangung des akademischen Grades
Doktor der Ingenieurwissenschaften
der KIT-Fakultät für Maschinenbau
Karlsruher Institut für Technologie (KIT)

genehmigte
Dissertation
von

M.Sc. Boliang Yi

Tag der mündlichen Prüfung:	05. Oktober 2017
Hauptreferent:	Prof. Dr.-Ing. Christoph Stiller
Korreferent:	Prof. Dr.-Ing. Ulrich Konigorski

Abstract

The introduction of driver assistance systems like Anti-lock Braking System and Electronic Stability Control has increased safety of vehicle passengers on public roads. The automotive industry has boosted the development of Advanced Driver Assistance Systems in the last years. Collision avoidance systems like emergency braking assist systems have demonstrated their effectiveness in various studies and thus gained broad acceptance in society. To further increase the effectiveness of collision avoidance systems, the exploitation of the lateral free space by evasive maneuvers is being investigated in this thesis.

Effective avoidance systems drive the need for holistic integrated approaches to cover planning and control in one method. This thesis focuses on methods for integrated trajectory planning and vehicle dynamics control in collision avoidance scenarios by combined evasion and braking. Integrated methods allow for consistent model representation for both planning and control functionality and lead to a reduced number of design parameters in the overall system. This strongly reduces the tuning effort required to adapt both modules to each other and avoids to tune both modules separately. The proposed nonlinear method based on a model predictive approach plans collision-free trajectories taking into account environmental information of obstacles and the available maneuver space. The concept of terminal collision avoidance provides a solution for planning with obstacles detected outside the current prediction horizon.

Application of methods for collision avoidance in real vehicles require algorithms which can be efficiently solved on a real time computation platform. The successive linearization technique is used in this work to linearize nonlinear constraints of the optimization problem and to enable planning in real vehicles. The concept of robust tube based model predictive control is exploited to consider for stability and constraint satisfaction with linearization errors. The performance and effectiveness of the proposed algorithm is demonstrated in a simulation environment and in a real vehicle application successfully avoiding the collision in characteristic scenarios and showing strong conformance for both environments.

The design concepts investigated in this thesis are promising to accelerate the realization of future design principles and thereby contribute to the overall development of automated driving technologies.

Kurzfassung

Die Einführung erster Fahrassistenzsysteme wie dem Anti-Blockier-System und dem Elektronischen-Stabilitäts-Programm hat erheblich zur Fahrsicherheit auf den Straßen beigetragen. Seitdem wurde in der Automobilindustrie die Entwicklung neuer Fahrassistenzsysteme stark vorangetrieben. Kollisionsvermeidungsysteme wie das Notbremssystem genießen hohe Akzeptanz und haben in Studien eine hohe Effektivität nachgewiesen. Um die Effektivität von Kollisionsvermeidungssystemen weiter auszubauen, wird in dieser Arbeit das Potenzial der Ausnutzung des verfügbaren lateralen Manöverraums durch Kollisionsausweichsysteme untersucht.

Effektive Kollisionsvermeidungssysteme benötigen holistische integrierte Ansätze, um Planung und Regelung in einer einzigen Methode zu vereinen. Der Fokus der Arbeit liegt daher in der Untersuchung von integrierten Planungs- und Regelungsmethoden für Kollisionsvermeidung-Manöver durch Ausweichen und Bremsen. Integrierte Methoden nutzen ein konsistentes physikalisches Model für die Planung und die Regelungsaufgabe, welches zu einer reduzierten Anzahl an Auslegungsparameter im Gesamtsystem führen. Diese Eigenschaft vermeidet die klassische separate Einstellung beider Module und reduziert damit den Anpassungsaufwand im Gesamtsystem. Die untersuchte Methodik basiert auf einem modelprädiktiven Regelungsansatz unter Berücksichtigung eines Fahrdynamikmodells, der erfassten potentiellen Kollisionsobjekte, der vorhandenen Fahrbahnbegrenzung sowie den Fahrdynamik- und Aktor-Limitierungen. Eine Endbeschränkung zur Kollisionsvermeidung ermöglicht die kollisionsfreie Planung mit potentiellen Hindernissen außerhalb des Prädiktionshorizonts.

Die Anwendung von Methoden zur Kollisionsvermeidung im Versuchsfahrzeug erfordert Algorithmen, die auf einer echtzeitfähigen Plattform effizient gelöst werden können. In dieser Arbeit wird daher das Verfahren der sukzessiven Linearisierung eingesetzt, um vorhandene Nichtlinearitäten in den Nebenbedingungen zu linearisieren und damit die Umsetzung der Planung in einem Versuchsfahrzeug zu ermöglichen. Das Konzept der robusten modelprädiktiven Regelung wird genutzt, um die Stabilität sowie die Einhaltung von Beschränkungen unter dem Einfluss von Linearisierungsfehlern zu garantieren. Die Performance und Effektivität der vorgeschlagenen Methoden wurden in einer Simulationsumgebung sowie in einem Versuchsfahrzeug getestet und bewertet. Die Ergebnisse zeigen die erfolgreiche Vermeidung von Kollisionen in allen charackteristischen Szenarien sowie eine hohe Übereinstimmung der Messergebnisse in Simulation und im Versuchsfahrzeug.

Die erarbeiteten Konzepte haben ein großes Potenzial, die Realisierung von zukünftigen Planungsprinzipien zu beschleunigen und damit zu der Entwicklung von automatisiertem Fahren beizutragen.

Acknowledgements

This dissertation is the result of my PhD study at Adam Opel AG, supervised by Prof. Dr.-Ing. Christoph Stiller from Karlsruhe Institute of Technology.
First of all, I would like to thank Prof. Stiller for his scientific advice and guidance during my thesis time. His trendsetting suggestions have been an inspirational source during the PhD time, as well as during the works on the research project. Furthermore, I would like to thank Prof. Konigorski from the Technical University of Darmstadt for serving as co-referent. His advice greatly influenced my research.
The research work was conducted within the framework of the research project UR:BAN - Urbaner Raum: Benutzergerechte Assistenzsysteme und Netz-management - funded by the German Federal Ministry of Economics and Energy (BMWi) in the context of the third traffic research program of the German government at Adam Opel AG. Special thanks are dedicated to my industrial supervisors during the PhD time, Frank Bonarens and Nikolas Wagner. Frank combined his extensive engineering experience from manufacturing and project leading skills with his enthusiasm for new technologies to support me as a mentor technically as well as personally during my PhD time. His patient guidance and readiness in all situations encouraged me in difficult circumstances. It is hard to imagine a better industrial supervisor than Frank. Nik enabled me in gaining first work experiences during my PhD time. His activities for the Opel PhD program integrated us well into the organization and represented our interests in all matters. Thanks Frank and Nik for your help over the last years.
I would like to thank my colleagues for the enjoyable collaborations during my time at Opel. I want to especially name Jens Ferdinand, Norbert Simm, Ulrich

Eberle, Steffen Knapp, Bernd Büchs, Harald Berninger, Björn Frank, Oliver Maier, Nikolai Moshchuk, Markus Bauer, Tobias Rückelt, Carsten Büttner, Thomas Streubel, Lena Rittger, Jens Heine, Gerald Schmidt, Stefan Berger, Timo Derichs, Rami Zarife, Jeremias Schucker, David Augustin, Bernhard Wandtner, Sven Hallerbach, Christina Kass, Maximilian Harr, Florian Jomrich and Fabian Fürst. I want to thank Denis Stein for support and help in the layout of this manuscript.

This thesis is based on the support of many students, who participated in the project as well as in the research work. I appreciate having worked with all students during this interesting and challenging time, namely Haoming Wu, Philipp Hoffmann, Christian Pfeiffer, Sebastian Engel, Alexander Schulzke, Holger Stübing, Fernando Gomes Papi, Florian Schindler, Stefan Gottschling, Theodor Rauch, Kay Hansel and Hermann Blondel Ngueko Kameni.

Finally I want to thank my parents. They taught me what is important in life and how to reach my goals. I can count on them when I need advice and in order to gather strength for my next steps. They are my backbone in every important step and shaped me how I am.

Wiesbaden, April 2017 *Boliang Yi*

Contents

1 Introduction **1**
1.1 Motivation 1
1.2 Aim of the Thesis 4
1.3 Outline 7

2 Concepts and Methods for Collision Avoidance **9**
2.1 Classical Concept of Collision Avoidance Systems 9
2.2 Trajectory Planning 10
2.3 Vehicle Dynamics Control 13
2.4 Integrated Planning and Control 15
2.5 Selection of Methods and Concepts 17

3 Fundamentals of Optimal Planning and Control **19**
3.1 Vehicle Dynamics Modeling 19
3.1.1 Lateral Dynamics Model 19
3.1.2 Longitudinal Dynamics Model 27
3.1.3 Combined Vehicle Dynamics Model 27
3.1.4 Actuator and State Limitations 29
3.1.5 Model Validation 30
3.2 Model Predictive Control 33
3.2.1 Nominal Model Predictive Control 34
3.2.2 Robust Model Predictive Control 38
3.3 Model Predictive Tracking 43
3.3.1 Trajectory Planning 44
3.3.2 Vehicle Dynamics Model 46

3.3.3 Cost Function 48
3.3.4 Constraints 50
3.3.5 Terminal Cost 53
3.3.6 MPC Formulation 54

4 Nonlinear Integrated Planning and Control 57
4.1 Design Goals 58
4.2 Cost Function 60
4.3 Obstacles and Road Boundaries 61
4.4 Terminal Collision Avoidance 64
4.5 Soft Constraints 68
4.6 MPC Formulation 69
4.7 Simulation Results 70
4.7.1 Metrics 70
4.7.2 Scenarios 71
4.7.3 Typical Scenarios 73
4.7.4 Terminal Collision Avoidance 82
4.8 Discussion 86

5 Fast Integrated Planning and Control 89
5.1 Successive linearization of nonlinear constraints 90
5.1.1 Vehicle Dynamics Model 91
5.1.2 Collision Avoidance Constraints 95
5.1.3 Terminal Collision Avoidance 100
5.1.4 Acceleration Circle 101
5.2 Compensation of Linearization Errors 103
5.3 MPC Formulation 107
5.4 Results 108
5.4.1 Simulation of Typical Scenarios 109
5.4.2 Simulation with Terminal Collision Avoidance 113
5.4.3 Real Vehicle Application 116
5.5 Discussion 117

6 Conclusion and Outlook 123

References 127

Publications by the author 137

Supervised theses 139

Notation and Symbols

General Notation

Scalars	Regular (greek)	$a, b, A, B, \sigma, \lambda$
Vectors	Bold (greek) lower case	$\mathbf{a}, \mathbf{b}, \mathbf{A}, \mathbf{B}, \boldsymbol{\sigma}, \boldsymbol{\lambda}$
Matrices	Bold upper case	$\mathbf{A}, \mathbf{B}, \mathbf{C}$
Sets	Bold blackboard upper case	$\mathbb{A}, \mathbb{B}, \mathbb{C}$

Acronym

ABS	Anti-lock Braking System
ADAS	Advanced Driver Assistance System
COG	Centre Of Gravity
ESC	Electronic Stability Control
EPS	Electronic Power Steering
IEEE	Institute of Electrical and Electronics Engineers
MPC	Model Predictive Control
RRT	Rapidly exploring Random Trees
SAE	Society of Automation Engineers
TCA	Terminal Collision Avoidance

Symbols

a	Slope of sigmoidal trajectory
$a_{(\mathrm{x}\,/\,\mathrm{y}\,/\,\mathrm{z})}$	Acceleration
$a_{\mathrm{x,cmd}}$	Longitudinal deceleration command

$a_{\mathrm{y,max}}$	Maximum planned lateral acceleration
a_{max}	Deceleration command limit
c	Half of longitudinal maneuver distance in sigmoidal trajectory
$c_{\alpha(\mathrm{f\,/\,r})}$	Cornering stiffness
$c_{\alpha(\mathrm{f\,/\,r}),0}$	Linearised cornering stiffness
d_{st}	steering damping coefficient
h_{cog}	Height of COG
i	Prediction time step
k	Control time step
l	Wheel base
l_{veh}	Overall length of vehicle
$l(\mathbf{x},\mathbf{u})$	Stage cost
$l_{(\mathrm{f\,/\,r})}$	Distance COG to axis
$l_{(\mathrm{f\,/\,r}),\mathrm{veh}}$	Distance COG to chassis edge
m	mass
$r_{\mathrm{T}}, r_{a_{\mathrm{x}}}, q_{\mathrm{y}}, q_{\varphi}, q_{\mathrm{v}}$	Tuning parameters of MPC on state and command signals
s_{obs}	Slack variable
t_{c}	Potential collision time
t_{cmp}	Computation time
t_{pred}	Prediction sampling time
t_{del}	Time constant in longitudinal dynamics
v	Velocity
w	Width of ego vehicle
$w_{\mathrm{obs(l\,/\,r)}}$	Obstacle width on corresponding lane
x	Longitudinal position
$x_{\mathrm{obs(l\,/\,r)}}$	Longitudinal position of obstacle on corresponding lane
$y_{\mathrm{obs(l\,/\,r)}}$	Lateral position of obstacle on corresponding lane

$x_{\mathrm{E}}, y_{\mathrm{E}}, t_{\mathrm{E}}$	Evasion maneuver distances and time
$x_{\mathrm{B}}, t_{\mathrm{B}}$	Braking maneuver distance and time
y	Lateral position
y_{Tol}	Tolerance parameter in sigmoidal trajectory
$y_{\max}$	Maximum lateral position
$y_{\mathrm{L}}, y_{\mathrm{R}}$	Left and right boundary of drivable area
Δv_{red}	Relative velocity reduction
$\alpha_{(\mathrm{f}\,/\,\mathrm{r})}$	Slip angle
$\alpha_{\mathrm{r,max}}$	Maximum rear slip angle
β	Side slip angle
$\beta_{\min}, \beta_{\max}$	Minimum and maximum side slip angle
δ	Steering angle
μ	Friction coefficient
μ_{y}	Lateral friction coefficient
$\mu_{\mathrm{y,max}}$	Assumed maximum lateral friction coefficient
φ	Yaw angle
$\dot{\varphi}$	Yaw rate
$\dot{\varphi}_{\min}, \dot{\varphi}_{\max}$	Minimum and Maximum yaw rate
θ	Vector of parametric uncertainties
$B/C/D/E$	Magic formula tire coefficient
D_{f}	Self alignment parameter
$B_{(\mathrm{l}\,/\,\mathrm{r})}$	Nearest point of obstacle on vehicle edge
$F_{(\mathrm{x}\,/\,\mathrm{y}\,/\,\mathrm{z})(\mathrm{f}\,/\,\mathrm{r})}$	Force on tire
$F_{(\mathrm{x}\,/\,\mathrm{y}\,/\,\mathrm{z})(\mathrm{f}\,/\,\mathrm{r}),\mathrm{max}}$	Maximum force on tire
$F_{\mathrm{y}(\mathrm{f}\,/\,\mathrm{r}),0}$	Force bias constant
J_{st}	Steering system inertia
J_{z}	Yaw inertia
M	Maximum lateral maneuver distance in sigmoidal trajectory
N	Number of prediction time steps

R	Curve radius
T	Total assisted torque
T_z	Self alignment torque
T_{max}	Maximum total assistant torque
T_{HW}	Driver hand torque
T_{OT}	Overlay torque
V	Cost function of finite horizon control problem
V_∞	Cost function of infinite horizon control problem
V_f	Terminal cost
$\mathbf{e}$	Error vector
$\mathbf{f}$	State transition function
$\mathbf{k}$	Linear feedback vector
$\mathbf{u}$	Input vector
$\mathbf{u}_{min}, \mathbf{u}_{max}$	Minimum and maximum input vector
$\mathbf{w}$	Additive uncertainty
$\mathbf{x}$	State vector
$\mathbf{x}^+$	Consecutive state vector
$\mathbf{x}_0$	Initial state vector
$\mathbf{z}$	Nominal state vector
$\mathbf{A}, \mathbf{B}, \mathbf{C}, \mathbf{E}$	State matrices of discrete state space model
$\mathbf{A}_c, \mathbf{B}_c, \mathbf{E}_c$	State matrices of continuous state space model
$\mathbf{P}$	Solution of Ricatti equation
$\mathbf{P}_f$	Terminal weight
$\mathbf{Q}$	Weight matrix of state vector in cost function
$\mathbf{R}$	Weight matrix of input vector in cost function
$\mathbb{C}$	Collision avoidance set
$\mathbb{C}_q$	Linearized collision avoidance set
$\mathbb{G}_B$	Terminal collision avoidance set of a braking maneuver
$\mathbb{G}_E$	Terminal collision avoidance set of an evasive maneuver

$\mathbb{L}$	Boundary constraint set
$\mathbb{S}$	Stability envelope set
$\mathbb{S}_{\mathrm{i}}$	Set of additive uncertainties for the predicted state *i*
$\mathbb{U}$	Input constraint
$\mathbb{V}$	Nominal input constraint
$\mathbb{V}$	Set of possible state vectors
$\mathbb{V}_{\mathrm{q}}$	Linearized set of possible state vectors
$\mathbb{W}$	Set of possible additive uncertainty vectors
$\mathbb{X}$	State constraint
$\mathbb{X}_{\mathrm{f}}$	Terminal constraint
$\mathbb{X}_{\mathrm{ol}}, \mathbb{X}_{\mathrm{or}}$	Longitudinal range of left and right obstacle
$\mathbb{Z}$	Nominal state constraint
Θ	Set of vectors for parametric uncertainties

where $_{(\mathrm{x}\,/\,\mathrm{y}\,/\,\mathrm{z})}$ denotes the vehicle dynamics variable in $_{\mathrm{x}}$ = longitudinal direction/ $_{\mathrm{y}}$ = lateral direction/ $_{\mathrm{z}}$ = vertical direction; $_{(\mathrm{f}\,/\,\mathrm{r})}$ denotes the $_{\mathrm{f}}$ = front axis/ $_{\mathrm{r}}$ = rear axis and $_{(\mathrm{l}\,/\,\mathrm{r})}$ denotes the $_{\mathrm{l}}$ = left direction/ $_{\mathrm{r}}$ = right direction. Furthermore, $_{\mathrm{traj}}$ = trajectory and $_{0}$ = current value at control time.

1 Introduction

1.1 Motivation

The introduction of driver assistance systems in the 1990's has greatly contributed to increased safety on the road [31][12]. Anti-lock Braking Systems (ABS) and Electronic Stability Control (ESC) assist drivers in controlling their vehicle on slippery roads by stabilzing the vehicle dynamics [75]. According to the natural driver studies in [17] Advanced Driver Assistance Systems (ADAS) have the potential to reduce the impact of negative driver-related factors in pre-crash situations. To further decrease the number of fatalities, car manufacturers have further boosted the development of ADAS technology to help the driver handling difficult situations.
The number of ADAS features has increased tremendously in the last years. In order to classify the systems and get a deeper understanding of the similarities as well as differences the Society of Automotive Engineers (SAE) has clustered the levels of driving automation as documented in the international standard J3016 [64]. The levels of automation provided by the SAE take four main criteria into account. A partially automated system in level 2 intervenes actively into the vehicle dynamics driving task. With level 3, the system being capable to monitor the surroundings of the vehicle takes over the driving task temporarily. The driver is responsible to take back the driving task in a short time frame when adverted by the system. A level 4 highly automated system provides a safe fallback solution during the driving task if needed e.g. by slowing down and swerving to the emergency lane. At level 5 the system is fully automated and has the capability to handle all driving modes. This differentiation of

automation levels provides an orientation for the classification of ADAS systems, but does not contain the complete range of ADAS features [26].
Gasser provides an extension which is based on the classification of the functionalities into three main operation types [24][23][25] . The operation type A consists of informing and warning functions taking only indirect influence on the vehicle control via the driver [25]. Examples of this type are traffic sign recognition or lane departure warning systems. Operation type B comprises continuously automating functions as described in the SAE standard J3016 like adaptive cruise control and traffic-jam-assist systems. Operation type C stands for intervening emergency functions, which take immediate control over the vehicle in near-accident situations that cannot be controlled by the driver. One example of this operation type is the emergency braking system, which can intervene independently of the driver intervention after detection of a critical situation by a sensing system. Recent studies [30][32][34] have shown the effectiveness of emergency braking systems with various system implementations. The study conducted by the Insurance Institute for Highway Safety in [32] showed a significant reduction of accident rates, expressed by lower claim frequency in all crash types of up to 51 % in personal injury claims and 22 % in first party claims.

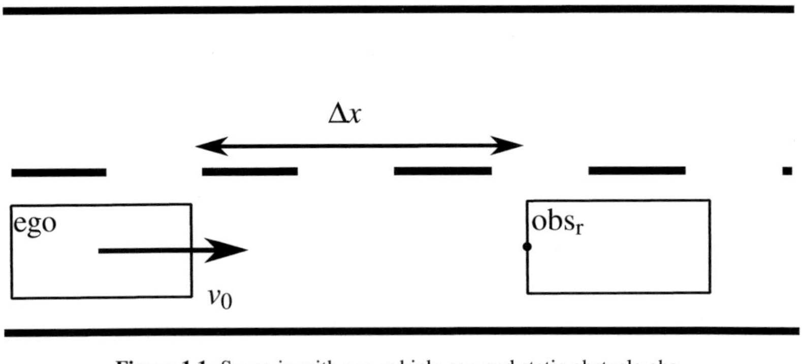

Figure 1.1: Scenario with ego vehicle ego and static obstacle obs_r

To further expand the effectiveness of collision avoidance systems recent studies [53] have investigated the exploitation of the lateral free space by evasive maneuvers.

Figure 1.1 illustrates a scenario with an ego vehicle and a static obstacle obs_r on the right lane. The ego vehicle drives with an initial velocity of v_0 and the rear edge of the static obstacle is located at the distance Δx in front of it. Different strategies of the ego vehicle shall be compared to avoid a collision with the obstacle.

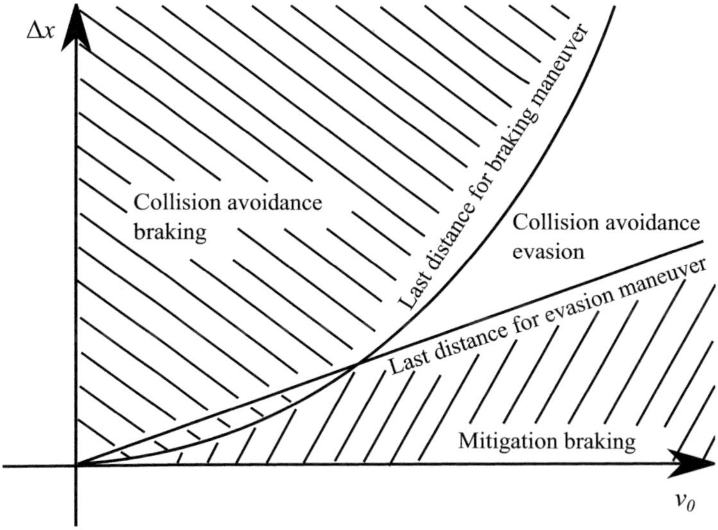

Figure 1.2: Areas of braking, evasion and mitigation braking maneuver in the distance and velocity diagram

The last-distance-to-brake and last-distance-to-evade curves are depicted in Figure 1.2 in the distance-velocity diagram. Areas above the last-distance-to-brake curve represent scenarios where a braking maneuver can avoid the collision, whereas in scenarios above the last-distance-to-evade curve collision-free evasion maneuvers are possible. Scenarios below both curves are maneuvers where a collision cannot be avoided and mitigation braking must be applied to reduce the collision energy. The figure illustrates that evasive maneuvers are more effective at higher velocity in avoiding the collision in this single obstacle scenario. To assist the driver in critical situations only, collision avoidance systems generally are designed to interact at the latest point in time at which the collision can be avoided by the system.

1.2 Aim of the Thesis

Evasive collision avoidance maneuvers have been introduced in Section 1.1 as a research field providing opportunities for further increasing the positive effect of collision avoidance systems. One of the main challenges of evasive collision avoidance maneuvers is the design of the trajectory planning and the vehicle dynamics control module for this task. The trajectory planning module plans the trajectory to avoid collision with potential obstacles on the road and the vehicle dynamics control module controls the vehicle according to this trajectory. This thesis investigates methods for trajectory planning as well as vehicle dynamics control for collision avoidance systems in urban environments. Key requirements need to be considered to select appropriate concepts and to design the planning and control modules as set down in the following paragraphs.

Guarantee of Collision Freeness

The main goal of collision avoidance systems is to avoid any collision with potential obstacles. Therefore the first requirement for a planning module is to plan collision-free trajectories and ensure safety of the vehicle with a specified safety margin.

Consideration of environmental information

Furthermore, the planning module should be able to take environmental information into account and consider the drivable area given by the road boundaries or the curbstones.

Representation of vehicle dynamics

Design of methods for safety critical maneuvers require a sufficient representation of the required vehicle dynamics for the given task. Planning of maneuvers with an adequate representation of the dynamics can limit the deviation between the planned maneuver and the driven maneuver.

Consideration of Actuator Limitations

Actuators cannot intervene with infinite power. Actuator limitations in braking and steering module constrain the applicable torque and deceleration values due to the design of the electrical and mechanical components.

Computational Complexity

Methods for planning and control of vehicle dynamics maneuvers need to be computationally efficient and realizable by algorithms that can be solved on a real time platform. Sophisticated algorithms that cannot be realized in vehicles cannot contribute to the safety of vehicles on the road. Due to the increase of computational capabilities from year to year the investigation of methods with a high computational burden are still interesting since chances are high that these algorithms can be applied in the near future.

Customization of Performance

Methods for planning and control of vehicle dynamics maneuvers should be able to meet customizable goals like customer experience (e.g. level and type of intervention) and to consider strategies for driver handover. These goals are different for each maneuver and must be defined appropriately to consider the characteristic of the maneuver.

Minimization of Tuning Effort

Customization of maneuver characteristics is usually associated with high tuning efforts. Tuning efforts hinder the methods from being applied to the vehicle

within a short development time. An ideal method should minimize the amount of tuning variables and the tuning effort necessary in the algorithm.
With the main requirements defined above the question adressed in this thesis is formulated as follows.

How can trajectory planning and vehicle dynamics control strategies be designed to

- guarantee collision-freeness with potential obstacles,
- consider complex environmental information,
- represent vehicle dynamics characteristics,
- incorporate actuator limitations,
- enable real time application,
- allow for customization and to reduce tuning effort?

In order to execute successful evasive collision avoidance maneuvers several key challenges need to be overcome. Precise perception of all relevant information in the environment is essential for planning and decision making in critical scenarios. The performance of the perception system is based on the neat cooperation of key technologies such as hardware design of the sensors and algorithmic processing modules such as sensor fusion, object tracking and object classification. In this thesis the environment perception system is assumed to be given and to reflect all relevant information. Design of driver assistance systems need to take the interaction and behaviour of the driver into account. Good system design enables effective cooperation between drivers and systems during the maneuver. This can be achieved by a suitable design of human machine interfaces and hardware as well as software modules. In this thesis, no driver intervention is assumed as is defined for highly automated vehicles [64].

The precise knowledge of the friction coefficient is fundamental for reliable planning of maneuvers in critical situation. At low friction values, the planned maneuver cannot achieve high acceleration dynamics. Therefore, the estimation of the friction coefficient has been investigated in several research works. In this thesis, the friction coefficient is assumed to correspond to the coefficient when driving on dry roads.

Planning and control in driver assistance systems is based on the interaction of high level and low level control systems. Low level control systems control the actuator dynamics to achieve the command input of a high level control system and to compensate for vehicle dynamics instabilities. High level control systems control the vehicle dynamics to achieve a goal in a specified maneuver. This thesis will focus on the design of the high level planning and control system and assume full functionality of the low level control system without additional interventions of ABS and ESC systems.

1.3 Outline

This thesis will focus on concepts and methods for trajectory planning and vehicle dynamics control in evasive collision avoidance systems and is structured as follows:

Chapter 2: Concepts and Methods for Collision Avoidance

Chapter 2 provides an overview of concepts and methods about trajectory planning and vehicle dynamics control suitable for evasive collision avoidance systems. Moreover this chapter will highlight the difference between the classical concepts with separated planning and control modules as well as the integrated planning and control concept discussed in this thesis.

Chapter 3: Fundamentals of Optimal Planning and Control

In Chapter 3 the needed vehicle dynamics models used for the steering and braking system as well as the vehicle dynamics will be introduced. The theoretical background of Model Predictive Control (MPC) as a finite time constrained optimal control problem as well as robust tube based MPC will be explained. Issues such as stability, robustness and constraint satisfaction are discussed. The applicability of the MPC theory will be shown on the classical trajectory tracking problem in a simulation environment.

Chapter 4: Nonlinear Integrated Planning and Control

A nonlinear model predictive control scheme integrating the vehicle dynamics planning and control together with combined longitudinal and lateral dynamics for collision avoidance maneuvers will be developed in Chapter 4. A nonlinear formulation of the optimization problem with nonlinear constraints for the vehicle dynamic model, the environmental constraints as well as vehicle dynamics limitations will be presented. In addition the collision avoidance system uses the terminal collision avoidance constraint to find a safe trajectory with obstacles outside of the prediction horizon. Characteristic scenarios with single and multiple obstacles will be used to demonstrate the performance of the method in simulation.

Chapter 5: Fast Integrated Planning and Control

The implementation of the nonlinear control problem on a real vehicle in a real time environment is critical due to the high computation time required by the solver. To overcome this issue successive linearization techniques will be presented in Chapter 5 to achieve convex quadratic problems that can be solved using standard optimization solvers. Robust MPC theory is employed to cope with linearization errors and to guarantee stability as well as constraint satisfaction with uncertainties. In addition, simulation and experimental results shall demonstrate the applicability of the algorithm in predefined scenarios.

Chapter 6: Conclusion and Outlook

Chapter 6 concludes this thesis with a summary and gives an outlook for future work.

2 Concepts and Methods for Collision Avoidance

Chapter 2 will first discuss the structure and the components of the classical collision avoidance concept. Furthermore, state of the art methods for the investigated modules, trajectory planning and vehicle dynamics control, will be presented. Subsequently, this section will discuss methods of the integrated planning and control concept in contrast to the classical concept. Finally, the concept and method investigated in this thesis will be motivated, considering the requirements and the key research question formulated in Section 1.2.

2.1 Classical Concept of Collision Avoidance Systems

Figure 2.1 illustrates the classical concept for realizing a collision avoidance system. The environment representation module processes the information of the sensor system and generates an appropriate representation, for example in the form of grids [33][73] or stixels [59][60]. A decision making module then determines the proper action in case of imminent collision risks and triggers the maneuver at the latest point in time unless the driver started to resolve the critical situation. The set of actions may consist of braking, evasion and mitigation maneuvers as discussed in Section 1.1. When the collision can be avoided by an evasive maneuver, the trajectory planning module calculates a drivable collision-free trajectory, considering a safety margin to the obstacles. Further, the vehicle dynamics control module activates the actuators to follow this trajectory. Finally, a handover strategy ensures safe transition to the driver.

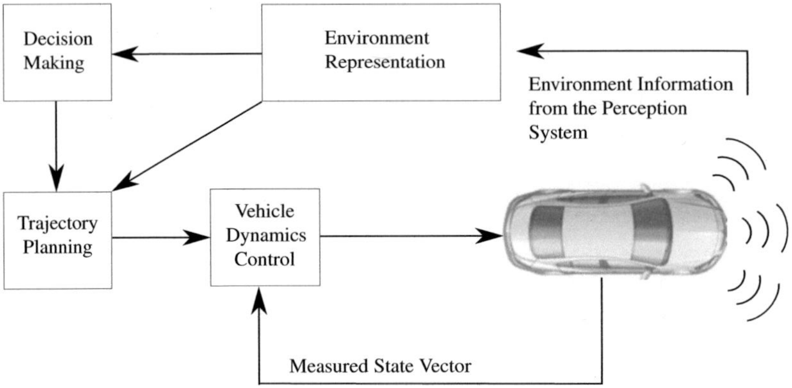

Figure 2.1: Classical concept of collision avoidance systems

The trajectory planning and the vehicle dynamics control module are separated in this classical concept. This implies that the vehicle dynamics control module tracks the trajectory calculated by the planning module and does not consider obstacles explicitly. Furthermore, both modules need to be tuned which requires a lot of tuning efforts.

2.2 Trajectory Planning

Several planning methods have been proposed for evasive maneuvers. Geometrical primitives have been used in many works such as [29][53] [67][68] [71] to calculate trajectories with low computational effort. In the following, properties of geometrical primitives are discussed based on the example of circular shaped and sigmoidal trajectories.

Circular shaped trajectories have been investigated in the context of collision avoidance [29][53][67]. The corresponding equations are composed two circular segments with continuous position and orientation. The discontinuous curvature of the circular shaped trajectory cannot be followed exactly considering actuator limitations of the vehicle.

The build-up time of the dynamics plays an essential role in critical maneuvers. Stählin and Schorn show how the sigmoidal equation can be adapted to consider build-up time of vehicle and different vehicle speeds [68][71]. Planning with geometrical primitives are suitable to handle simple collision avoidance maneuvers or overtaking maneuvers in static scenarios with single obstacles. However, their design does not offer the necessary flexibility to appropriately react under arbitrary environmental conditions with multiple static and dynamic road participants, as discussed in Section 1.2.

Sampling based approaches such as rapidly exploring random trees (RRT) have been proposed by Kuwata [43] and Farinella [21]. These methods sample a map randomly and connect these samples in a tree when their sequential connection is collision-free. The method is able to consider non holonomic constraints such as a vehicle dynamics model and additional environmental information. Although collision-free trajectories can be found with sampling based methods, the solution does not take performance metrics into account. Consequently, the resulting performance metric, in general, is random which does not fulfil the requirement of customization.

Potential field approaches have been investigated by Krogh [42] and Khatib [39]. The basic idea is that environmental obstacles as well as road boundaries shape a gradient field which repel the trajectory of the vehicle. Ideally the vehicle should follow the direction of the negative gradient to prevent collision with other obstacles and keep the vehicle on the center of the lane. Problems arise when overlapping influences of environmental information result in low gradients and local minima such that the potential fields may lead to a collision with an obstacle. Complex environmental information provide qualitative contribution in the potential field approach, however the quantitative contribution of the potential field is difficult to design, in general, to guarantee collision-freeness.

Optimization based methods build on a general framework in which relevant components such as environmental information and vehicle dynamics characteristics as well as desired driving properties can properly be formulated. The components of the framework can be featured with elements of different complexities that is suitable for the requirements of the considered maneuver. Optimal trajectory planning methods have been proposed by many researchers in different setups [78][79][86] which vary in component complexity as well as in their computational effort.

Simple setups of optimization based methods which rely on a point mass model or kinematic model with a given start and end position can be solved analytically and in real time without the need of complex solver algorithms [78]. However, the applicability is limited since environmental information such as obstacle position or geometry cannot be considered with fast analytic solutions.

In complex scenarios like multiple obstacles and arbitrary road shapes analytical solutions cannot be given explicitly. A solution can be provided by efficient optimization solvers which are available for mathematically standardized formulations. Ziegler [86] designs a quadratic program with a linear model and convex sets for environmental representations. The shown test results verify the performance for autonomous driving as demonstrated in the Bertha Benz drive [87]. However, the applicability of this research work is restricted to simple vehicle models and low environmental complexities.

Nonlinear optimal problems for trajectory planning provide the opportunity to consider complex vehicle dynamics characteristics such as coupling effects between lateral and longitudinal dynamics [28]. Yet no efficient solvers are available for general nonlinear optimization problems which prevents the algorithms from being applied in real time on real vehicles. The trajectory planning methods discussed are evaluated with respect to the requirements formulated according to Section 1.2 in Table 2.1 considering their suitability in collision avoidance tasks.

Table 2.1: Methods of trajectory planning for collision avoidance maneuvers

Planning method	Geometrical	Sampling based	Linear Optimal	Nonlinear Optimal	Potential field
Guaranteed Avoidance	+	+	+	+	-
Environmental Information	-	+	+	+	+
Vehicle Dynamics	-	+	o	+	o
Actuator Limits	-	+	+	+	o
Computational Complexity	+	+	o	-	o
Customization of Performance	-	-	+	+	o

2.3 Vehicle Dynamics Control

Vehicle dynamics control has also been addressed in previous publications. In early investigations by Ackermann [1] robust control methods have been proposed to control a bus on a given path. The method has been demonstrated in a simulation environment. The strength of this approach is the robustness to parameter deviations such as mass changes. A characteristic of robust control methods is its tradeoff betweeen robustness and performance. In [84] the authors have investigated robust control for a collision avoidance application in comparison with other control approaches. It can provide theoretical guarantees for robustness and requires less tuning effort, but it also shows a significant lack of performance for collision avoidance applications.

Linear controllers with feed forward components have been suggested in [37] [41][68]. The feedforward component is calculated based on the sigmoidal design of the trajectory and a simple kinematic single track model. To consider different parameter setups in different speed ranges, Schorn [68] suggests to pair the controller with a gain scheduling approach. Performance in different speed ranges can be achieved by proper tuning in each speed range separately. The controller has been demonstrated in the vehicle for trajectories up to 6 m/s^2.

Longitudinal dynamics control is achieved by calculating the remaining friction capability. Given a suitable trajectory, this approach provides an effective way for collision avoidance control. The tuning effort of linear controllers required for complex systems is enormous as explained in [2][70][76]. In addition actuator limits cannot be considered and the vehicle dynamics characteristics is used only by the feedforward component but not by the feedback component.
Input-output linearization has been considered by Werling [77] and Kranz [40]. The advantage of these approaches is that stability and performance of the control system can be easily analyzed by applying methods from linear control theory on nonlinear systems. Yet, these control methods are not suitable for maneuvers at the limits of the actuators, when executing evasive maneuvers as is required in this thesis.
Katrionik [35] proposes a MPC scheme for combined vehicle dynamics control of the lateral and longitudinal dynamics. The approach considers a single track model with a nonlinear tire model as well as a PT1 characteristic for the longitudinal dynamics. To guarantee vehicle dynamics stability, the maximum tire slip angles are constrained. The performance has been demonstrated in a simulation environment and in a real vehicle. The focus of this thesis is to provide a method for vehicle state estimation and sensor calibration as well as vehicle dynamics control with low cost sensors. Model predictive control is a promising approach for collision avoidance maneuvers due to the explicit consideration of actuator and state limitations. It can handle a rather complex dynamics model and adapt to model errors by repetitive planning of future interventions. An advantage of control algorithms that consider complex models is a reduced tuning effort as the model information can be exploited.
All works mentioned in this section assume a drivable and safe trajectory to be given such that the algorithm achieves good performance by following it. The control methods presented in this section are evaluated in Table 2.2 based on the predefined requirements in Section 1.2.

Table 2.2: Methods of vehicle dynamics control for collision avoidance maneuvers

Control method	Robust Control	Linear Control	Input-Output Linearization	Model Predictive Control
Vehicle Dynamics	o	-	o	+
Actuator Limits	o	-	-	+
Computational Complexity	o	+	+	-
Customization of Performance	-	+	o	+
Tuning Effort	+	-	o	+

2.4 Integrated Planning and Control

State of the art work on trajectory planning and vehicle dynamics control has been presented in previous sections. Due to the separation of both modules high tuning efforts are required to adapt both modules to each other and to cover the functionality in all relevant scenarios, as the planning method generally does not take the control properties into account. On top of that, the missing interaction between the trajectory planning and the control modules implies that the overall system, without an explicit replanning module, is most suitable for static scenarios, but not appropriate for dynamically changing scenarios. By integrating both modules, the interaction between both modules can be realized with reduced tuning effort. Further, the integrated concept allows to handle scenarios with new obstacles entering the environment during the maneuver by adapting the trajectory to updated environmental information in each planning step.

This section reviews collision avoidance methods that integrate the trajectory planning and the vehicle dynamics control module into one single module as shown in Figure 2.2 in contrast to the classical concept shown in Figure 2.1.

Keller and Bauer proposed methods for integrated planning and control combining potential fields and optimal control. Keller [38] has chosen an algorithm with rough discretization for the steering wheel angle and assumed the angle to stay constant over the complete planning to realize the algorithm in real time. In contrast, Bauer [4] calculated a command sequence for the maneuver based on optimal potential field approach. As described in Section 2.3, potential field approaches cannot guarantee collision-freeness due to local minima and low gradients.

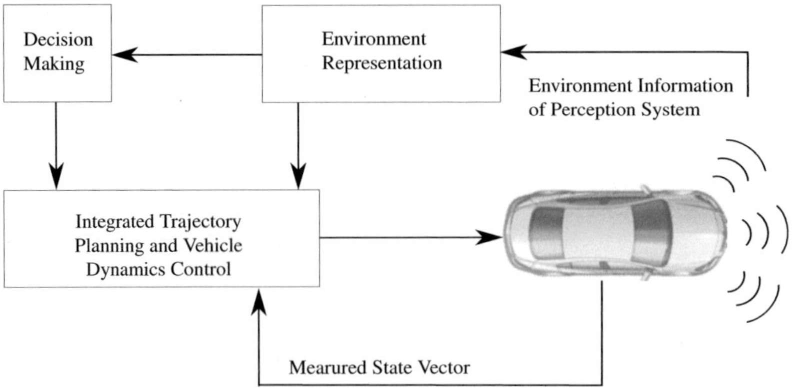

Figure 2.2: Integrated planning and control concept of collision avoidance systems

Erlien [19] investigated an integrated planning and control method with optimal control where inequality constraints are used to avoid collisions with obstacles within the prediction horizon. Similarly to [4] the investigated MPC takes all environmental information into account and further can guarantee collision-free trajectories with obstacles within the horizon. However the method does not consider braking intervention in the optimal control problem and cannot take into account obstacles outside the prediction horizon.
Table 2.3 shows an evaluation of the discussed approaches based on requirements presented in Section 1.2. All methods share the MPC's characteristic of low tuning effort, explicit consideration of actuator limits and vehicle dynamics model information. Further, the mentioned methods plan the maneuver separately for lateral and longitudinal dynamics, which shows deficiencies compared to planning of combined dynamics for the maneuver [67].

Differences arise in handling of obstacles, as potential field based approach cannot guarantee collision-freeness, whereas the work in [19] can provide a guarantee by using hard inequality constraints. However collision avoidance on obstacles that are outside the prediction horizon have not been considered. The method in [38] excells in terms of simplicity such that the computation time is negligible, but the control horizon is not representative for complex scenarios.

2.5 Selection of Methods and Concepts

Methods for trajectory planning, vehicle dynamics control as well as for the integrated planning and control concept have been presented in previous sections. A method that integrates both modules can plan a trajectory in each consecutive step and adapt to new environmental information instead of tracking a fixed trajectory as explained in Section 2.4. Further integration of planning and control functionality simplifies the interaction between both modules, as less parameters need to be tuned and the same vehicle dynamics model is shared. Due to these reasons the integrated planning and control concept will be further exploited in this thesis.

MPC as an optimality based method (Section 2.2 and 2.3) is a method that gives a good base to fulfill all requirements as defined in Section 1.2. In particular nonlinear optimal methods offer a high potential to meet the challenge of evasive collision avoidance systems. The main weakness of this approach is the computational burden required to solve the resulting optimization problem. The methods presented in Section 2.4 are all based on MPC. Among these methods, the method proposed by Erlien[19] can achieve collision avoidance due to inequality constraints, whereas the potential field based approaches cannot guarantee collision-freeness, as discussed in Section 2.2.

Therefore this thesis will investigate the potential of optimality based methods with collision avoidance constraints to solve the research question formulated in Section 1.2. Furthermore, this thesis shall take into account braking interventions in the optimization problem and investigate solutions to avoid collision with obstacles outside of the prediction horizon.

Table 2.3: Methods of integrated vehicle dynamics planning and control for collision avoidance maneuvers

Integrated method	Keller [38]	Bauer [3]	Erlien [19]
Guaranteed Avoidance	-	-	o
Environmental Information	+	+	+
Vehicle Dynamics	o	o	o
Actuator Limits	-	+	+
Computational Complexity	+	o	o
Customization of Performance	o	+	+
Tuning Effort	+	+	+

3 Fundamentals of Optimal Planning and Control

This chapter explains the theoretical background on vehicle dynamics modeling and model predictive control that is required for integrated vehicle dynamics planning and control in the Chapters 4 and 5. Furthermore this chapter will introduce fundamental terminologies, the general optimal framework for trajectory planning and vehicle dynamics control as well as the notation used in this thesis.

3.1 Vehicle Dynamics Modeling

The vehicle dynamics model used in this thesis is composed of the lateral and the longitudinal dynamics as illustrated in Figure 3.1. The lateral dynamics model consists further of the lateral vehicle model, the steering dynamics model and the tire model. The model is complemented by actuator and state limitations.

3.1.1 Lateral Dynamics Model

Lateral Vehicle Model

The vehicle model provides information about the dependancy between basic vehicle dynamics states such as the yaw rate $\dot{\varphi}$, the velocity v, the steering wheel angle δ and the side slip angle β. In this thesis, the single track model by Riekert and Schunck [63] is taken as a starting point and extended in subsequent sections.

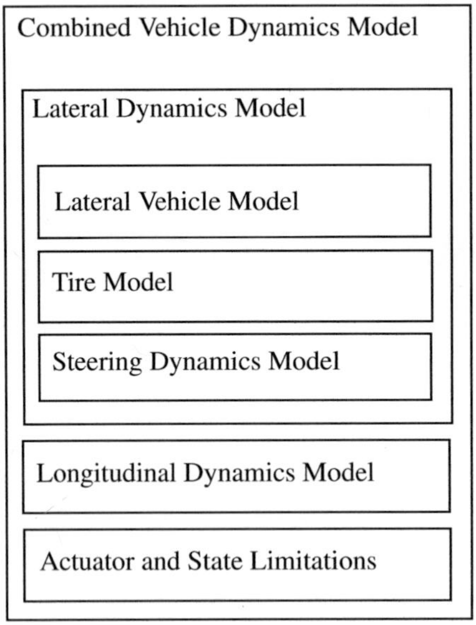

Figure 3.1: Components of the Combined Vehicle Dynamics Model

In Figure 3.2 the single-track model and the real vehicle used in this thesis are illustrated. The single-track model is a simplified approach to represent the lateral vehicle dynamics, which (in its original version) is described by two states corresponding to the yaw rate and the side slip angle. The single-track model models the axes of a vehicle with a single front tire and a single rear tire. The mass m is located at the center of gravity (COG), which is positioned at the distance l_{f} from the front axis and the distance l_{r} from the rear axis. The COG moves with the speed v and the side slip angle β to the longitudinal orientation of the vehicle. The vehicle rotates with the inertia J_z and yaw rate $\dot{\varphi}$ around the vertical axis z. The front and rear edge of the vehicle chassis, which need to be taken into account in collision avoidance maneuvers, are located at the longitudinal distances $l_{\mathrm{f,veh}}$ and $l_{\mathrm{r,veh}}$ from the COG. δ corresponds to the steering wheel angle, i_{L} is the steering ratio and α_{f} as well as α_{r} are the slip angles of the front and rear wheel. The slip angle is the angle between the tire orientation and the moving direction of the vehicle [49]. Using the simplification for small angles, the angles can be calculated by

$$\alpha_{\mathrm{f}} = \frac{\delta}{i_{\mathrm{L}}} - \beta - \frac{l_{\mathrm{f}}\dot{\varphi}}{v} \tag{3.1}$$

$$\alpha_{\mathrm{r}} = -\beta + \frac{l_{\mathrm{r}}\dot{\varphi}}{v} \; . \tag{3.2}$$

The lateral and longitudinal tire forces of the front and rear tires are F_{yf}, F_{xf}, F_{yr} and F_{xr} respectively. Setting up the equilibrium of forces for the dynamics of the lateral movement as well as the equilibrium of moments for the yaw dynamics results in

$$m\frac{v^2}{R}cos\beta - F_{\mathrm{yr}} - F_{\mathrm{xf}}sin\delta - F_{\mathrm{yf}}cos\delta = 0 \tag{3.3}$$

$$J_{\mathrm{z}}\ddot{\varphi} - (F_{\mathrm{yf}}cos\delta + F_{\mathrm{xr}}sin\delta)l_{\mathrm{f}} + F_{\mathrm{yr}}l_{\mathrm{r}} = 0 \; . \tag{3.4}$$

In addition, the states of the single track model are extended by the lateral position with the dynamics

$$\dot{y} = v\sin(\varphi + \beta) \; . \tag{3.5}$$

By applying the small angle simplifications and neglecting the effect of longitudinal forces, the nonlinear continuous state transition for the lateral dynamics can be given by

$$\dot{\mathbf{x}}_{\mathrm{lat}} = \mathbf{f}_{\mathrm{lat}}(\mathbf{x}_{\mathrm{lat}}, \mathbf{u}_{\mathrm{lat}}) = \begin{bmatrix} \frac{F_{\mathrm{yf}}+F_{\mathrm{yr}}}{mv} - \dot{\varphi} \\ \frac{F_{\mathrm{yf}}l_{\mathrm{f}}-F_{\mathrm{yr}}l_{\mathrm{r}}}{J_{\mathrm{z}}} \\ \dot{\varphi} \\ v(\varphi+\beta) \end{bmatrix} \tag{3.6}$$

with the state vector $\mathbf{x}_{\mathrm{lat}} = [\beta,\ \dot{\varphi},\ \varphi,\ y]^T$ and the system input $\mathbf{u}_{\mathrm{lat}} = \delta$.

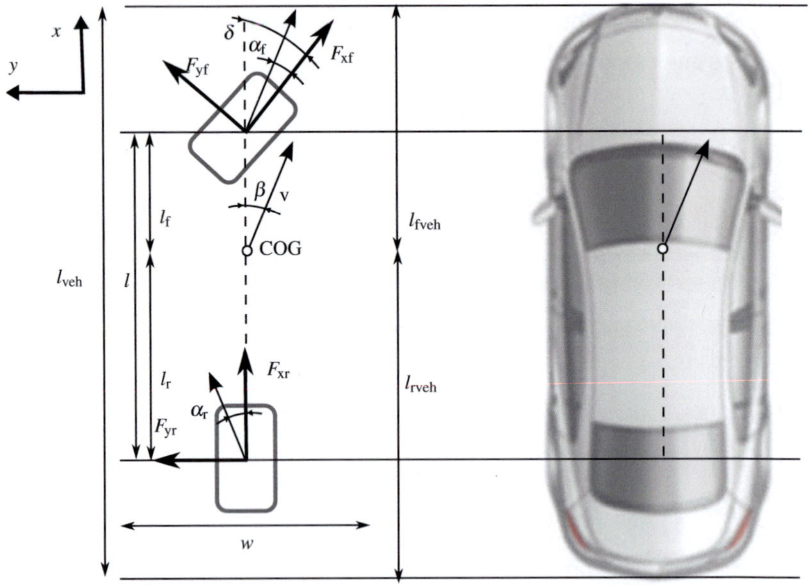

Figure 3.2: Illustration of single-track model and real vehicle

Tire Model

Tires are essential in the execution of vehicle dynamics maneuvers, since they determine the amount of force that can be transmitted to the ground. A basic tire model assumes the tire forces to be linear dependant on the slip angle α with a cornering stiffness c_α given in

$$F_y = c_\alpha \alpha \ . \tag{3.7}$$

In the equation above, the index which reveals the position of the tire is omitted for generalization purpose. Although (3.7) is a good approximation for the tire model at maneuvers with low slip angles, it does not represent the nonlinear tire behaviour of maneuvers with high slip angles[49][56][68]. To improve the model performance, the so called Magic Formula [56] is used to represent the tire behaviour also for high slip angles. This simple tire model has a limited number of parameters and represents the tire forces at different slip angles α and wheel loads F_z. The structure of the tire model is

$$F_y(\alpha) = F_z \mu_y(\alpha) \tag{3.8}$$

$$\mu_y(\alpha) = D \sin\{C \arctan[B\alpha - E(B\alpha - \arctan(B\alpha))]\} \quad . \tag{3.9}$$

In (3.8) and (3.9), parameter B determines the stiffness, parameter C the shape, parameter D the peak and parameter E the curvature of the function and parameter μ_y the lateral friction coefficient [56].
This thesis neglects coupling effects of the longitudinal and the lateral tire forces. A suitable representation for the friction circle will be presented in Section 3.1.4 to compensate for this weakness.

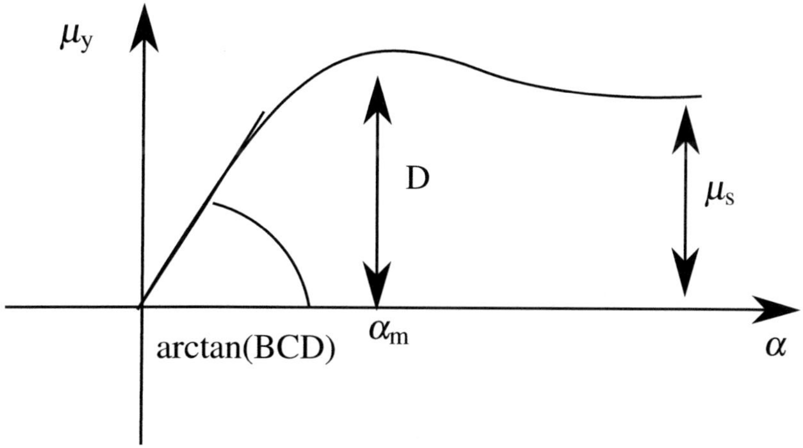

Figure 3.3: Pacejka's tire model [56] for firction coefficient

In Figure 3.3 the function for the friction coefficient is illustrated together with the relevant parameters, where μ_s is the stationary value of the friction coefficient for big slip angles and α_m is the slip angle at the peak of the friction coefficient. The tire parameters in (3.9) can be calculated according to [56] by

$$D = \mu_y(\alpha_m) \tag{3.10}$$

$$C = 1 + \left(1 - \frac{2}{\pi} \arcsin \frac{\mu_s}{D}\right) \tag{3.11}$$

$$B = [\frac{d\mu_y}{d\alpha}|_{\alpha=0}]/CD \tag{3.12}$$

$$E = \frac{B\alpha_m - \tan\{\pi/(2C)\}}{B\alpha_m - \arctan\{B\alpha_m\}} \; . \tag{3.13}$$

The wheel load F_z in equation (3.8) at the front and rear wheel is approximated according to [68] by

$$F_{zf} = m \cdot \left(\frac{l_r}{l} \cdot g + \frac{h_{cog}}{l} \cdot a_x\right) \tag{3.14}$$

$$F_{zr} = m \cdot \left(\frac{l_f}{l} \cdot g - \frac{h_{cog}}{l} \cdot a_x\right) \tag{3.15}$$

with the height of the COG h_{cog}.

To integrate the tire model into the state space equation in Section 3.1.3, the nonlinear tire equation is linearized according to [15] at the operation point given by the slip angle α_f and α_r. The linearized equation for the front and rear tires is then given by

$$F_{yf} = c_{\alpha f,0}\alpha_f + F_{yf,0} \tag{3.16}$$

$$F_{yr} = c_{\alpha r,0}\alpha_r + F_{yr,0} \tag{3.17}$$

and illustrated in Figure 3.4. In the equation above, the tire model is described by the cornering stiffnesses $c_{\alpha f,0}$ and $c_{\alpha r,0}$ as well as the bias values $F_{yf,0}$ and $F_{yr,0}$ for a slip angle α_0.

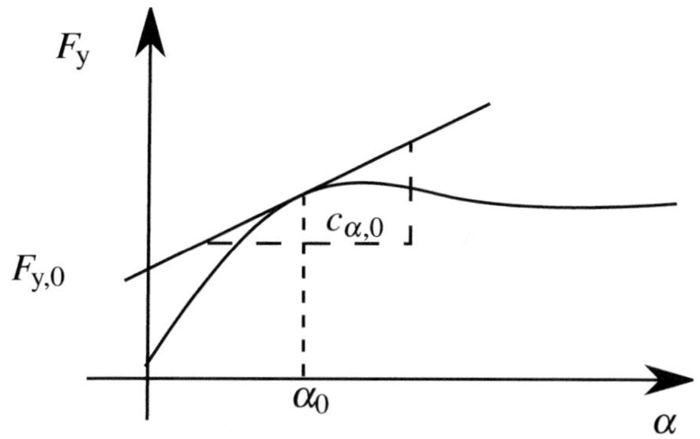

Figure 3.4: Approximation of nonlinear tire todel by a linear equation

Steering model

In order to support the driver during the collision avoidance maneuver, the real vehicles is equipped with an Electrical Power Steering (EPS) system, which can apply a steering torque on the steering system by an electrical motor on the steering rack.

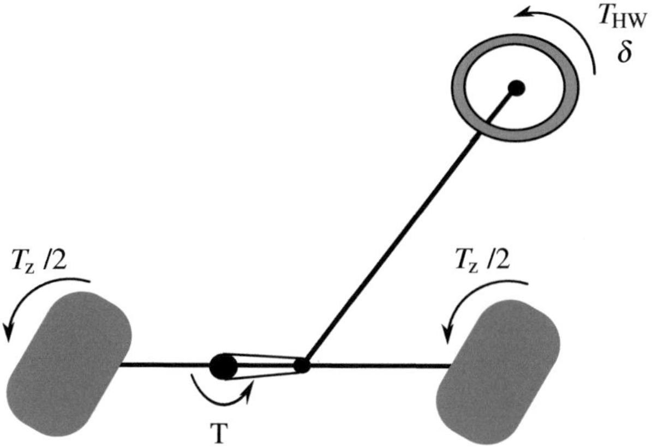

Figure 3.5: Mechanical structure of steering system

Moshchuk [52] provides a simple steering model structure for the mechanical components which will be used in this thesis. As illustrated in Figure 3.5, the electrical motor on the rack amplifies the driver torque T_{HW} by the total assisted torque T and counteracts the self aligning torque of the tires T_{z}. The dynamics equation of the steering system is given by

$$J_{\mathrm{st}}\ddot{\delta}+d_{\mathrm{st}}\dot{\delta}=T+T_{\mathrm{z}}+T_{\mathrm{HW}} \tag{3.18}$$

with the steering inertia J_{st}, the damping parameter d_{st} and the self alignment torque T_{z} which is the torque caused by tire friction on the road. The self alignment torque is modeled linearly by using the front slip angle α_{f} and the self alignment parameter D_{f} with

$$T_{\mathrm{z}}=-2D_{\mathrm{f}}\alpha_{\mathrm{f}}\ . \tag{3.19}$$

In the software of the EPS system, boost curves are implemented to translate the driver torque to an equivalent total assisted torque T on the steering system. The translation ratio is mainly dependant on the driver torque T_{HW} and the vehicle velocity v. In order to command a required total assisted torque T in the real vehicle, the torque T need to be translated into a driver equivalent torque using the boost curve and finally send to the EPS software. As the specific implementation depends on the design of the steering system manufacturer the specific algorithm will not be explained in this section. In the following, this thesis assumes the translation algorithm to be given.
Note that the steering system is modeled based on mechanical components and the boost curves only. The electrical power steering module of a state of the art vehicle is comprised of additional features such as friction compensation, active wheel return and arbitrage modules which may further influence the steering system behaviour during maneuver control. To keep the simplicity of the model, these investigations are not the focus of this thesis. More details may

be retrieved from [83], where the author has investigated the effects of these features for vehicle dynamics control using local linear model trees [54].

3.1.2 Longitudinal Dynamics Model

The electronic brake control module (EBCM) of the real vehicle contains a low level control module, which controls the hydraulic brake pressure in the braking pipe to achieve a desired deceleration command value. Instead of a complex physical model of the brake system, the behaviour of the overall low level control unit is represented by a double integrator with a time delay characteristic. Here, effects of the ABS and the ESC systems are neglected. The vehicle dynamics model for the longitudinal dynamics is given by

$$\dot{\mathbf{x}}_{\mathrm{ln}} = \mathbf{f}_{\mathrm{ln}}(\mathbf{x}_{\mathrm{ln}}, \mathbf{u}_{\mathrm{ln}}) = \begin{bmatrix} v\cos(\varphi+\beta) \\ a_{\mathrm{x}} \\ -\frac{1}{t_{\mathrm{del}}}a_{\mathrm{x}} + \frac{1}{t_{\mathrm{del}}}a_{\mathrm{x,cmd}} \end{bmatrix}. \tag{3.20}$$

Let t_{del} denote the time delay constant and $\mathbf{x}_{\mathrm{ln}} = [x,\ v,\ a_{\mathrm{x}}]^T$ denote the longitudinal state vector of the ego vehicle with the longitudinal position x, the velocity v as well as the longitudinal acceleration a_{x}. The input of the system $\mathbf{u}_{\mathrm{ln}}$, the deceleration command $a_{\mathrm{x,cmd}}$, should be distinguished from the measured acceleration a_{x}, which deviates due to control characteristic of the low level controller.

3.1.3 Combined Vehicle Dynamics Model

This section integrates all presented vehicle dynamics components into one combined vehicle dynamics model. The combined nonlinear vehicle dynamics model is given below. The state vector and the input vector are given by $\mathbf{x} = [\beta,\ \dot{\varphi},\ \varphi,\ y,\ \dot{\delta},\ \delta,\ x,\ v,\ a_{\mathrm{x}}]^T$ and $\mathbf{u} = [T,\ a_{\mathrm{x,cmd}}]^T$.

$$\alpha_f = \frac{\delta}{i_L} - \beta - \frac{l_f \dot{\varphi}}{v}$$

$$\alpha_r = -\beta + \frac{l_r \dot{\varphi}}{v}$$

$$F_{yf}(\alpha_f) = F_{zf} D \sin\{C \arctan[B\alpha_f - E(B\alpha_f - \arctan(B\alpha_f))]\}$$

$$F_{yr}(\alpha_r) = F_{zr} D \sin\{C \arctan[B\alpha_r - E(B\alpha_r - \arctan(B\alpha_r))]\}$$

$$F_{zf} = m \cdot \left(\frac{l_r}{l} \cdot g + \frac{h_{cog}}{l} \cdot a_x \right)$$

$$F_{zr} = m \cdot \left(\frac{l_f}{l} \cdot g - \frac{h_{cog}}{l} \cdot a_x \right)$$

$$T_z = -2 D_f \alpha_f$$

$$\dot{\mathbf{x}} = \mathbf{f}(\mathbf{x}, \mathbf{u}) = \begin{bmatrix} \frac{F_{yf} + F_{yr}}{mv} - \dot{\varphi} \\ \frac{F_{yf} l_f - F_{yr} l_r}{J_z} \\ \dot{\varphi} \\ v \sin(\varphi + \beta) \\ \frac{T + T_z + T_{HW} - d_{st} \dot{\delta}}{J_{st}} \\ \dot{\delta} \\ v \cos(\varphi + \beta) \\ a_x \\ -\frac{1}{t_{del}} a_x + \frac{1}{t_{del}} a_{x,cmd} \end{bmatrix} \tag{3.21}$$

3.1.4 Actuator and State Limitations

The equation (3.21) describes a detailed model for the vehicle dynamics. The validity of this physical model is limited due to actuator and friction limitations. The EPS unit assists the driver by translating the driver torque T_{HW} with a boost curve to the total assisted torque T. In the frame of this thesis, the torque T will be used to perform the evasive collision avoidance maneuver without intervention of the driver. The limitations of the demanded total assisted torque can be simplified to a maximum steering assist torque T_{max} with

$$-T_{\text{max}} \leq T \leq T_{\text{max}} . \tag{3.22}$$

Furthermore, the EBCM unit introduced in Section 3.1.2 controls the brake pressure in the brake pipe to achieve a given deceleration value. Due to limitations and properties of the actuator, the performance of a braking intervention is limited as well. The maximum deceleration value of the EBCM is $a_{\text{x,cmd,max}} = -9.81\ \text{m}/s^2$ in the chosen real vehicle, such that the limit is given by

$$a_{\text{x,cmd,max}} \leq a_{\text{x,cmd}} \leq 0 . \tag{3.23}$$

An acceleration circle represents the combined effect of the friction limitation on each wheel. Figure 3.6 shows the relevant left half of the acceleration circle in the g-g diagram from [5], as only negative acceleration (braking) shall be applied in longitudinal direction during the maneuver. The acceleration circle is given in

$$\sqrt{a_{\text{x}}^2 + a_{\text{y}}^2} \leq a_{\text{max}} . \tag{3.24}$$

The lateral acceleration is not represented in the vehicle dynamics states, but can be calculated according to [49] by

$$a_{\text{y}} = v(\dot{\varphi} + \dot{\beta}) . \tag{3.25}$$

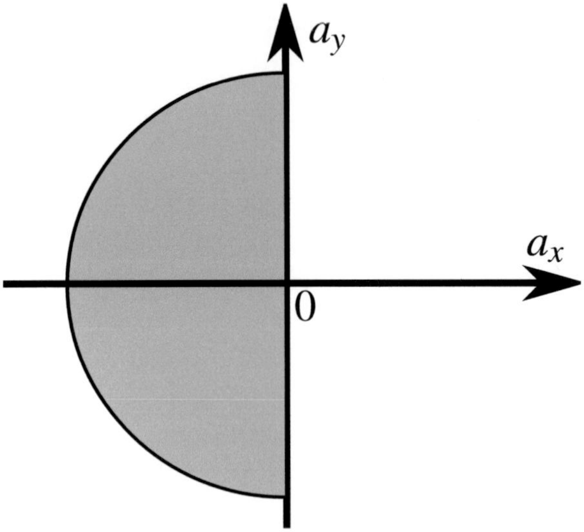

Figure 3.6: Acceleration circle in g-g diagram with shaded area for maneuver dynamics with combined braking and steering [6]

3.1.5 Model Validation

In this section, relevant parameters of the real vehicle, an Opel Insignia, are given based on the vehicle dynamics model presented in previous section. The basic model parameters are listed in Table 3.1.
Subsequently, the model is evaluated using test drives by executing an evasive maneuver in the real vehicle on the proving ground. Modelled signals are generated using the model structure and parameters as presented in this chapter with commanded steering torque T and deceleration $a_{\mathrm{x,cmd}}$ as input.
Figure 3.7 illustrates a comparison of the modelled signals with the measured signals for the positions x and y, the accelerations a_{x} and a_{y}, the steering wheel angle δ, the velocity v, the yaw rate $\dot{\varphi}$ and the yaw angle φ. The maneuver is performed at approximately 50 km/h, a typical urban velocity.
During the maneuver lateral accelerations of up to 6 m/s^2 are achieved, which corresponds to maximum values of evasive collision avoidance maneuvers [68].

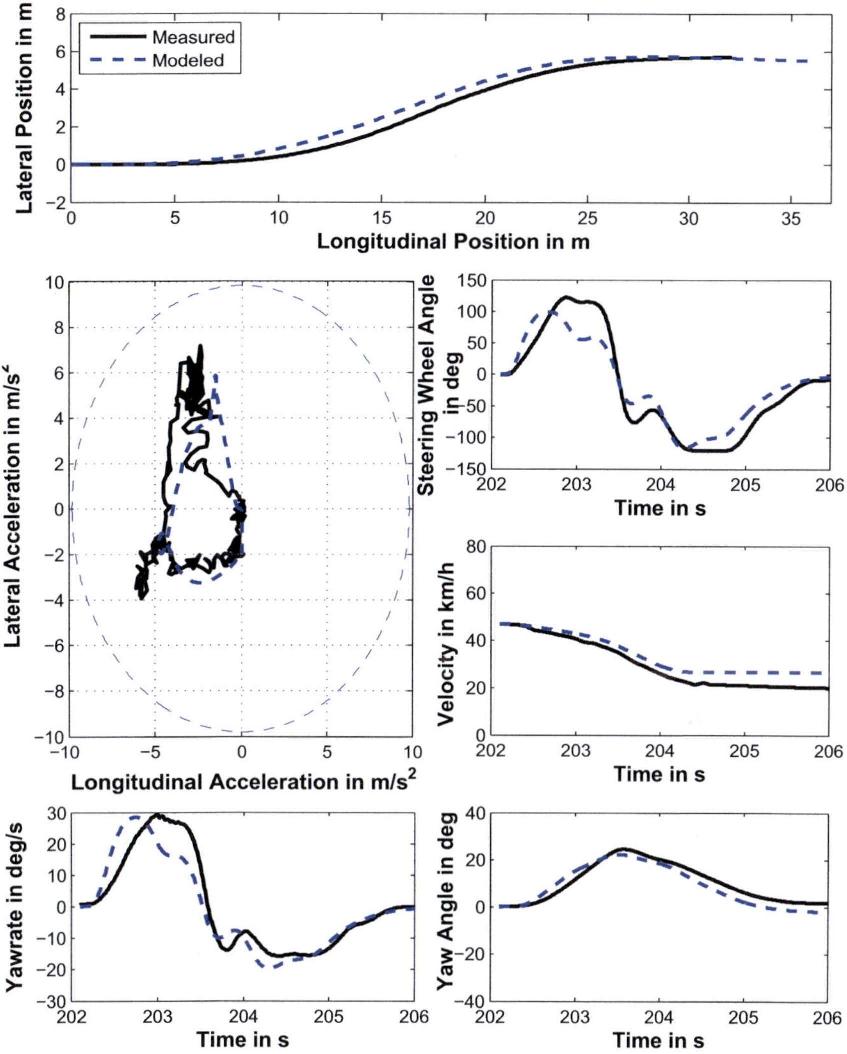

Figure 3.7: Model Validation on a maneuver with 50 km/h and lateral acceleration up to 6 m/s^2

The signal curves in Figure 3.7 show the main curve characteristics, such that the combined vehicle dynamics model can represent the considered maneuvers. Table 3.2 shows the model deviations when commanded signals in a test drive are applied to the vehicle dynamics model in open loop simulation. The model deviations can be attributed to deviations in model structure, parameter uncertainties and environmental changes. Because of these effects, this model can not be used in open loop simulation for localization tasks without feedback information. In a closed loop setup, effect of uncertainties and deviations can be reduced due to replanning.

Table 3.1: Measured and estimated parameters of the vehicle dynamics model in Section 3.1.3

Parameter	Symbol	Value	Unit
Vehicle mass	m	2050	kg
Wheel base	l	2.74	m
Track width	w	1.585	m
Distance to front axis	l_f	1.227	m
Distance to ground	h_cog	0.548	m
Inertia of Vehicle	J_z	3500	kgm^2
Tire parameters	$B/C/D/E$	19.56/0.44/ 2.05/-0.70	1/rad /(none)/ (none)/(none)
Steering inertia	J_st	0.57	kgm^2
Steering ratio	i_L	16	(none)
Steering damping	d_st	2.54	Nms/rad
Steering self alignment torque	D_f	460	Nm/rad

Figure 3.8: Opel Insignia Sports Tourer

Table 3.2: Model deviation for test drive in Figure 3.7 in open loop

Maximum lateral deviation	0.7 m
Maximum longitudinal deviation	7.2 m
Maximum acceleration deviation	1.3 m/s^2

3.2 Model Predictive Control

Optimality based methods like model predictive control (MPC) have gained growing application in a broad range of industrial fields. Since its first successfull application in 1978 by Richalet [62], the method has been applied in various industrial applications like: fluid catalytic cracking, batch reactors, regenerators loadings, pressure controls, temperature control etc. [16][22][45][47][55][65]. Due to increased computation power and faster optimization algorithms in recent years, MPC can now be applied on highly dynamical problems and systems such as adaptive cruise control [50], DC/DC converters[7], traction control [9] or autonomous vehicle steering [51].

The basic principle of MPC is to solve a finite horizon optimal control problem online starting from an initial state of a given discretized plant model [61]. The result of the method is a finite input vector sequence, where the first input vector is applied. The advantage compared to other control methods is its ability to consider input and state constraints as well as model equations over a prediction horizon [10].
This section explains the theoretical background of the nominal and the robust MPC by giving a brief extract of the work in [61].

3.2.1 Nominal Model Predictive Control

The ideal regulating problem that the nominal MPC tries to solve is to find a solution for the following optimization problem:

$$\begin{aligned} \min\ V_\infty &= \sum_{i=0}^{\infty} l(\mathbf{x}(i,k),\mathbf{u}(i,k)) \\ \text{s.t. } \mathbf{x}(i+1,k) &= f(\mathbf{x}(i,k),\mathbf{u}(i,k)),\ \ \mathbf{x}(0,k)=\mathbf{x}_0 \\ \mathbf{u}(i,k) &\in \mathbb{U},\ \ \mathbf{x}(i,k)\in\mathbb{X} \end{aligned} \tag{3.26}$$

The control problem above has been designed in [61] to steer the state to the origin under consideration of the state transition equation $\mathbf{x}(i+1,k) = f(\mathbf{x}(i,k),\mathbf{u}(i,k))$, the initial state $\mathbf{x}_0$, the input constraint set $\mathbb{U}$ and the state constraint set $\mathbb{X}$. In the considered problem, $\mathbf{x}(i,k)$ and $\mathbf{u}(i,k)$ are the discrete state and input vector. Here i is the index for the prediction step and k is the index for the control time step. If the target state is not the origin, modifications [61] can be made to use the generalized optimization problem in (3.26). The input constraint set $\mathbb{U}$ and state constraint set $\mathbb{X}$ are exploited to consider for limitations of actuators and requirements of the control problem. The cost function V_∞ in (3.26) consists of stage costs $l(\mathbf{x},\mathbf{u})$ for predictive time steps of an infinite time horizon. The consideration of an infinite horizon has advantages in terms of closed loop properties compared to the consideration of a

finite horizon [61]. Yet, the considered horizon is infinite and the cost function is not necessarily a convex function of $\mathbf{u}$ which makes the problem hard to solve numerically for generalized settings of optimization problems.
In [61] this infinite time optimal control problem is replaced by two sub-problems, one representing a finite horizon optimal control problem and one describing an infinite horizon optimal control problem [61]. The equations of the resulting quasi-infinite control problem are given by

$$\begin{aligned} \min\ V &= \sum_{i=0}^{N-1} l(\mathbf{x}(i,k),\mathbf{u}(i,k)) + V_{\mathrm{f}}(\mathbf{x}(N)) \\ \text{s.t.}\ \ & \mathbf{x}(i+1,k) = \mathbf{f}(\mathbf{x}(i,k),\mathbf{u}(i,k)) \\ & \mathbf{u}(i,k) \in \mathbb{U},\ \ \mathbf{x}(i,k) \in \mathbb{X},\ \ \forall i = 0,\cdots,N \\ & \mathbf{x}(N,k) \in \mathbb{X}_{\mathrm{f}}\ . \end{aligned} \tag{3.27}$$

with the terminal set $\mathbb{X}_{\mathrm{f}}$ and the terminal penalty V_{f}. The solution calculated with the finite time interval control problem should be equivalent to the infinite time optimal control problem by forcing the terminal state of the finite horizon to enter the terminal set $\mathbb{X}_{\mathrm{f}}$. Within the terminal set, the method should be able to guarantee stability and recursive feasibility with a linear feedback vector $\mathbf{k}$ and the terminal weight $\mathbf{P}_{\mathrm{f}}$, such that

$$V_{\mathrm{f}}(\mathbf{x}) = \mathbf{x}^T \mathbf{P}_{\mathrm{f}} \mathbf{x} \tag{3.28}$$

$$\mathbf{k} = -\mathbf{R}^{-1} \mathbf{B} \mathbf{P}_{\mathrm{f}}\ . \tag{3.29}$$

Here $\mathbf{B}$ is the input matrix of the state space model and $\mathbf{R}$ is the input weight of the control problem.
Rawling [61] shows that an arbitrary optimization problem does not necessarily have a solution. In order to guarantee the existence of a solution according to the extreme value theorem of Weierstrass in [36], the cost function needs to be continuous and the set constraints should be compact. This leads to the following assumptions for the existence of a solution according to [61].

Assumption 1: *(Continuity of system and cost)*
The functions $\mathbf{f}: \mathbb{X} \times \mathbb{U} \to \mathbb{R}^n, l: \mathbb{X} \times \mathbb{U} \to \mathbb{R}_{\geq 0}$ *and* $V_f: \mathbb{X}_f \to \mathbb{R}_{\geq 0}$ *are continuous and satisfy* $\mathbf{f}(\vec{0},\vec{0}) = 0, l(\vec{0},\vec{0}) = 0$ *as well as* $V_f(\vec{0},\vec{0}) = 0$

The notation used should be understood as follows, $\mathbf{f}: \mathbb{X} \times \mathbb{U} \to \mathbb{R}^n$ describes a function **f**, which maps a state vector given in the admissible state constraint set $\mathbb{X}$ and the input vector in the input constraint set $\mathbb{U}$ to a vector of real numbers of dimension n and $\mathbb{R}_{\geq 0}$ is the set of positive real numbers.

Assumption 2: *(Properties of constraint sets)*
The sets $\mathbb{X}$ *and* $\mathbb{X}_f$ *are closed,* $\mathbb{X}_f$ *is a subset of* $\mathbb{X}$ *and* $\mathbb{U}$ *is compact; each set contains the origin.*

Proposition 1: *(Existence of a solution to an optimal control problem)*
Suppose assumptions 1 and 2 hold. Then
a. The function V is continuous in $\mathbb{X}$ *and* $\mathbb{U}$.
b. For each $\mathbf{x} \in \mathbb{X}$ *a solution to the problem in (3.27) exists.*

To ensure asymptotic stability of the MPC controller, the Lyapunov theorem is deployed. This implies that the origin is asymptotically stable with a region of attraction $\mathbb{X}$ for the autonomous system $\mathbf{x}^+ = \mathbf{f}(\mathbf{x})$ if there exists a Lyapunov function V, a positive definite invariant set $\mathbb{X}$, two $\mathcal{K}_\infty$ functions $\gamma_1(\cdot)$ and $\gamma_2(\cdot)$, and a positive definite function $\gamma_3(\cdot)$ satisfying

$$V(\mathbf{x}) \geq \gamma_1(|\mathbf{x}|) \tag{3.30}$$

$$V(\mathbf{x}) \leq \gamma_2(|\mathbf{x}|) \tag{3.31}$$

$$V(f(\mathbf{x})) \leq V(\mathbf{x}) - \gamma_3(|\mathbf{x}|) \tag{3.32}$$

for all $\mathbf{x} \in \mathbb{X}$. The properties of the functions and sets are defined below.

Definition 1: *(Properties of functions and sets)*
a. A function γ *is a* $\mathcal{K}_\infty$ *function, if* $\gamma: \mathbb{R} \to \mathbb{R}_{\geq 0}$ *is continuous, strictly increasing, satisfies* $\gamma(0) = 0$ *and is unbounded.*

b. A function γ is a positive definite function, if it is continuous and positive everywhere except at the origin.
c. A set $\mathbb{X}$ is positive invariant for the system $\mathbf{x}^+ = \mathbf{f}(\mathbf{x})$ if $\mathbf{x} \in \mathbb{X}$ implies $\mathbf{f}(\mathbf{x}) \in \mathbb{X}$.
d. A set $\mathbb{X}$ is a region of attraction for the system $\mathbf{x}^+ = \mathbf{f}(\mathbf{x})$, if the states in the set $\mathbb{X}$ evolve to zero, $|\mathbf{x}(k)|_{\mathbb{X}} \to 0$, as $k \to \inf$.

A standard approach is to consider the value function of the infinite horizon optimal control problem as a Lyapunov function. For the quasi-infinite optimal control problem this implies several design concepts that need to be satisfied to guarantee asymptotic stability according to the Lyapunov theorem.

Assumption 3: *(Basic stability assumptions)*

$$\min_{u \in \mathbb{U}} \{V_f(\mathbf{f}(\mathbf{x},\mathbf{u}) + l(\mathbf{x},\mathbf{u})) | \mathbf{f}(\mathbf{x},\mathbf{u}) \in \mathbb{X}_f\} \leq V_f(\mathbf{x}), \quad \forall \mathbf{x} \in \mathbb{X}_f$$

Assumption 4: *(Implied invariance) assumptions*
The set $\mathbb{X}_f$ is control invariant for the system $\mathbf{x}^+ = \mathbf{f}(\mathbf{x},\mathbf{u})$

Further to guarantee (3.30) and (3.31), bounds need to be considered for the stage cost and the terminal cost.

Assumption 5: *(Bounds on stage and terminal cost)*
The stage cost l and the terminal cost V_f satisfy

$$l(\mathbf{x},\mathbf{u}) \geq \gamma_l(|\mathbf{x}|) \ \forall \mathbf{x} \in \mathscr{X}_N, \forall \mathbf{u} \in \mathbb{U} \tag{3.33}$$

$$V_f(\mathbf{x}) \leq \gamma_2(|\mathbf{x}|) \ \forall \mathbf{x} \in \mathbb{X}_f \tag{3.34}$$

in which γ_l and γ_2 are $\mathscr{K}_\infty$ functions.

It can be derived that (3.33) guarantees (3.30) for the Lyapunov function if $\mathbb{X}_\mathrm{f}$ contains the origin in its interior [61]. The conditions defined above thus guarantee Lyapunov stability [61] summarized in the following proposition. The terminal set $\mathbb{X}_\mathrm{f}$ can be chosen to be the maximal control invariant set, which corresponds to the inclusion of the set $\mathscr{W} \subseteq \{\mathbf{x} \in \mathbb{X} | \mathbf{k}\mathbf{x} \in \mathbb{U}\}$ and the invariant set in which $\mathbf{x} \in \mathbb{X}$ implies $\mathbf{f}(\mathbf{x}) \in \mathbb{X}$.

Proposition 2: *(Optimal value function properties)*
Suppose that Assumptions 1, 2, 3, 4 and 5 are satisfied. Then there exist $\mathscr{K}_\infty$ functions $\gamma_1(\cdot)$ and $\gamma_2(\cdot)$ such that $V(\cdot)$ has the following properties

$$V(\mathbf{x}) \geq \gamma_1(|\mathbf{x}|) \ \forall \mathbf{x} \in \mathbb{X}$$
$$V(\mathbf{x}) \leq \gamma_2(|\mathbf{x}|) \ \forall \mathbf{x} \in \mathbb{X}_f$$
$$V(f(\mathbf{x}, \mathbf{u})) \leq V(\mathbf{x}) - \gamma_3(|\mathbf{x}|) \ \forall \mathbf{x} \in \mathbb{X}$$

Finally the stability of the MPC can be ensured by Theorem 1.

Theorem 1: *(MPC stability)*
Suppose that Assumptions 1, 2, 3, 4 and 5 are satisfied and that $\mathbb{X}_f$ contains the origin in its interior. Then the origin is asymptotically stable with a region of attraction $\mathbb{X}$ for the system $\mathbf{x}^+ = \mathbf{f}(\mathbf{x}, \mathbf{u})$.

For proofs and detailed explanations for this section please refer to [61]. This section presented requirements formulated for linear time invariant (LTI) systems. Consider that the assumptions and propositions formulated above can be equivalently used to design for control of linear time varying (LTV) systems, as is presented in Chapter 5.

3.2.2 Robust Model Predictive Control

The robust design of controllers is important in the presence of uncertainties causing a deviation between predicted behaviour based on the nominal system and actual behavior. Two different types of uncertainties have been adressed in [61]: additive and parametric uncertainties. Parametric uncertainties can directly represent effects of parameters in sophisticated models, such as mass or inertia, whereas additive uncertainties can be easily considered mathematically as a further summand in the equation and can be converted from the parametric uncertainties.

Consider the linear system with parametric uncertainties in the vector θ in

$$\mathbf{x}^{+} = \mathbf{f}(\mathbf{x}, \mathbf{u}, \theta) = \mathbf{A}(\theta)\mathbf{x} + \mathbf{B}(\theta)\mathbf{u} \tag{3.35}$$

where $\theta \in \Theta$ and Θ being the uncertainty set.
This system with parametric uncertainties can be reformulated in a form with additive uncertainties

$$\mathbf{x}^{+} = \mathbf{f}(\mathbf{x}, \mathbf{u}, \theta) = \mathbf{A}\mathbf{x} + \mathbf{B}\mathbf{u} + \mathbf{w} \tag{3.36}$$

$$\mathbf{w} \in \mathbb{W}, \quad \mathbb{W} := \{(\mathbf{A}(\theta) - \mathbf{A})\mathbf{x} + (\mathbf{B}(\theta) - \mathbf{B})\mathbf{u} | \theta \in \Theta, (\mathbf{x}, \mathbf{u}) \in \mathbb{X} \times \mathbb{U}\},$$

where the matrices **A** and **B** are the nominal state space matrices defined at nominal parameters in θ_0.
To treat the additive uncertainties, [48] and [61] proposed tube based MPC as a computationally efficient method for this task. The feedback policy proposed by tube based MPC is

$$\mathbf{u} = \mathbf{v} + \mathbf{k}(\mathbf{x} - \mathbf{z}) \tag{3.37}$$

Here **v** and **z** are the input and state vector of a nominal system given by

$$\mathbf{z}^{+} = \mathbf{A}\mathbf{z} + \mathbf{B}\mathbf{v}, \tag{3.38}$$

k is a linear feedback vector calculated offline to stabilize the system and minimize the deviation between the actual state **x** and the nominal state **z**. The feedback policy of the tube based MPC in (3.37) is composed of two components. The open loop control **v** controls the nominal system in (3.38) whereas the feedback control with linear feedback vector **k** minimizes the deviation between the nominal system states and the actual system states.
With the control policy given, the overall system dynamics is

$$\mathbf{x}^{+} = \mathbf{A}\mathbf{x} + \mathbf{B}\mathbf{v} + \mathbf{B}\mathbf{k}\mathbf{e} + \mathbf{w} \tag{3.39}$$

where $\mathbf{e}$ is the error between the actual state $\mathbf{x}$ and the nominal state $\mathbf{z}$ with

$$\mathbf{e} = \mathbf{x} - \mathbf{z} \tag{3.40}$$

and the error dynamics

$$\mathbf{e}^{+} = \mathbf{A}_{\mathrm{k}}\mathbf{e} + \mathbf{w} \tag{3.41}$$

$$\mathbf{A}_{\mathrm{k}} = \mathbf{A} + \mathbf{B}\mathbf{k}\,. \tag{3.42}$$

The error can be calculated with the closed loop state matrix $\mathbf{A}_{\mathrm{k}}$ using

$$\mathbf{e}(i) = \mathbf{A}_{\mathrm{k}}^{i}\mathbf{e}_0 + \sum_{j=0}^{i-1} \mathbf{A}_{\mathrm{k}}^{j}\mathbf{w}(j) \tag{3.43}$$

with $\mathbf{e}(i) \in \mathbb{S}_{\mathrm{i}}$ where $\mathbb{S}_{\mathrm{i}}$ is composed of the uncertainties and state transition matrices up to the prediction time step i.

$$\mathbb{S}_{\mathrm{i}} = \sum_{j=0}^{i-1} \mathbf{A}_{\mathrm{k}}^{j}\mathbb{W} = \mathbb{W} \oplus \mathbf{A}_{\mathrm{k}}\mathbb{W} \oplus \ldots \oplus \mathbf{A}_{\mathrm{k}}^{i-1}\mathbb{W} \tag{3.44}$$

Control with tube based MPC can be viewed as control of tubes rather than control of trajectories. The feedback policy enables control under uncertainty and can be explained by controlling a tube around the nominal trajectory, in which each trajectory in this tube corresponds to a particular realization of the uncertainty.

To satisfy the constraints $\mathbb{X}$ and $\mathbb{U}$ in the real system, suitable tightened constraint sets $\mathbb{Z}$ and $\mathbb{V}$ need to be considered for the open loop control with the nominal system formulated by

$$
\begin{aligned}
&\mathbf{z}(i,k) \in \mathbb{Z} := \mathbb{X} \ominus \mathbb{S} \\
&\mathbf{v}(i,k) \in \mathbb{V} := \mathbb{U} \ominus \mathbf{k}\mathbb{S} \\
&\mathbf{x}(0,k) \in \{\mathbf{z}(0,k)\} \;\; \mathbf{e}(i,k) \in \mathbb{S}_{\mathrm{i}} \\
&\mathbf{z}(N,k) \in \mathbb{Z}_{\mathrm{f}} \;\; \mathbb{Z}_{\mathrm{f}} \subseteq \mathbb{Z}
\end{aligned}
$$

where the set operations are defined by

Definition 2: *(Set algebra)*
(a) Set addition $\mathbb{A} \oplus \mathbb{B} := \{\mathbf{a}+\mathbf{b} | \mathbf{a} \in \mathbb{A}, b \in \mathbb{B}\}$.
(b) Set substraction $\mathbb{A} \ominus \mathbb{B} := \{\mathbf{x} \in \mathbb{R}^n | \{\mathbf{x}\} \oplus \mathbb{B} \subseteq \mathbb{A}\}$.
(c) Set multiplication: Let $\mathbf{K} \in \mathbb{R}^{m \times n}$. *Then* $\mathbf{K}\mathbb{A} := \{\mathbf{K}\mathbf{a} | \mathbf{a} \in \mathbb{A}\}$.

The tightened constraints guarantee satisfaction of constraints for $\mathbf{x}$ and $\mathbf{u}$ with the feedback control law in presence of uncertainties. Calculation of the set $\mathbb{Z}$ and $\mathbb{V}$ is in general a difficult task. [61] shows how the set can be calculated with minimal computational burden achieving a tradeoff between conservatism and simplicity.
Let the state constraints be given by

$$
\mathbb{X} := \{\mathbf{x} \in \mathbb{R}^n | \mathbf{c}_{\mathrm{j}}\mathbf{x} \leq d_{\mathrm{j}}\}
$$

with the number of the constraint given by the index j then the tightened state constraints corresponding to the nominal states are given by

$$
\mathbb{Z} := \{\mathbf{z} \in \mathbb{R}^n | \mathbf{c}_{\mathrm{j}}\mathbf{z}(i,k) \leq d_{\mathrm{j}} - \max\{\mathbf{c}_{\mathrm{j}}\mathbf{e}(i) | \mathbf{e}(i) \in S_{\mathrm{i}}\}\}
$$

where the maximum error in the set S_i with respect to the vector $\mathbf{c}_\mathrm{j}$ is calculated for each constraint equation separately. Similarly, if the input constraint of the actual input is given by

$$\mathbb{U} := \{\mathbf{u} \in \mathbb{R}^m | \mathbf{a}_\mathrm{j}\mathbf{u} \leq b_\mathrm{j}\}$$

the tightened input constraint can be calculated by

$$\mathbb{V} := \{\mathbf{v} \in \mathbb{R}^m | \mathbf{a}_\mathrm{j}\mathbf{v} \leq b - \max\{\mathbf{a}_\mathrm{j}\mathbf{k} \sum_{i=0}^{N-1} \mathbf{A}_\mathrm{k}^i \mathbf{w} | \mathbf{w} \in \mathbb{W}\}\} \ .$$

To guarantee stability of the tube based controller, the definitions and theorems stated in Chapter 3 need to be reformulated in consideration of the uncertainty. The basic stability assumption with uncertainties can be reformulated as below.

Assumption 6: *(Basic stability assumption; robust case)*
(a) For all $\mathbf{x} \in \mathbb{X}_f$

$$\min_{\mathbf{u}\in\mathbb{U}} \max_{\mathbf{w}\in\mathbb{W}} [\Delta V_f + l](\mathbf{x},\mathbf{u},\mathbf{w}) \leq 0$$

with $\quad \Delta V_f(\mathbf{x},\mathbf{u},\mathbf{w}) = V_f(\mathbf{f}(\mathbf{x},\mathbf{u},\mathbf{w})) - V_f(\mathbf{x},\mathbf{u},\mathbf{w})$
(b) $\mathbb{X}_f \subseteq \mathbb{X}$

Assumption 6 requires the reduction of the loss function under any uncertainty $\mathbf{w} \in \mathbb{W}$, which makes the loss function a local Lyapunov function in $\mathbb{X}_\mathrm{f}$. This implies properties for the terminal set which are equivalent to Assumption 4 for the nominal MPC in Chapter 3.

Assumption 7: *(Implied stability assumption; robust case)*
The set $\mathbb{X}_f$ *is robust control invariant for* $\mathbf{x}^+ = \mathbf{f}(\mathbf{x},\mathbf{u},\mathbf{w}), \mathbf{w} \in \mathbb{W}$

The term robust control invariant is defined by

Definition 3: *(Robust control invariance)*
A set $\mathbb{X} \subseteq \mathbb{R}^n$ *is robust control invariant for* $\mathbf{x}^+ = \mathbf{f}(\mathbf{x}, \mathbf{u}, \mathbf{w}), \mathbf{w} \in \mathbb{W}$ *if, for every* $\mathbf{x} \in \mathbb{X}$*, there exists a* $\mathbf{u} \in \mathbb{U}$ *such that* $\mathbf{f}(\mathbf{x}, \mathbf{u}, \mathbf{w}) \in \mathbb{X}$

Satisfaction of the Assumptions 1, 2, 5, 6 and 7 with bounded disturbances ensures asymptotic stability to a robust control invariant set discussed in [61]. Finally, limitation on the uncertainty set need to be formulated as unlimited disturbances cannot be compensated and stability as well as constraint satisfaction can not be guaranteed. The characteristic of the uncertainty is described in the following assumptions.

Assumption 8: *(Restricted disturbance for constraint satisfaction)*
$S \subseteq \mathbb{X}, \mathbf{k}S \subseteq \mathbb{U}$.

Assumption 9: *(Compact convex disturbance set)*
The compact convex set $\mathbb{W}$ *contains the origin in its interior.*

The Assumptions 8 and 9 underline that the compensation of uncertainty is limited by the input and the state constraint. If the effect of the uncertainties reaches the input limits, the performance and stability of the system can no longer be guaranteed. In this thesis, the Assumptions 8 and 9 are assumed to be satisfied, such that the uncertainty can be compensated by the control.
Similarly to Section 3.2.1 the assumptions and propositions formulated above can be equivalently used to design for control of linear time varying systems, as is used in Chapter 5.

3.3 Model Predictive Tracking

In the last sections, the theoretical background of the MPC necessary for guaranteeing the existence and the stability of the solution have been briefly presented. In this section these results will be applied to the simple control problem of trajectory tracking for a vehicle. Similar approaches have been presented in works like [35][15][20][6]. The structure of the MPC in the classical control concept is presented in Figure 3.9. The MPC for the tracking problem makes use of a preplanned trajectory and the state vector information of the plant vehicle. The core of the MPC method is the optimization algorithm, which comprehends several modules such as the vehicle dynamics model, the cost function, the state and input constraints as well as the terminal cost. After solving the MPC optimization problem the first input vector is applied to the actors of the vehicle.

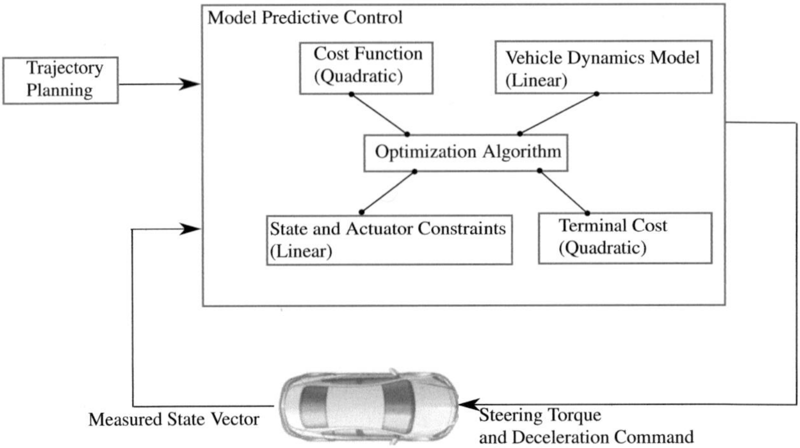

Figure 3.9: MPC structure in Classical Control Concept

3.3.1 Trajectory Planning

The trajectory planning method exploited in this chapter for the lateral dynamics is the sigmoidal trajectory presented by Stählin [71] and Schorn [68].
The equation of the sigmoidal trajectory is

$$y(x) = \frac{M}{1 + e^{-a(x-c)}} \tag{3.45}$$

where M is the lateral maneuver distance, c is half of the longitudinal maneuver distance and a is the slope at the middle of the maneuver. Since the sigmoidal reaches its both ends in positive and negative infinity only, y_{Tol} is introduced, such that $y(0) = y_{\text{Tol}}$ and $y(2c) = M - y_{\text{Tol}}$. The parameters of the sigmoidal trajectory are illustrated in Figure 3.10. Given a maximum lateral acceleration $a_{\text{y,max}}$, the lateral maneuver distance M and the tolerance constant y_{Tol}, the remaining parameters c and a have been derived in [71] to

$$c = \frac{1}{a_{\text{y}}} ln(\frac{M - y_{\text{Tol}}}{y_{\text{Tol}}})$$

$$a = \frac{(p+1)^2 \sqrt{-pM(a_{\text{y,max}} pM - v^2 p^2 + v^2) a_{\text{y,max}}}}{pM(a_{\text{y,max}} pM - v^2 p^2 + v^2)}$$

with the parameters

$$p = \sqrt{2} \frac{s_2}{\sqrt{s_1}} \left(6v^2 + 2a_{\text{y,max}} M + \frac{4a_{\text{y,max}}^2 M^2}{3v^2} \right) + \frac{2a_{\text{y,max}} M}{3v^2} + 1$$

$$s_1 = 9v^4 + 3a_{\text{y,max}} M v^2 + 2a_{\text{y,max}}^2 M^2$$

$$s_2 = \cos \left(\frac{1}{3} \arctan \left(\frac{3v^2 \sqrt{num}}{den} \right) \right)$$

$$num = 81v^8 + 27v^4 a_{\text{y,max}}^2 M^2 + 4a_{\text{y,max}}^3 M^3 + 3a_{\text{y,max}}^4 M^4$$

$$den = 27v^6 + 9v^2 a_{\text{y,max}}^2 M^2 + 4a_{\text{y,max}}^3 M^3 + 27v^4 a_{\text{y,max}} M \ .$$

The trajectory for the longitudinal dynamics is generated using the approach of [35]. Katriniok distinguishes between several maneuver phases and plans different constant deceleration values for different phases. Similarly to [35], the trajectory in this section is created by planning a constant deceleration of -5 m/s^2 from t = 0 s to 0.8 s and -1.5 m/s^2 from t = 0.8 s to 1.3 s. From t = 1.3 s to 2.2 s no braking is planned. Finally for t > 2.2 s, a braking maneuver of -8 m/s^2 is planned until standstill.

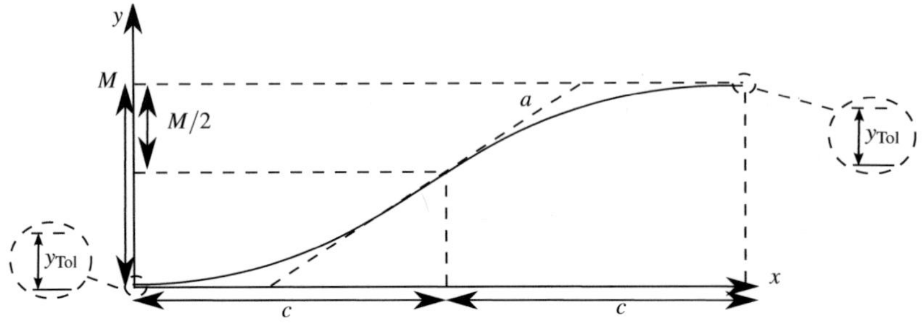

Figure 3.10: Sigmoidal Trajectory [71]

3.3.2 Vehicle Dynamics Model

The state transition model represents the system characteristics in the considered control process, i.e. the combined braking and steering dynamics behaviour of the vehicle in the collision avoidance maneuver. Though the combined vehicle dynamics model can represent the model behaviour for these maneuvers, a control algorithm based on the nonlinear model in (3.21) cannot be implemented in the real vehicle due to computational limitations. In the equation below the affine state space matrices with the system matrix $\mathbf{A}_c$, the input matrix $\mathbf{B}_c$ and the bias vector $\mathbf{E}_c$ are calculated by taking the derivates with respect to the state and the input vector of (3.21) respectively at the operation point (measured state) $\mathbf{x}_0$ and the last input vector $\mathbf{u}_0$.

$$\mathbf{A}_c = \frac{\partial \mathbf{f}(\mathbf{x}_0, \mathbf{u}_0)}{\partial \mathbf{x}} \tag{3.46}$$

$$\mathbf{B}_c = \frac{\partial \mathbf{f}(\mathbf{x}_0, \mathbf{u}_0)}{\partial \mathbf{u}} \tag{3.47}$$

$$\mathbf{E}_c = \mathbf{f}(\mathbf{x}_0, \mathbf{u}_0) - \mathbf{A}_c\mathbf{x}_0 - \mathbf{B}_c\mathbf{u}_0 \ . \tag{3.48}$$

The affine continuous state space model and the corresponding matrices are given by

$$\dot{\mathbf{x}} = \mathbf{A}_c \cdot \mathbf{x} + \mathbf{B}_c \cdot \mathbf{u} + \mathbf{E}_c \tag{3.49}$$

with

$$\mathbf{A}_\mathrm{c} = \begin{bmatrix} A_{11} & A_{12} & 0 & 0 & 0 & \dfrac{c_{\alpha\mathrm{f},0}}{m v_0 i_\mathrm{L}} & 0 & A_{18} & 0 \\ A_{21} & A_{22} & 0 & 0 & 0 & A_{26} & 0 & A_{28} & 0 \\ 0 & 1 & 0 & 0 & 0 & 0 & 0 & 0 & 0 \\ v_0 & 0 & v_0 & 0 & 0 & 0 & 0 & 0 & 0 \\ \dfrac{2D_\mathrm{f}}{J_\mathrm{st}} & \dfrac{2l_f D_\mathrm{f}}{v_0 J_\mathrm{st}} & 0 & 0 & -\dfrac{-d_\mathrm{st}}{J_\mathrm{st}} & \dfrac{-2D_\mathrm{f}}{i_\mathrm{L} J_\mathrm{st}} & 0 & A_{58} & 0 \\ 0 & 0 & 0 & 0 & 1 & 0 & 0 & 0 & 0 \\ 0 & 0 & 0 & 0 & 0 & 0 & 1 & 0 & 0 \\ 0 & 0 & 0 & 0 & 0 & 0 & 0 & 1 & 0 \\ 0 & 0 & 0 & 0 & 0 & 0 & 0 & 0 & -\dfrac{1}{t_\mathrm{del}} \end{bmatrix};$$

$$\mathbf{B}_\mathrm{c} = \begin{bmatrix} 0 & 0 \\ 0 & 0 \\ 0 & 0 \\ 0 & 0 \\ \dfrac{1}{J_\mathrm{st}} & 0 \\ 0 & 0 \\ 0 & 0 \\ 0 & 0 \\ 0 & \dfrac{1}{t_\mathrm{del}} \end{bmatrix}; \mathbf{E}_\mathrm{c} = \begin{bmatrix} \dfrac{F_{\mathrm{yf},0} + F_{\mathrm{yr},0}}{m v_0} - A_{18} v_0 \\ \dfrac{F_{\mathrm{yf},0} l_\mathrm{f} + F_{\mathrm{yr},0} l_\mathrm{r}}{J_\mathrm{z}} - A_{28} v_0 \\ 0 \\ 0 \\ -A_{58} v_0 \\ 0 \\ 0 \\ 0 \\ 0 \end{bmatrix};$$

$$A_{11} = -\frac{c_{\alpha_{\mathrm{f},0}} + c_{\alpha_{\mathrm{r},0}}}{mv_0}; A_{12} = \frac{c_{\alpha_{\mathrm{r},0}} l_{\mathrm{r}} - c_{\alpha_{\mathrm{f},0}} l_{\mathrm{f}}}{mv_0^2} - 1;$$

$$A_{18} = \frac{c_{\alpha_{\mathrm{f},0}} + c_{\alpha_{\mathrm{r},0}}}{mv_0^2} \beta_0 - 2\frac{c_{\alpha_{\mathrm{r},0}} l_{\mathrm{r}} - c_{\alpha_{\mathrm{f},0}} l_f}{mv_0^3} \dot{\varphi}_0 - \frac{c_{\alpha_{\mathrm{f},0}}}{i_{\mathrm{L}} mv_0^2} \delta_0;$$

$$A_{21} = \frac{c_{\alpha_{\mathrm{r},0}} l_{\mathrm{r}} - c_{\alpha_{\mathrm{f},0}} l_{\mathrm{f}}}{J_{\mathrm{z}}}; A_{22} = -\frac{c_{\alpha_{\mathrm{r},0}} l_{\mathrm{r}}^2 + c_{\alpha_{\mathrm{f},0}} l_f^2}{J_{\mathrm{z}} v_0};$$

$$A_{26} = -\frac{c_{\alpha_{\mathrm{r},0}} l_{\mathrm{r}}^2 + c_{\alpha_{\mathrm{f},0}} l_{\mathrm{f}}^2}{J_{\mathrm{z}} v_0}; A_{28} = \frac{c_{\alpha_{\mathrm{f},0}} l_{\mathrm{f}}^2 + c_{\alpha_{\mathrm{r},0}} l_{\mathrm{r}}^2}{J_{\mathrm{z}} v_0^2} \dot{\varphi}_0;$$

$$A_{58} = \frac{-2 l_f D_{\mathrm{f}}}{J_{\mathrm{st}} v_0^2} \dot{\varphi}_0;$$

Here v_0, δ_0, $\dot{\varphi}_0$ and β_0 denote the measured velocity, steering wheel angle, yaw rate as well as the side slip angle and $c_{\alpha_{\mathrm{f},0}}$, $c_{\alpha_{\mathrm{r},0}}$, $F_{\mathrm{yf},0}$ and $F_{\mathrm{yr},0}$ denote parameters of the linearized tire model in (3.16). In the model predictive control design, the discretized state space equation

$$\mathbf{x}_{k+1} = \mathbf{A}\mathbf{x}_k + \mathbf{B}\mathbf{u}_k + \mathbf{E} \tag{3.50}$$

is used. The state vector is given by $\mathbf{x}^T = [\beta, \dot{\varphi}, \varphi, y, \dot{\delta}, \delta, x, v, a_{\mathrm{x}}]$ and the input vector by $\mathbf{u}^T = [T, a_{\mathrm{x,cmd}}]$. The state transition equation is continuous for the state and the input vector, so that it satisfies Assumption 1. Note that the affine state space equation in (3.50) needs to be reformulated in section 3.3.5 to calculate the penalty matrix based on a linear state space model.

3.3.3 Cost Function

A central component of MPC methods is the selection of a suitable cost function, which follows the design goal of the control problem. In this section we choose the cost function to minimize the deviation of the vehicle states from a preplanned trajectory. The trajectory shall be followed explicitly for the lateral position, the velocity as well as the yaw angle of the vehicle dynamics state. Furthermore, we need to consider weights for the control signals, the steering torque and the commanded deceleration, to avoid permanent planning of maneuvers at the actuator limits. The cost function is given by

$$\begin{aligned} J = \sum_{i=0}^{N-1} & q_y(y(i,k) - y_{\text{traj}}(i,k))^2 + q_\varphi(\varphi(i,k) - \varphi_{\text{traj}}(i,k))^2 \\ & + q_v(v(i,k) - v_{\text{traj}}(i,k))^2 + r_{\text{T}} T^2(i,k) + r_{\text{a}_\text{x}} a^2_{x,cmd}(i,k) \\ = & (\mathbf{x}(i,k) - \mathbf{y}_{\text{traj}}(i,k))^T \mathbf{Q}(\mathbf{x}(i,k) - \mathbf{y}_{\text{traj}}(i,k)) + \mathbf{u}(i,k)'\mathbf{R}\mathbf{u}(i,k) \end{aligned} \tag{3.51}$$

$$\text{with } \mathbf{Q} = \begin{bmatrix} 0 & 0 & 0 & 0 & 0 & 0 & 0 & 0 & 0 \\ 0 & 0 & 0 & 0 & 0 & 0 & 0 & 0 & 0 \\ 0 & 0 & q_\varphi & 0 & 0 & 0 & 0 & 0 & 0 \\ 0 & 0 & 0 & q_y & 0 & 0 & 0 & 0 & 0 \\ 0 & 0 & 0 & 0 & 0 & 0 & 0 & 0 & 0 \\ 0 & 0 & 0 & 0 & 0 & 0 & 0 & 0 & 0 \\ 0 & 0 & 0 & 0 & 0 & 0 & 0 & 0 & 0 \\ 0 & 0 & 0 & 0 & 0 & 0 & 0 & q_\text{v} & 0 \\ 0 & 0 & 0 & 0 & 0 & 0 & 0 & 0 & 0 \end{bmatrix} \quad \mathbf{R} = \begin{bmatrix} r_\text{T} & 0 \\ 0 & r_{\text{a}_\text{x}} \end{bmatrix}$$

The chosen cost function is continuous for the state as well as the input vector and the matrices **P** and **Q** are chosen to be symmetric positive semidefinite such that the Assumptions 1, 3 and 5 are satisfied. The trajectory vector $\mathbf{y}_{\text{traj}}$ is composed of the trajectory values for the yaw angle q_φ, the lateral position q_y and the velocity q_v with

$$\mathbf{y}_{\text{traj}}(i,k) = [0, 0, \varphi_{\text{traj}}(i,k), y_{\text{traj}}(i,k), 0, 0, 0, v_{\text{traj}}(i,k), 0]^T.$$

3.3.4 Constraints

One of the main advantages of model predictive control is the ability to consider state and input constraints in the control design. In this section, constraints are used to ensure stability when controlling the vehicle at high yaw rate φ and side slip angle β as well as to consider actuator limitations for the input vector $\mathbf{u}$.

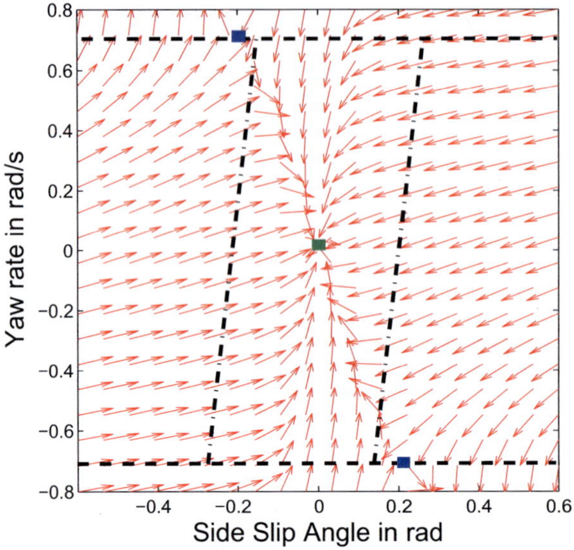

Figure 3.11: Phase plane map for v =50km/h and δ =3 °

Stable Handling Envelopes

Stable handling envelopes can guarantee for vehicle dynamics stability when integrated in MPC methods for maneuvers with high yaw rate and side slip angle. This was addressed by Beal [6] by presenting an invariant set in a phase plane, spanned by the yaw rate and the side slip angle axis. Beal shows that the limit of instability in the phase plane map for a vehicle dynamics model with nonlinear tires and different steering wheel angle as well as velocity values can be considered by a maximum yaw rate and a maximum side slip angle limit.

A phase plane map is exemplarily shown in Figure 3.11 for a single track model with constant velocity and steering wheel angle. The figure illustrates the snapshot dynamics of the model at different states (arrows), the stable equilibrium point (marked point near the origin), to which states in the stable set evolve to, and the unstable equilibrium points (marked points at +0.7 rad/s and -0.7 rad/s), which separates the stable states from the unstable states. The unstable states are characterized by states moving away from the origin. Beal [5] explains that the stable equilibrium points move to higher yaw rate values with increasing steering wheel angle, whereas the unstable equilibrium points move to lower slip angles at constant yaw rate value with increasing steering wheel angle. The yaw rate value of the unstable equilibrium point is

$$\dot{\varphi}_{max} = \begin{cases} \frac{F_{\mathrm{yr,max}}(1+\frac{l_r}{l_f})}{mv} & \text{for } F_{\mathrm{yf,max}} < \frac{l_\mathrm{r}}{l_\mathrm{f}} F_{\mathrm{yr,max}} \\ \frac{F_{\mathrm{yf,max}}(1+\frac{l_\mathrm{f}}{l_\mathrm{r}})}{mv} & \text{for } F_{\mathrm{yf,max}} < \frac{l_\mathrm{r}}{l_\mathrm{f}} F_{\mathrm{yr,max}} \end{cases} . \tag{3.52}$$

for different maximum lateral forces. These forces are dependent on the tire parameter, slip angle and load of each axis. The conditional equation can be rewritten with (3.14) in terms of the longitudinal acceleration.

$$\begin{aligned} & F_{\mathrm{yf,max}} - \frac{l_\mathrm{r}}{l_\mathrm{f}} F_{\mathrm{yr,max}} \\ &= \mu_{\mathrm{y,max}} m (\frac{l_\mathrm{r}}{l} g + \frac{h_{\mathrm{cog}}}{l} a_\mathrm{x}) - \frac{l_\mathrm{r}}{l_\mathrm{f}} \mu_{\mathrm{y,max}} m (\frac{l_\mathrm{f}}{l} g - \frac{h_{\mathrm{cog}}}{l} a_\mathrm{x}) \\ &= \mu_{\mathrm{y,max}} m (\frac{h_{\mathrm{cog}}}{l} + \frac{l_\mathrm{r}}{l_\mathrm{f}} \frac{h_{\mathrm{cog}}}{l}) a_\mathrm{x} \\ &= \mu_{\mathrm{y,max}} m \frac{h_{\mathrm{cog}}}{l_\mathrm{f}} a_\mathrm{x} \end{aligned}$$

For the considered maneuver the longitudinal acceleration a_x is negative for the duration of the maneuver, so that the second case in (3.52) applies.
The limit for the maximum yaw rate can thus be calculated by

$$\begin{aligned}\dot{\varphi}_\mathrm{max} &= \frac{F_\mathrm{yf,max}(1+\frac{l_\mathrm{f}}{l_\mathrm{r}})}{mv} \\ &= \frac{\mu_\mathrm{y,max}m(\frac{l_\mathrm{r}}{l}g+\frac{h_\mathrm{cog}}{l}a_\mathrm{x})(1+\frac{l_\mathrm{f}}{l_\mathrm{r}})}{mv} \\ &= \frac{\mu_\mathrm{y,max}(g+\frac{h_\mathrm{cog}}{l_\mathrm{r}r}a_\mathrm{x})}{v} \ .\end{aligned}$$

The maximum side slip angle β_max is

$$\beta_\mathrm{max} = -\alpha_\mathrm{r,max} + \frac{l_\mathrm{r}\dot{\varphi}}{v} \tag{3.53}$$

with $\alpha_\mathrm{r,max}$ being the maximum allowed slip angle for the rear tire. The stablility envelope can then be given by

$$-\dot{\varphi}_\mathrm{max} \leq \dot{\varphi} \leq \dot{\varphi}_\mathrm{max} \tag{3.54}$$

$$-\beta_\mathrm{max} \leq \beta \leq \beta_\mathrm{max} \ . \tag{3.55}$$

The admissible set for the stability is therefore

$$\mathbb{S} = \{\mathbf{x} \in \mathbb{R}^9 \mid \dot{\varphi} \in [-\dot{\varphi}_\mathrm{max}, \dot{\varphi}_\mathrm{max}], \beta \in [-\beta_\mathrm{max}, \beta_\mathrm{max}]\} \ . \tag{3.56}$$

For the case that the maximum lateral force on the front and rear tires are achieved simultaneously and a maximum slip angle of $\alpha_\mathrm{r,max} = 12°$ should not be exceeded, the maximum yaw rate and side slip angle limits are illustrated in Figure 3.11. In this chapter the state constraint set $\mathbb{X}$ comprises of the set $\mathbb{S}$ only, such that $\mathbb{X} = \mathbb{S}$ holds.

Actuator Limitations

The input vector $\mathbf{u}$ is limited by an admissible actuator range, i.e. the maximum and minimum possible steering wheel torque and the maximum longitudinal deceleration command.

The set of admissible actuator limits is thus given by

$$\mathbb{U} = \{\mathbf{u} \in \mathbb{R}^2 \mid T \in [-T_{\max}, T_{\max}], a_{\mathrm{x,cmd}} \in [a_{\mathrm{x,cmd,max}}, 0]\} \ . \tag{3.57}$$

Note that the set $\mathbb{U}$ is compact and $\mathbb{X}$ is closed which satisfies Assumption 2.

3.3.5 Terminal Cost

As explained in Section 3.2.1 the quasi-infinite optimal control problem is composed of two sub-problems, the finite optimal control problem and the infinite optimal control problem. The infinite control problem is designed to guarantee stability for consecutive control step $i > N$. Therefore a penalty matrix $\mathbf{P}_{\mathrm{f}}$ needs to be calculated to stabilize the system for the time steps $i > N$ in a wide range of vehicle dynamics. To calculate the penalty matrix the affine state space equation in (3.50) will be rewritten into the linear state space equation

$$\begin{bmatrix} \mathbf{x}_{k+1} \\ 1 \end{bmatrix} = \begin{bmatrix} \mathbf{A} & \mathbf{E} \\ 0 & 1 \end{bmatrix} \begin{bmatrix} \mathbf{x}_k \\ 1 \end{bmatrix} + \begin{bmatrix} \mathbf{B} \\ 0 \end{bmatrix} \mathbf{u}_k \tag{3.58}$$

Based on this linear state space model, the theory of linear quadratic control and Lyapunov inequality [11] will be exploited with

$$\begin{bmatrix} \mathbf{A}(\theta) & \mathbf{E}(\theta) \\ 0 & 1 \end{bmatrix}^T \mathbf{P}_{\mathrm{f}} \begin{bmatrix} \mathbf{A}(\theta) & \mathbf{E}(\theta) \\ 0 & 1 \end{bmatrix} - \mathbf{P}_{\mathrm{f}} \leq \begin{bmatrix} -\mathbf{Q} & 0 \\ 0 & 0 \end{bmatrix} \tag{3.59}$$

Here $\mathbf{A}(\theta)$ denotes the dependance of the state space matrix from the parameter vector $\theta \in \Theta$ with

$$\begin{aligned}\Theta := \{\theta \in \mathbb{R}^3 | v \in [v_{\min}, v_{\max}], \alpha_{\mathrm{r}} \in [-\alpha_{\mathrm{r,max}}, \alpha_{\mathrm{r,max}}], \\ \alpha_{\mathrm{f}} \in [-\alpha_{\mathrm{f,max}}, \alpha_{\mathrm{f,max}}]\}\end{aligned} \tag{3.60}$$

The parameter set Θ considers the change of velocity and slip angles for a wide range of vehicle dynamics maneuvers. The matrix $\mathbf{P}_{\mathrm{f}}$ is calculated offline with solvers for linear matrix inequalities [11][44] with the parameter vector $\theta \in \Theta$.

3.3.6 MPC Formulation

The overall MPC method proposed in this chapter is

$$\begin{aligned}\min_{\mathbf{u}} J = \sum_{i=0}^{N-1} (\mathbf{x}(i,k) - \mathbf{y}_{\mathrm{traj}}(i,k))'\mathbf{Q}(\mathbf{x}(i,k) - \mathbf{y}_{\mathrm{traj}}(i,k)) \\ + \mathbf{u}(i,k)'\mathbf{R}\mathbf{u}(i,k) \\ + [\mathbf{x}^T(N,k) - \mathbf{y}_{\mathrm{traj}}^T(N,k)\ 1]^T \mathbf{P}_{\mathrm{f}} [\mathbf{x}^T(N,k) - \mathbf{y}_{\mathrm{traj}}^T(N,k)\ 1]^T \\ \mathbf{x}(i+1,k) = \mathbf{A}\mathbf{x}(i,k) + \mathbf{B}\mathbf{u}(i,k) + \mathbf{E},\ \forall i = 0, \cdots N \\ \mathbf{u} \in \mathbb{U},\ \mathbf{x} \in \mathbb{X}\,.\end{aligned} \tag{3.61}$$

The components of the proposed tracking method satisfy the Assumptions 1, 2, 3, 4 and 5. Theorem 1 can thus provide stability guarantee for the presented control algorithm. Figure 3.12 illustrates the performance of this simple tracking method on a scenario with a static obstacle. The sigmoidal trajectory and the MPC are set up with the parameters given in Table 3.3. The top picture illustrates the position of the planned trajectory (thick dashed line) and the driven trajectory of the COG (thick solid line) as well as of the four vehicle corners (thin dashed lines line). The method shows lateral deviations of approximately y_{dev} = 0.65m which results from an aggressive parameterization of the trajectory planner. The maneuver is triggered at $x = 4$m. At ① the maximum steering torque and the maximum deceleration command values are applied. The counter steering phase starts at ② where the deceleration command is reduced to zero. In the stabilization phase at ③ maximum deceleration command is applied to reduce the vehicle velocity. The deviation between the planned trajectory and the driven trajectory indicates that proper tuning of the geometrical parameters in the planning method on the control characteristic is very important to ensure collision avoidance in the maneuver. Furthermore, both steering torque and deceleration command are not planned according to the most recent obstacle position but according to the curvature of the trajectory which is fixed during maneuver execution.

Table 3.3: Parameter setup for trajectory planning and vehicle dynamics control module

Parameter	Symbol	Value	Unit
Lateral maneuver distance	M	2.1	m
Tolerance Value	y_{Tol}	0.01	m
Maximum lateral acceleration	$a_{\mathrm{y,max}}$	9.81	$\mathrm{m/s^2}$
Penalty value on lateral position	q_{y}	1000	$\mathrm{1/m^2}$
Penalty value on yaw angle	q_{φ}	10	$\mathrm{1/rad^2}$
Penalty value on velocity	q_{v}	50	$\mathrm{s^2/m^2}$
Penalty value on steering torque	r_{T}	0.01	$\mathrm{1/Nm^2}$
Penalty value on deceleration	$r_{\mathrm{a_x}}$	0.01	$\mathrm{s^4/m^2}$
Maximum steering assist torque	T_{max}	50	Nm
Maximum rear slip angle	$\alpha_{\mathrm{r,max}}$	12	°
Maximum lateral friction	$\mu_{\mathrm{y,max}}$	1	(none)
Maximum deceleration value	$a_{\mathrm{x,cmd,max}}$	−9.81	$\mathrm{m/s^2}$

In Chapter 4 and 5, methods for the integrated planning and control concept will be investigated, where the obstacle is considered by the method directly and no tuning effort is needed between the planning and the control module.

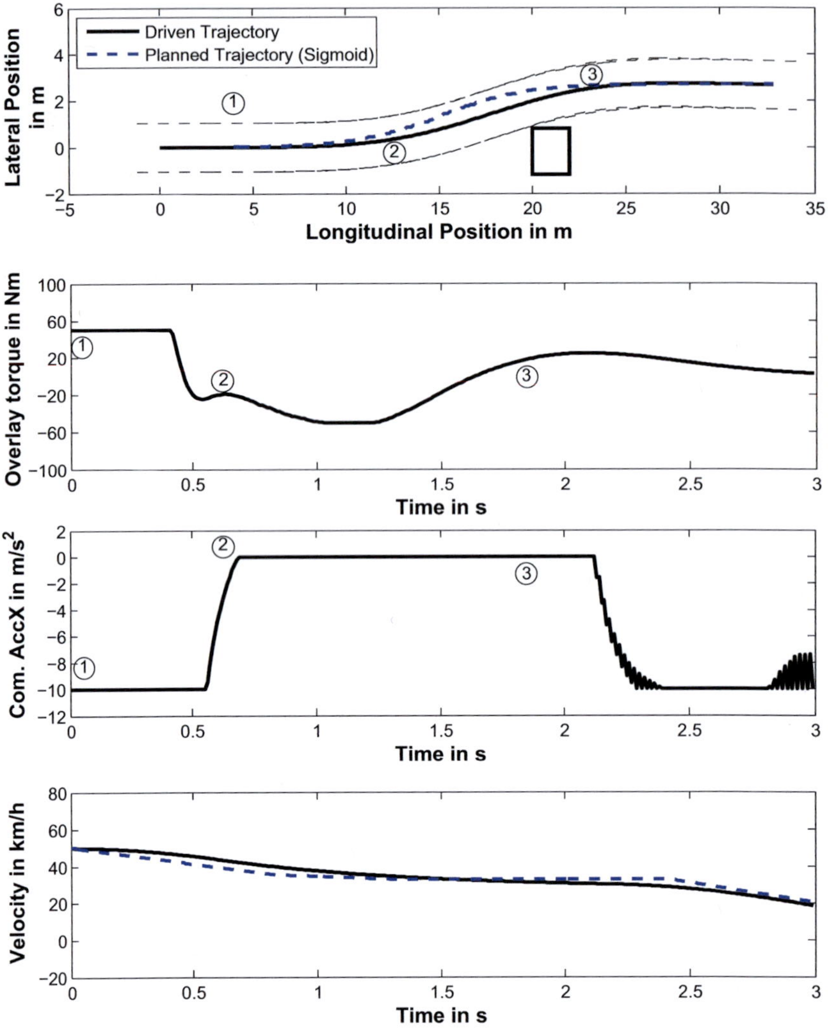

Figure 3.12: Tracking on predefined trajectory in single obstacle scenario

4 Nonlinear Integrated Planning and Control

The classical planning and control concept based on a trajectory, calculated with geometrical primitives, and a MPC approach as described in Section 3.3 is suitable to avoid a collision in simple scenarios, but cannot consider environmental information in complex scenarios with multiple obstacles and road boundaries. The integrated planning and control concept according to Section 2.4 can overcome the weaknesses of the classical concept. This chapter describes a nonlinear integrated planning and control method based on a nonlinear MPC structure. Preliminary work of this chapter has been published at the IEEE Intelligent Vehicles Symposium 2016 [83].

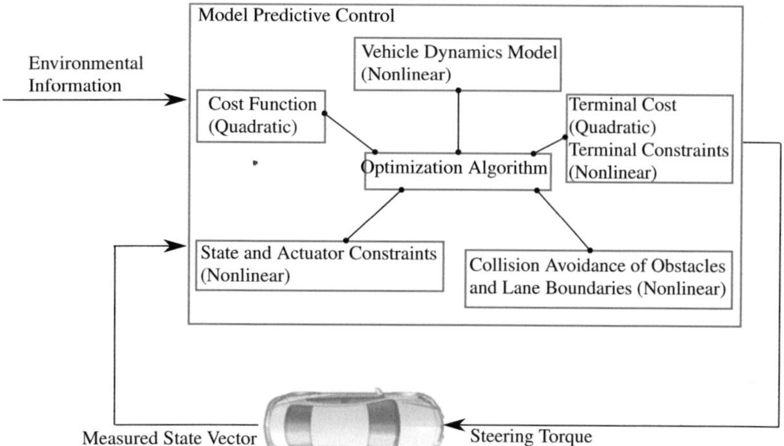

Figure 4.1: MPC structure of Nonlinear Integrated Planning and Control Method

The structure of the integrated planning and control method is illustrated in Figure 4.1. Instead of the preplanned trajectory as part of the classical concept, environmental information such as road boundaries and obstacle information serve as direct input for the proposed method. This information is used in the collision avoidance constraint to predict collision-free trajectories. The state and actuator constraints, the terminal cost as well as the vehicle dynamics model are taken from Chapter 3.

The remainder of this chapter is structured as follows. Section 4.1 introduces design goals for the desired planning characteristic of the proposed method. The cost function and the collision avoidance constraints required to consider the design goals and the environmental information are explained in Sections 4.2 and 4.3. The problem of obstacles outside of the prediction horizon is discussed and the terminal collision avoidance constraint is formulated for specific scenarios in Section 4.4. In Section 4.5, soft constraints are introduced to ensure smooth calculation of the result. The overall nonlinear MPC formulation is summarized in Section 4.6. The performance of the integrated method in a simulation environment is demonstrated in Section 4.7.

4.1 Design Goals

Key requirements to design the trajectory planning and the vehicle dynamics control module have been introduced in Section 1.2. By selecting the integrated planning and control concept based on the MPC method with collision avoidance constraints, several requirements can be fulfilled with reasonable effort as will be explained in this chapter. After establishing a proper implementation in this chapter to fulfill all given requirements independent on computational load, Chapter 5 will serve to deduce a time efficient method.

The optimization goal will be designed through a cost function as will be shown in the next section. The cost function enables the customization of the collision avoidance maneuver performance. This customization defines the characteristic of optimality and determines how the performance of the method shall be later evaluated in the simulation environment and in the real vehicle.

The design goals for collision avoidance maneuvers impact constraints and cost function. Design goals like guaranteed collision-freeness, consideration of environmental information, representation of vehicle dynamics and actuator limitations directly lead to the formulation of constraints. The performance of the optimization algorithm is thus indirectly influenced by those design goals. Two design goals have been chosen to be directly considered in the cost function. This concept can be easily modified with respect to performance by changing the states being considered in the cost function.

An emergency maneuver in urban environments should always be designed to reduce the velocity until the critical situation is dissolved. This supports the item of uncertain information as discussed in Section 1.2, leading to a reduced potential collision energy in the case that obstacles did not move dynamically as predicted. By reducing the velocity during the evasive maneuver, the prerequisite to do the best for avoiding the collision with an obstacle not detected early enough is given.

Similarly the planning and control method should aim to reduce the final yaw angle of the maneuver. By following this design goal, larger yaw angles not needed for the maneuvver are reduced. A potential threat of the maneuvering vehicle to other road participants like oncoming traffic can also be reduced by considering the yaw angle in the cost function. This indirectly supports the minimization of the lateral displacement. Furthermore, both criteria support better coorporation with the driver after being exposed to an emergency maneuver.

The key requirements and the design goals shape the properties of the integrated planning and control method investigated in this thesis. By selection of the concept and method based on the key requirements already build the fundament to cope with the objectives of evasion maneuvers. In addition, the described design goals shall further shape the desired driving characteristics of the planned maneuver and will be used to develop the required metrics in Section 4.7.1.

4.2 Cost Function

While the cost function for the trajectory tracking method in Section 3.3 serves to reduce deviations to a preplanned trajectory, the function for the integrated planning and control method directly considers the design goals from Section 4.1, i.e. it is designed to reduce the velocity and the final yaw angle deviation to the orientation of the road boundaries. In scenarios with straight lanes the yaw angle deviation can be substituted by the yaw angle. In addition, command signals are weighted in the cost function to prevent the method from permanent planning of interventions at the limits of the actuators. In addition to the desired control characteristic in (4.1), the cost function needs to be appended with a terminal cost V_f which can ensure stability of the control characteristic. The cost function is thus given by

$$J = \sum_{i=1}^{N-1} \mathbf{x}(i)^T \mathbf{Q}\mathbf{x}(i) + \mathbf{u}(i)^T \mathbf{R}\mathbf{u}(i) + [\mathbf{x}(N)^T \;\; 1]\mathbf{P}_\mathrm{f}[\mathbf{x}(N)^T \;\; 1]^T \tag{4.1}$$

$$\mathbf{Q} = \begin{bmatrix} 0 & 0 & 0 & 0 & 0 & 0 & 0 & 0 & 0 \\ 0 & 0 & 0 & 0 & 0 & 0 & 0 & 0 & 0 \\ 0 & 0 & q_\varphi & 0 & 0 & 0 & 0 & 0 & 0 \\ 0 & 0 & 0 & 0 & 0 & 0 & 0 & 0 & 0 \\ 0 & 0 & 0 & 0 & 0 & 0 & 0 & 0 & 0 \\ 0 & 0 & 0 & 0 & 0 & 0 & 0 & 0 & 0 \\ 0 & 0 & 0 & 0 & 0 & 0 & 0 & 0 & 0 \\ 0 & 0 & 0 & 0 & 0 & 0 & 0 & q_\mathrm{v} & 0 \\ 0 & 0 & 0 & 0 & 0 & 0 & 0 & 0 & 0 \end{bmatrix} \quad \mathbf{R} = \begin{bmatrix} r_\mathrm{T} & 0 \\ 0 & r_{\mathrm{a_x}} \end{bmatrix}$$

The chosen cost function is continuous for the state vector $\mathbf{x}$ as well as the command vector $\mathbf{u}$ and the matrices $\mathbf{Q}$ as well as $\mathbf{R}$ are chosen to be symmetric positive semi-definite to satisfy Assumptions 1, 3 and 5 from Section 3.2.1. The penalty matrix $\mathbf{P}_\mathrm{f}$ is calculated to stabilize the system with (3.59). Note that the

MPC theory, explained in Section 3.2, designed methods to steer the whole state vector $\mathbf{x}$ to the origin (zero), whereas the proposed method is designed to steer individual states (velocity and yaw angle) to zero.

4.3 Obstacles and Road Boundaries

To explain the design of the proposed constraints in this section, consider a scenario with two straight lanes and a static obstacle on the right lane with the width w_{obsr} and the length l_{obsr}. Further the relative position from the obstacle rear edge to the origin is x_{obsr} and y_{obsr}. All road participants like pedestrians, cyclists or vehicles are enframed by a rectangle.

The proposed planning and control method considers potential collision obstacles and lane boundary information to plan a collision-free trajectory during the maneuver without exceeding of road boundaries.

The planned trajectory is calculated in relative position to the ego vehicle's COG at current time step i = 0. Figure 4.2 illustrates the scenario with the ego vehicle and the obstacle at current time step i = 0 (solid lines) and the predicted ego vehicle for a time step i (dotted lines). The longitudinal and lateral position of the front left, front right, rear left and rear right corners of the ego vehicle are marked in the figure and can be denoted by $(x_{\text{fl}}, y_{\text{fl}})$, $(x_{\text{fr}}, y_{\text{fr}})$, $(x_{\text{rl}}, y_{\text{rl}})$ and $(x_{\text{rr}}, y_{\text{rr}})$. In this scenario the ego vehicle shall swerve to the left to avoid the collision. The evasion direction is predetermined by a higher order decision making module (e.g. [8]). Let B_{r} be defined as the intersection point of the right edge of the ego vehicle rectangle at time step i with the elongated rear edge of the obstacle vehicle.

The lateral position of the point B_{r} for the time step i, y_{Br}, is then calculated based on the lateral position y of the COG and the yaw angle φ at time step i in

$$y_{\text{Br}} = y - \frac{w}{2\cos\varphi} + \Delta x \tan\varphi \ . \tag{4.2}$$

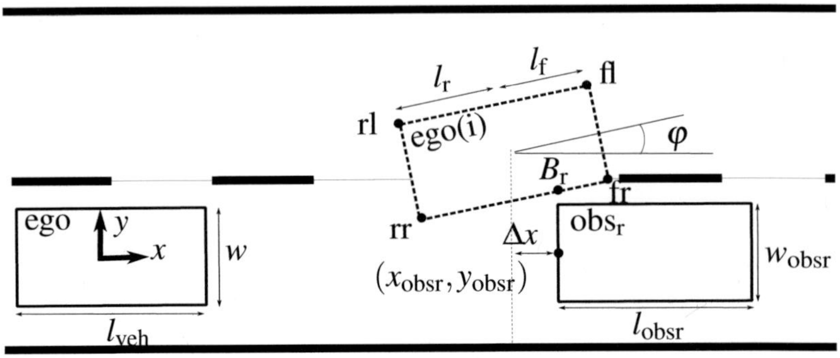

Figure 4.2: Illustration of collision avoidance constraints for an evasive collision avoidance maneuver to the left

The collision avoidance constraint can be expressed as

$$\begin{aligned} y_{\mathrm{Br}} &\geq y_{\mathrm{obsr}} + w_{\mathrm{obsr}}/2 \\ \forall(x_{\mathrm{fr}} &> x_{\mathrm{obsr}}) \cap (x_{\mathrm{rr}} < x_{\mathrm{obsr}} + l_{\mathrm{obsr}}) \end{aligned} \tag{4.3}$$

for an evasive maneuver to the left. In the equation above, the constraint is only active when the dimension of the vehicle overlaps with the dimension of the obstacle. The longitudinal positions x_{fr} and x_{rr} are calculated with

$$\begin{aligned} x_{\mathrm{fr}} &= x + l_{\mathrm{f,veh}} \cos(\varphi) + \frac{w}{2} \sin(\varphi) \quad \text{and} \\ x_{\mathrm{rr}} &= x - l_{\mathrm{r,veh}} \cos(\varphi) + \frac{w}{2} \sin(\varphi). \end{aligned} \tag{4.4}$$

Similarly the constraint for a collision avoidance maneuver to the right as shown in Figure 4.3 can be treated analogously with an obstacle obs_{l} with the width w_{obsl}, the length l_{obsl} as well as the the distances x_{obsl} and y_{obsl}.
The set for the vehicle dynamics states to ensure collision avoidance is given by

$$\begin{aligned} \mathbb{C} = \{\mathbf{x} \in \mathbb{R}^9 |\ & y_{\mathrm{Br}} \geq y_{\mathrm{obsr}} + w_{obsr}/2, \forall x \in \mathbb{X}_{\mathrm{or}}, \\ & y_{\mathrm{Bl}} \leq y_{\mathrm{obsl}} - w_{\mathrm{obsl}}/2, \forall x \in \mathbb{X}_{\mathrm{ol}}\} \end{aligned} \tag{4.5}$$

$$\begin{aligned}
\text{with}\quad & \mathbb{X}_{\text{or}} := \{x \in \mathbb{R} | (x_{\text{fr}} > x_{\text{obsr}}) \cap (x_{\text{rr}} < x_{\text{obsr}} + l_{\text{obsr}})\} \\
& \mathbb{X}_{\text{ol}} := \{x \in \mathbb{R} | (x_{\text{fl}} > x_{\text{obsl}}) \cap (x_{\text{rl}} < x_{\text{obsl}} + l_{\text{obsl}})\} \\
& y_{\text{Bl}} = y + \frac{w}{2\cos\varphi} + \Delta x \tan\varphi \\
& x_{\text{fl}} = x + l_{\text{f,veh}} \cos(\varphi) - \frac{w}{2}\sin(\varphi) \\
& x_{\text{rl}} = x - l_{\text{r,veh}} \cos(\varphi) - \frac{w}{2}\sin(\varphi).
\end{aligned}$$

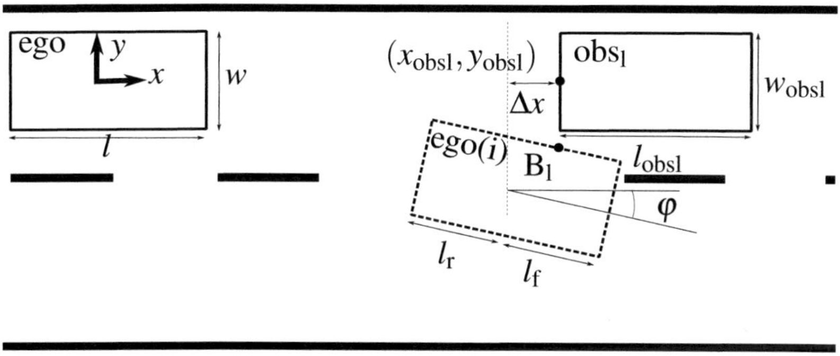

Figure 4.3: Illustration of collision avoidance constraints for an evasive collision avoidance maneuver to the right

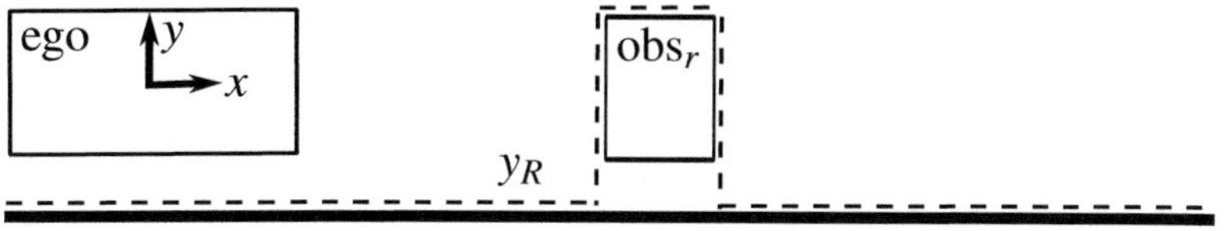

Figure 4.4: Illustration of boundary lines y_R and y_L to consider for road boundaries and restrictions due to obstacles

Furthermore, constraints for the planning method will be presented to consider position restrictions such as road boundaries and curb stones. Let the boundaries on the left and on the right side be given by y_{L} and y_{R} as shown in Figure 4.4.
To keep all corners of the vehicle within these boundaries, the following constraints shall be considered

$$\begin{aligned}
\min(y_{\mathrm{fr}}, y_{\mathrm{rr}}) &\geq y_{\mathrm{R}}(x) \\
\max(y_{\mathrm{fl}}, y_{\mathrm{rl}}) &\leq y_{\mathrm{L}}(x) \\
\text{with } y_{\mathrm{fr}} &= y + l_{\mathrm{f,veh}} \sin(\varphi) - \frac{w}{2}\cos(\varphi) \\
y_{\mathrm{rr}} &= y - l_{\mathrm{f,veh}} \sin(\varphi) - \frac{w}{2}\cos(\varphi) \\
y_{\mathrm{fl}} &= y + l_{\mathrm{f,veh}} \sin(\varphi) + \frac{w}{2}\cos(\varphi) \\
y_{\mathrm{rl}} &= y - l_{\mathrm{r,veh}} \sin(\varphi) + \frac{w}{2}\cos(\varphi)
\end{aligned} \tag{4.6}$$

The corresponding set to consider the road boundaries is defined in

$$\mathbb{L} = \{\mathbf{x} \in \mathbb{R}^9 \mid \min(y_{\mathrm{fr}}, y_{\mathrm{rr}}) \geq y_{\mathrm{R}}(x),\ \max(y_{\mathrm{fl}}, y_{\mathrm{rl}}) \leq y_{\mathrm{L}}(x)\}. \tag{4.7}$$

The constraints in the set $\mathbb{C}$ and $\mathbb{L}$ are closed such that Assumption 2 is satisfied.

4.4 Terminal Collision Avoidance

One aspect that has not been adressed in prior works, according to the knowledge of the author, is the guarantee of collision avoidance for obstacles outside of the prediction horizon. In practice, the length of the prediction horizon and the number of prediction steps are selected as a trade off between high coverage of relevant obstacles and low computational burden of the resulting optimization problem. Problematic scenarios can be found when the prediction horizon is not long enough by design to consider all relevant obstacles that have been detected. To enable planning of collision-free trajectories in these scenarios,

terminal collision avoidance constraint (TCA) introduced in this section can be exploited. The terminal collision avoidance constraints are implemented on the terminal prediction step of the prediction horizon to enable planning of safe trajectories with a simple point mass model [67].
First consider a scenario as illustrated in Figure 4.5 where the ego vehicle needs to perform an evasive maneuver to avoid collision with $\mathrm{obs_r}$ followed by a braking maneuver in order to avoid collision with vehicle obstacles in the traffic jam at a distance of x_{Tjam}.

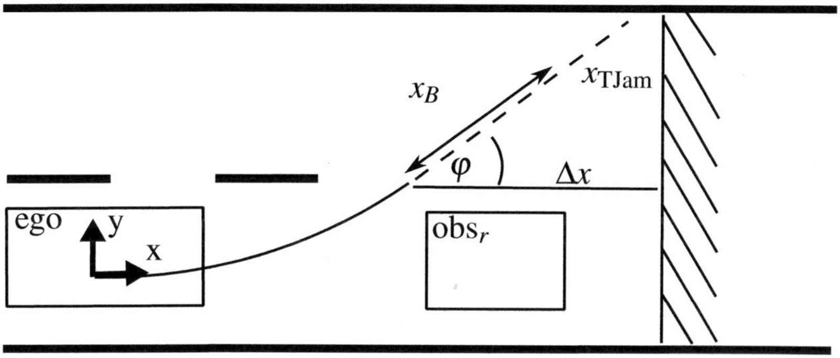

Figure 4.5: Illustration of terminal collision avoidance constraint for a braking maneuver

Figure 4.5 illustrates the predicted trajectory to avoid collision with the first obstacle, but due to the length of the prediction horizon and the chosen control parameters it may not be able to avoid collision with vehicle obstacles in the traffic jam. The terminal collision avoidance constraint in this section requires the vehicle dynamics state at the last prediction step to comply with collision avoidance requirements based on a point mass model.
Let the equation for the maneuver distance of a braking maneuver from [67] be given by

$$x_{\mathrm{B}} = -\frac{v_0^2}{2a_{\mathrm{x}}} \tag{4.8}$$

where v_0 is the initial velocity and a_x is the required longitudinal acceleration value, then the terminal collision avoidance constraint in this section requires the vehicle dynamics state for the last prediction step to satisfy

$$\Delta x(N)/\cos\varphi(N) \geq x_B(N) \equiv (x_{Tjam} - x(N)) \geq \frac{v(N)^2 \cos(\varphi(N))}{2a_x(N)}. \tag{4.9}$$

where $x(N), v(N), \varphi(N)$ and $a_x(N)$ correspond to predicted vehicle dynamics states and $\Delta x(N) = x_{Tjam} - x(N)$ is the remaining longitudinal distance to the traffic jam at the last prediction step of the prediction horizon N. To avoid division by zero in the equation for the optimization problem, $a_x(N)$ is substituted by $\min(a_x, -0.1)$ which has marginal influence on the overall performance. Further, consider Scenario 2 with two obstacles in Figure 4.6 and assume again the short prediction horizon for this scenario by design. The algorithm is not able to consider an avoidance maneuver with the second obstacle already at maneuver start without any additional constraints.

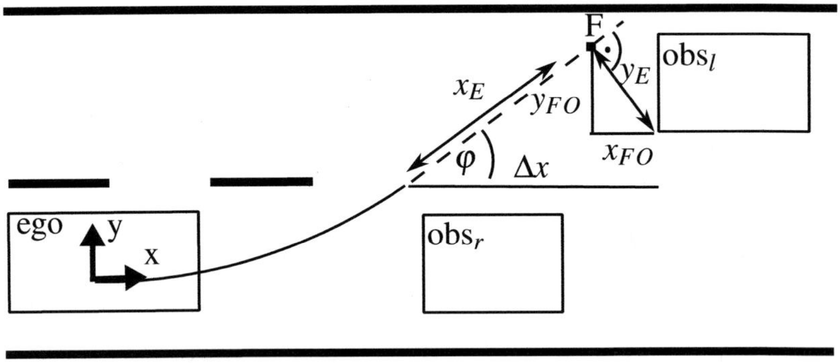

Figure 4.6: Illustration of terminal collision avoidance constraint for an evasion maneuver

Subsequently the terminal collision avoidance constraint for evasive maneuver will be developed.

Let the required maneuver distance for an evasive maneuver according to [67] be given as

$$x_{\mathrm{E}} = v_0 \sqrt{\frac{2y_{\mathrm{E}}}{a_{\mathrm{y}}}} \tag{4.10}$$

where y_E is the lateral maneuver distance and a_{y} is the assumed lateral acceleration, then the terminal collision avoidance constraint for evasive maneuver requires the maneuver distance to be smaller than the available longitudinal distance in the equation

$$\begin{aligned} &\Delta x(N) \geq x_{\mathrm{E}}(N) \\ \equiv\ &(x_{\mathrm{obsl}} - x(N)) \geq v(N) \sqrt{\frac{2y_{\mathrm{E}}(N)}{a_{\mathrm{y}}(N))}}\ . \end{aligned} \tag{4.11}$$

The required lateral maneuver distance in the orientation of the last prediction step y_{E} illustrated in Figure 4.6 is calculated with

$$y_{\mathrm{E}} = \sqrt{x_{\mathrm{FO}}^2 + y_{\mathrm{FO}}^2} \tag{4.12}$$

$$x_{\mathrm{FO}} = x_{\mathrm{obsl}} - x_{\mathrm{F}} \tag{4.13}$$

$$y_{\mathrm{FO}} = y_{\mathrm{F}} - y_{obsl} - w_{\mathrm{obsl}}/2\ . \tag{4.14}$$

Here point F is located on the straight line starting at the position of the last predicted time step N with the orientation of the corresponding yaw angle with the nearest distance to the critical corner G at obs_{l} (see Figure 4.6). The exact position can be determined by calculating the crossing point of the two lines

$$\begin{aligned} y_1(x) &= y(N) + \tan(\varphi(N))(x - x(N)) \quad \text{and} \\ y_2(x) &= y_{\mathrm{obsl}} - w_{\mathrm{obsl}}/2 - \tan(\varphi(N))(x - x_{\mathrm{obsl}}). \end{aligned} \tag{4.15}$$

The coordinates of the point F can then be calculated by

$$y_{\mathrm{F}} = (y(N) + \tan(\varphi(N))(x_{\mathrm{obsl}} - x(N)) + y_{\mathrm{obsl}} - w_{\mathrm{obsl}}/2)/2$$
$$x_{\mathrm{F}} = (y_{\mathrm{F}} - y(N))/\tan\varphi(N) + x(N)$$

Similarly to the terminal constraint for a subsequent braking maneuver, $a_{\mathrm{y}}(N)$ is substituted by $\min(a_{\mathrm{y}}(N), -0.1)$ in (4.11) to avoid division by zero in. The sets of the terminal collision avoidance constraints for a braking and an evasive maneuver are

$$\mathbb{G}_{\mathrm{B}} := \{\mathbf{x} \in \mathbb{R}^9 | (x_{\mathrm{obsl}} - x(N,k)) \geq \frac{v(N)^2 \cos^2(\varphi(N,k))}{2a_{\mathrm{x}}(N,k)}\} \tag{4.16}$$

and

$$\mathbb{G}_{\mathrm{E}} := \{\mathbf{x} \in \mathbb{R}^9 | (x_{\mathrm{obsl}} - x(N)) \geq v(N)\sqrt{\frac{2y_{\mathrm{E}}(N)}{a_{\mathrm{y}}(N))}}\}. \tag{4.17}$$

4.5 Soft Constraints

The introduction of soft constraints using slack variables is important for the practical application of the method with many numerical solvers. To ensure satisfaction of the collision avoidance constraints with numerical smoothness of the results according to [19], slack variables s_{obs} are added to the sets in $\mathbb{C}, \mathbb{L}, \mathbb{G}_{\mathrm{B}}$ and $\mathbb{G}_{\mathrm{E}}$ such that the modified sets are

$$\begin{aligned}
\mathbb{C} := \{\mathbf{x} \in \mathbb{R}^9 |\ & y_{\mathrm{Br}} + s_{\mathrm{obs}} \geq y_{\mathrm{obsr}} + w_{\mathrm{obsr}}/2, \forall x \in \mathbb{X}_{\mathrm{or}}, \\
& y_{\mathrm{Bl}} + w_{\mathrm{obsl}}/2 \leq y_{\mathrm{obsl}} + s_{\mathrm{obs}}, \forall x \in \mathbb{X}_{\mathrm{ol}}\}, \\
\mathbb{L} := \{\mathbf{x} \in \mathbb{R}^9 \mid\ & \min(y_{\mathrm{fr}}, y_{\mathrm{rr}}) + s_{\mathrm{obs}} \geq y_{\mathrm{R}}(x)\ , \\
& \max(y_{\mathrm{fl}}, y_{\mathrm{rl}}) \leq y_{\mathrm{L}}(x) + s_{\mathrm{obs}}\},
\end{aligned}$$

$$\mathbb{G}_{\text{B}} := \{\mathbf{x} \in \mathbb{R}^9 | (x_{\text{tjam}} - x(N,k)) + s_{\text{obs}} \geq \frac{v(N)^2 \cos^2(\varphi(N,k))}{2a_{\text{x}}(N,k)}\} \text{ and}$$

$$\mathbb{G}_{\text{E}} := \{\mathbf{x} \in \mathbb{R}^9 | (x_{\text{obsl}} - x(N)) + s_{\text{obs}} \geq v(N)\sqrt{\frac{2y_{\text{E}}(N)}{a_{\text{y}}(N)}}\} .$$

To minimize these slack values, the variables are considered by a modified cost function J_{s} and the cost function J defined in (4.1) in Section 4.2.

$$J_{\text{s}} = J + Q_{\text{s}} s_{\text{obs}}(i)^2 \tag{4.18}$$

The introduction of soft constraints may endanger the calculation of collision-free trajectories. The already introduced safety margins help to cope with minor constraint violations. The safety integrity with the soft constraints can be further ensured by an additional module double checking the trajectory after the optimization step prior to any activation of the planned maneuver.

4.6 MPC Formulation

The overall nonlinear optimization problem in this chapter is a nonconvex optimization problem and is formulated below.

$$\begin{aligned}
\min_{\mathbf{u}} \; J_{\text{s}} = & \sum_{i=1}^{N} \mathbf{x}(i,k)^T \mathbf{Q} \mathbf{x}(i,k) + \mathbf{u}(i,k)^T \mathbf{R} \mathbf{u}(i,k) + \cdots \qquad (4.19) \\
& [\mathbf{x}(N,k)^T \;\; 1] \mathbf{P}_{\text{f}} [\mathbf{x}^T(N,k) 1]^T + Q_{\text{s}} s_{\text{obs}}(i,k)^2 \\
\text{s.t.} \;\; & \mathbf{x}(i+1,k) = \mathbf{x}(i,k) + t_{\text{pred}} \mathbf{f}(\mathbf{x}(i,k), \mathbf{u}(i,k)) \\
& \mathbf{u}(i,k) \in \mathbb{U} \\
& \mathbf{x}(i,k) \in \mathbb{X} \\
& \mathbb{X} = \mathbb{S} \cap \mathbb{V} \cap \mathbb{C} \cap \mathbb{L} \;\; i = 1, \cdots, N \\
& \mathbf{k}\mathbf{x}(N,k) \in \mathbb{U} \\
& \mathbf{x}(N,k) \in \mathbb{G}_{B/E} \;\; \text{if TCA is applied}
\end{aligned}$$

Here t_{pred} is the discretization time between predicted time steps and $\mathbb{V}$ is the set described in

$$\mathbb{V} = \{\mathbf{x} \in \mathbb{R}^9 \mid v \geq 0,\ \sqrt{a_x{}^2 + a_y^2} \leq \mu g,\ a_x \leq 0\}\ . \tag{4.20}$$

The set $\mathbb{V}$ is closed, such that Assumption 2 is satisfied.

4.7 Simulation Results

After explaining the modules of the integrated trajectory planning and vehicle dynamics control method, the performance of the proposed method will be presented in a simulation environment with the vehicle dynamics model as presented in Section 3.1. The integrated trajectory planning and vehicle dynamics control method has been implemented in MATLAB Simulink on a computer with 8 GB RAM and Intel CORE i5 processor with 2.90 GHz.
First this section will present the considered metrics and scenarios to evaluate the integrated planning and control method for collision avoidance maneuvers. Finally the performance of the proposed method on the predefined metrics and scnearios will be shown in the simulation environment.

4.7.1 Metrics

The metrics defined in this section are derived from the design goals of Section 4.1. The first metric relates to the amount of **velocity reduction** at the longitudinal position of the potential collision. The relative velocity reduction is calculated by

$$\Delta v_{\text{red}} = \frac{v_0 - v(t_c)}{v_0}\ . \tag{4.21}$$

Here v_0 is the initial velocity of the ego vehicle at the start of the manuver intervention, whereas $v(t_c)$ is the velocity at the potential collision time - the point in time when the front edge of the vehicle bypasses the rear edge of the obstacle.

The second metric considers the **final yaw angle** φ_{fnl} of the vehicle at the end of the maneuver. According to Section 4.1, the vehicle shall reorientate the vehicle at the end of an ideal maneuver.

The third metric is the **computation time** t_{cmp} of the implemented algorithm per control step. The method presented in this chapter does not focus on the reduction of computation time. Therefore the computation time metric will not be used for evaluation in this chapter but shall be calculated prior to further modification in Chapter 5.

4.7.2 Scenarios

In this section typical scenarios in urban environments are presented to evaluate the performance of the proposed planning and control concept. The scenarios consider a straight road with two lanes. The overall width is 7 m with 3.5 m for each lane. All road participants like pedestrians, cyclists or vehicles are enframed by a rectangle. The ego vehicle with the width $w = 2$ m and the length $l_{\text{veh}} = 3.5$ m moves with the velocity $v_0 = 50$ km/h on the right lane. In the presented scenarios, the collision avoidance maneuvers will not start immediately but triggered after $x = 4$ m which corresponds to the latest point in time.

Scenario 1: Single Obstacle

In Scenario 1, the static obstacle with the width $w_{\text{obsr}} = 2$ m and the length $l_{\text{obsr}} = 3.5$ m, which represent a static vehicle, is located on the center of the right lane. The corresponding scenario is given in Figure 4.7.

Let the origin of the reference system be located at the COG of the ego vehicle, then the relative position of the obstacle rear edge to the origin shall be ($x_{\text{obsr}} = 20$ m, $y_{\text{obsr}} = 0$ m).

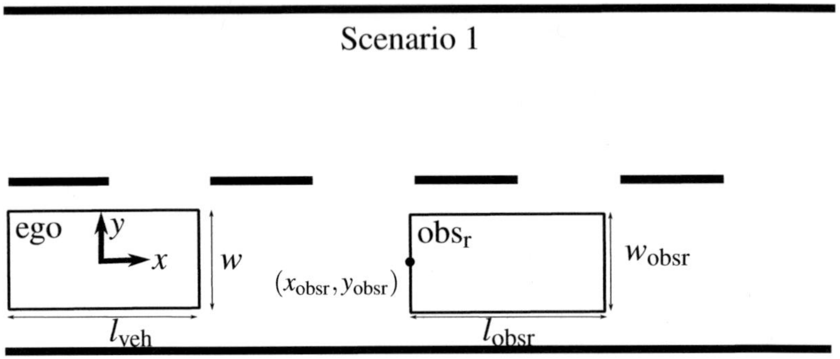

Figure 4.7: Scenario 1 with one static obstacle

Scenario 2: Multiple Obstacles

The second use case considers multiple obstacles representing another typical situation in urban environments. The first obstacle with the width $w_{obsr} = 2$ m and the length $l_{obsr} = 1$ m, representing a pedestrian or a cyclist, is located on the center of the right lane similarly to Scenario 1. The second obstacle (obs_l) is located on the left lane with the relative distances $x_{obsl} = 24$ m and $y_{obsl} = 4.5$ m to the COG of the ego vehicle and with the width $w_{obsl} = 2$ m and length $l_{obsl} = 3.5$ m. The gap between both obstacles is chosen such that it can be passed through by the ego vehicle without a collision. The scenario is shown in Figure 4.8.

Scenario 3: Single Obstacle with Traffic Jam

The last scenario shown in Figure 4.9 considers a single obstacle obs_r, with the width $w_{obsr} = 2$ m and the length $l_{obsr} = 3.5$ m on the right lane, and a traffic jam at the distance $x_{TJam} = 28$ m illustrated by a wall, such that the system needs to apply a braking maneuver after the first evasive maneuver in order to avoid a collision.

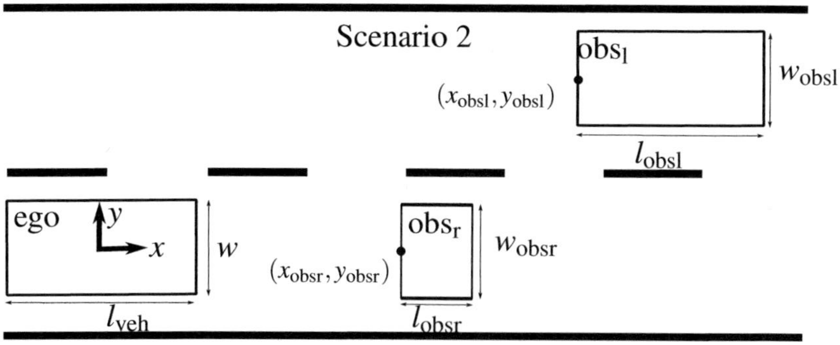

Figure 4.8: Scenario 2 with two static obstacles on both lanes

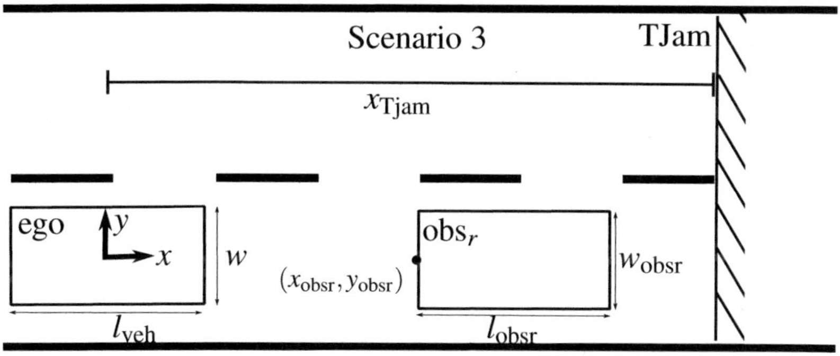

Figure 4.9: Scenario 3 with single obstacle and end of a traffic jam / wall

4.7.3 Typical Scenarios

The performance of the proposed method for the Scenarios 1 and 2 (Section 4.7.2) shall be demonstrated in this section.
The focus of this section is to discuss the maneuver performance with three different parameter setups for the proposed method. Parameter setup 1 for the proposed nonlinear MPC is shown in Table 4.1.

Table 4.1: Parameter setup 1 for the Nonlinear Integrated Planning and Control Method

Parameter	Symbol	Value	Unit
Prediction horizon length	N	15	(none)
Prediction horizon step size	dt_{pred}	0.14	s
Controller time step size	dt_{ctrl}	0.03	s
Penalty of slack variable	q_{s}	$3{\cdot}10^6$	$1/\text{m}^2$
Penalty of yaw angle	q_{φ}	$3{\cdot}10^3$	$1/\text{rad}^2$
Penalty of velocity	q_{v}	10	s^2/m^2
Penalty of steering torque command	r_{T}	10^{-1}	$1/\text{Nm}^2$
Penalty of deceleration command	r_{a_x}	10^{-1}	s^4/m^2
Limit for steering system	T_{max}	50	Nm
Limit for braking system	$a_{\text{x,cmd,max}}$	-9.81	m/s^2

Figure 4.10 shows the result for the single obstacle scenario. In the top picture, the driven trajectories of the vehicle with the COG (solid lines) and its four corners (front right, front left, rear right and rear left in dashed lines) are illustrated in the plane, spanned by the axes for the lateral and longitudinal position. The figure shows how the vehicle avoids a collision with the obstacle and brakes the vehicle to a standstill.

At maneuver start ①, the algorithm commands the maximum steering torque and maximum deceleration to avoid the collision. This may seem to violate the restrictions of the friction circle but due to the time delay characteristic of the vehicle dynamics model the friction limits can not be reached instantly. Subsequently at ②, the algorithm introduces the counter steering phase. When the ego vehicle with its front edge is passing the rear edge of the obstacle ③, the algorithm again commands a high steering torque value in the stabilization phase and prevents further rotation towards the obstacle. At ④, the algorithm stops the steering and the braking intervention and due to the characteristic of the longitudinal model the vehicle is decelerated to a standstill.

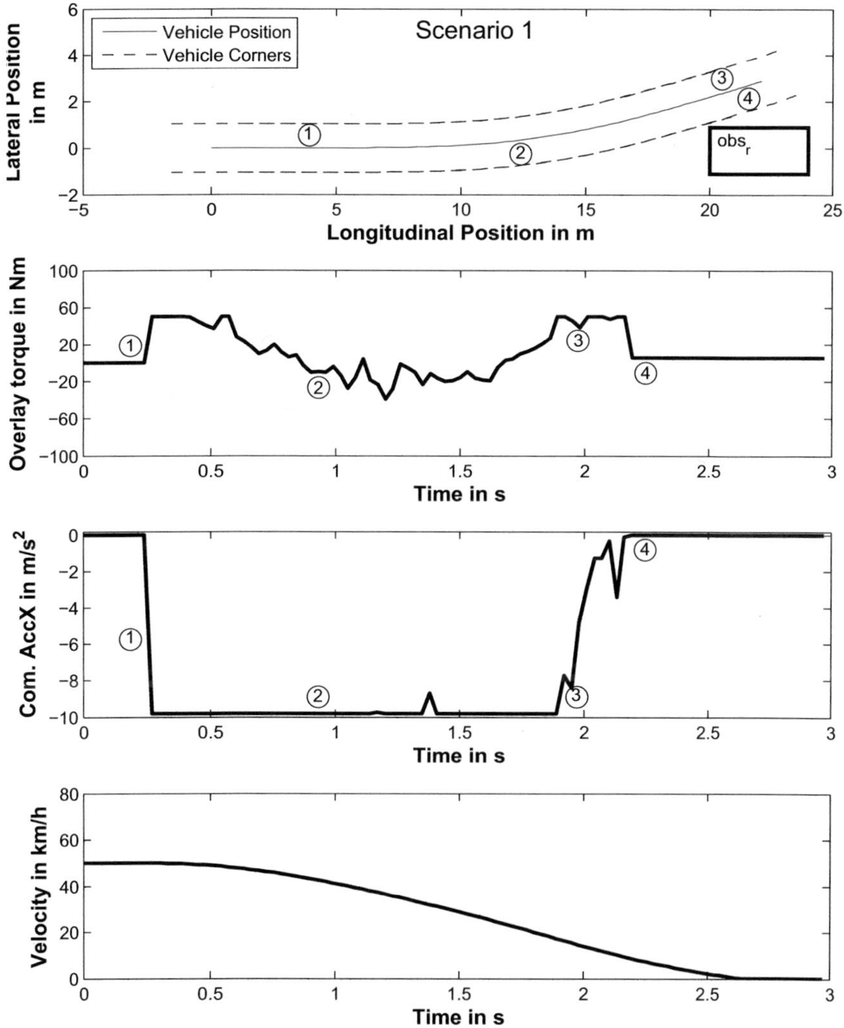

Figure 4.10: Simulation results of the nonlinear integrated trajectory planning and vehicle dynamics control method with parameter setup 1 in Scenario 1

To explain the main influence of different parameters on the maneuver performance, parameter setup 2 in Table 4.2 will be used. The parameter setup 2 is designed with a higher penalty on the yaw angle and a lower penalty on the velocity, which represents higher emphasis on the reorientation of the vehicle.

Table 4.2: Parameter setup 2 for Nonlinear Integrated Planning and Control Method

Parameter	Symbol	Value	Unit
Prediction horizon length	N	15	(none)
Prediction horizon step size	dt_{pred}	0.14	s
Controller time step size	dt_{ctrl}	0.03	s
Penalty of slack variable	q_{s}	$3{\cdot}10^7$	$1/\mathrm{m}^2$
Penalty of yaw angle	q_{φ}	10^4	$1/\mathrm{rad}^2$
Penalty of velocity	q_{v}	1	$\mathrm{s}^2/\mathrm{m}^2$
Penalty of steering torque command	r_{T}	10^{-1}	$1/\mathrm{Nm}^2$
Penalty of deceleration command	$r_{\mathrm{a_x}}$	10^{-1}	$\mathrm{s}^4/\mathrm{m}^2$
Limit of steering system	T_{max}	50	Nm
Limit of braking system	$a_{\mathrm{x,cmd,max}}$	-9.81	$\mathrm{m/s}^2$

Figure 4.11 shows the performance of the algorithm in the same scenario. Similarly to Figure 4.10 the maneuver is characterized by several time marks. A difference in the control characteristic can be found in the counter steering phase, between ② and ③, where the deceleration command is reduced to zero for a longer time period. This reduction of deceleration command ensures faster yaw angle reduction and avoids standstill of the vehicle without reorientation during this time period which can be explained by the high penalty on the yaw angle and the lower penalty on the velocity. The figure illustrates that zero velocity and yaw angle at the end of the maneuver at ④ is achieved.

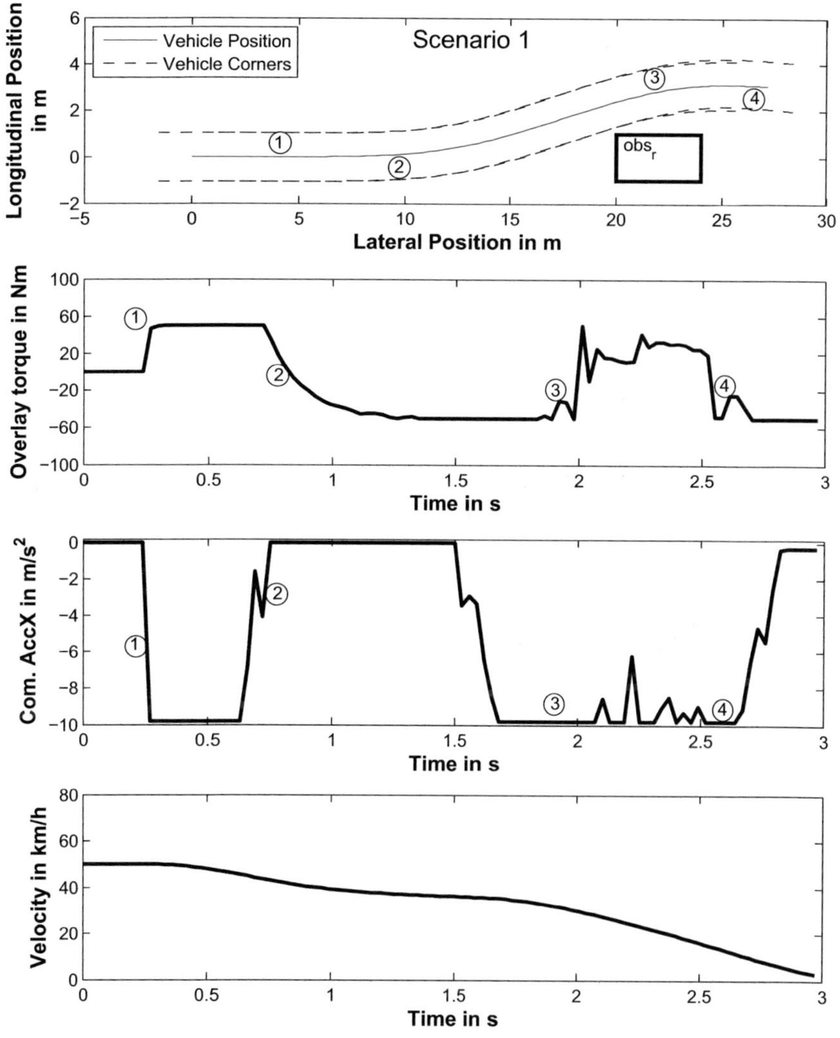

Figure 4.11: Simulation results of the nonlinear integrated trajectory planning and vehicle dynamics control method with parameter setup 2 in Scenario 1

In a last variation, the parameter setup 3 in Table 4.3 is used. This setup is very different compared to the setups 1 and 2. The step size of the prediction horizon is reduced by half, such that the algorithm has a shorter prediction horizon. The sampling time of the algorithm is increased to 100 ms which corresponds to a lower sampling frequency. The penalty on the velocity is decreased to zero and the penalty of both the command values are increased to high values.
Figure 4.12 shows the performance of the method with the parameter setup 3. Similarly to the previous figures, high steering torque and deceleration commands are applied at ①. However, the high command values are reduced after a short time and the counter steering phase is introduced at ② with a small deceleration command which can be attributed to high penalty on the command values. After passing the obstacle edge the command values are reduced continuously to zero.

Table 4.3: Parameter setup 3 for the Nonlinear Integrated Planning and Control Method

Parameter	Symbol	Value	Unit
Prediction horizon length	N	15	(none)
Prediction horizon step size	dt_{pred}	0.07	s
Controller time step size	dt_{ctrl}	0.1	s
Penalty of slack variable	q_s	$3 \cdot 10^7$	1/m^2
Penalty of yaw angle	q_φ	10^4	1/rad^2
Penalty of velocity	q_v	0	s^2/m^2
Penalty of steering torque command	r_T	10^0	1/Nm^2
Penalty of deceleration command	r_{a_x}	10^1	s^4/m^2
Limit of steering system	$T_{\max}$	50	Nm
Limit of braking system	$a_{x,\text{cmd},\max}$	-9.81	m/s^2

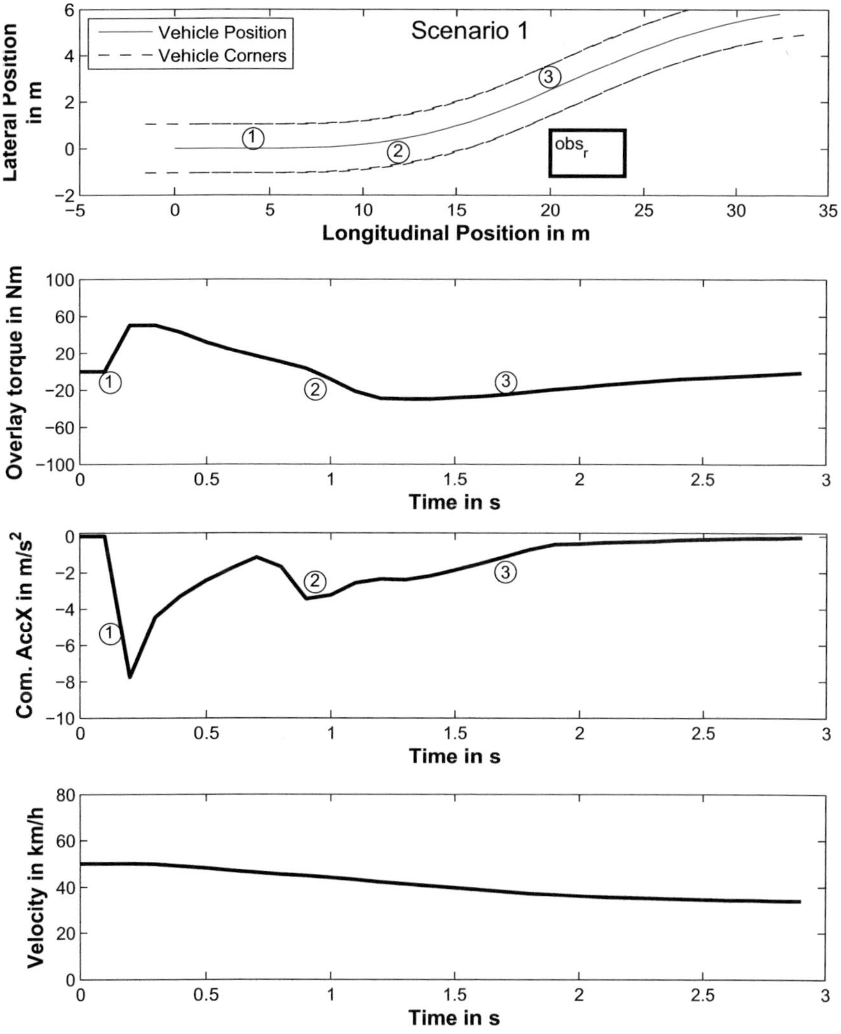

Figure 4.12: Simulation results of the nonlinear integrated trajectory planning and vehicle dynamics control method with parameter setup 3 in Scenario 1

The figure demonstrates that the vehicle avoids the collision with the obstacle and has a velocity of 35 km/h at the end of the maneuver. Overall, the algorithm applied low command values during the maneuver so that it does not brake the velocity to a standstill and reorientate the vehicle to zero yaw angle. This performance can as well be explained by high penalties on the command steering torque and deceleration command as well as zero penalty on the velocity in parameter setup 3.

The results with the three different parameter setups in Scenario 1 illustrate the influence of the key parameters on the maneuver performance. With parameter setup 1, the evasive collision avoidance maneuver applies nearly full braking throughout the evasive maneuver. The vehicle avoids a collision with the obstacle but does not reorientate the vehicle to a yaw angle of zero degree. By choosing a higher yaw angle penalty and lower velocity penalty (parameter setup 2), the vehicle is reorientated at the end of the maneuver but achieves slower reduction of the velocity during the evasion.

Higher penalty values on the applied command values (parameter setup 3) lead to an avoidance of the collision with limited braking intervention and without full reorientation.

For the purpose of this thesis, parameter setup 1 is further used due to higher reduction of velocity during the evasive maneuver. After demonstrating the method in Scenario 1 with a single obstacle, Scenario 2 with two obstacles shall be considered to further investigate the performance. In Figure 4.13 the results of the method with parameter setup 1 on Scenario 2 is shown.

Compared to the performance in Scenario 1 in Figure 4.10, the performance in the second scenario is characterized by a longer counter steering phase to avoid collision with the second obstacle. The braking characteristic is comparable to Scenario 1. The figure shows a peak reduction in deceleration command in the middle of the maneuver which helps the vehicle to achieve faster reorientation in a short time.

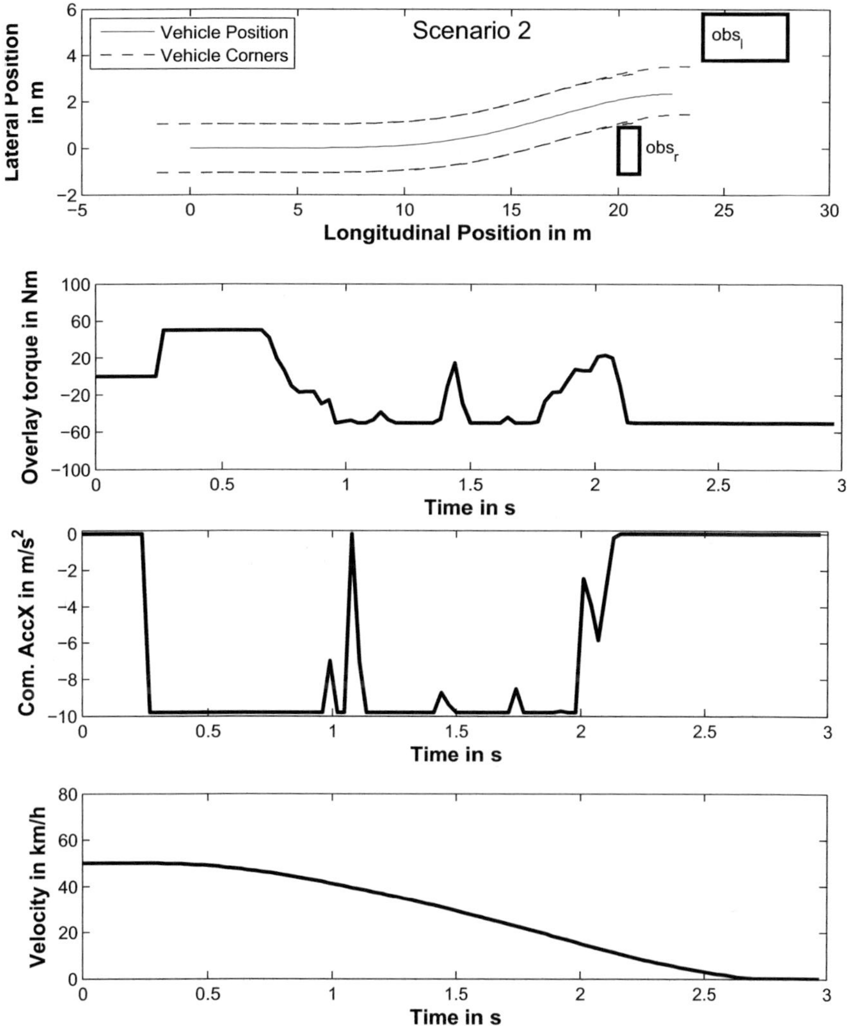

Figure 4.13: Simulation results of the nonlinear integrated trajectory planning and vehicle dynamics control method with parameter setup 1 in Scenario 2

4.7.4 Terminal Collision Avoidance

The terminal collision avoidance constraint introduced in Section 4.4 shall lead to trajectories avoiding a collision with obstacles outside of the prediction horizon.

In the first step of this section, the integrated method is applied without the terminal collision avoidance constraint in Scenario 2 with two obstacles while reusing parameter setup 3 (Table 4.3). The result is shown in Figure 4.14, where the ego vehicle drives around the first obstacle, but cannot avoid collision with the second obstacle, because the second obstacle is being considered in the planning horizon too late.

In a second step, the terminal collision avoidance constraint for a subsequent evasive maneuver shall be implemented and the method applied to Scenario 2 with the same parameter setup. The result in Figure 4.15 shows that collision with the obstacles has been successfully avoided by strong braking and steering at maneuver start followed by strong counter steering. It should be emphasized that this result could be achieved without extra tuning of the parameters.

In the third and final step, the terminal collision avoidance constraint for a braking maneuver in (4.16) can even be deployed to ensure collision avoidance in Scenario 3, which requires braking in front of a traffic jam as shown in Figure 4.16. This is achieved by considering the traffic jam from the beginning of the maneuver through the terminal collision avoidance constraint.

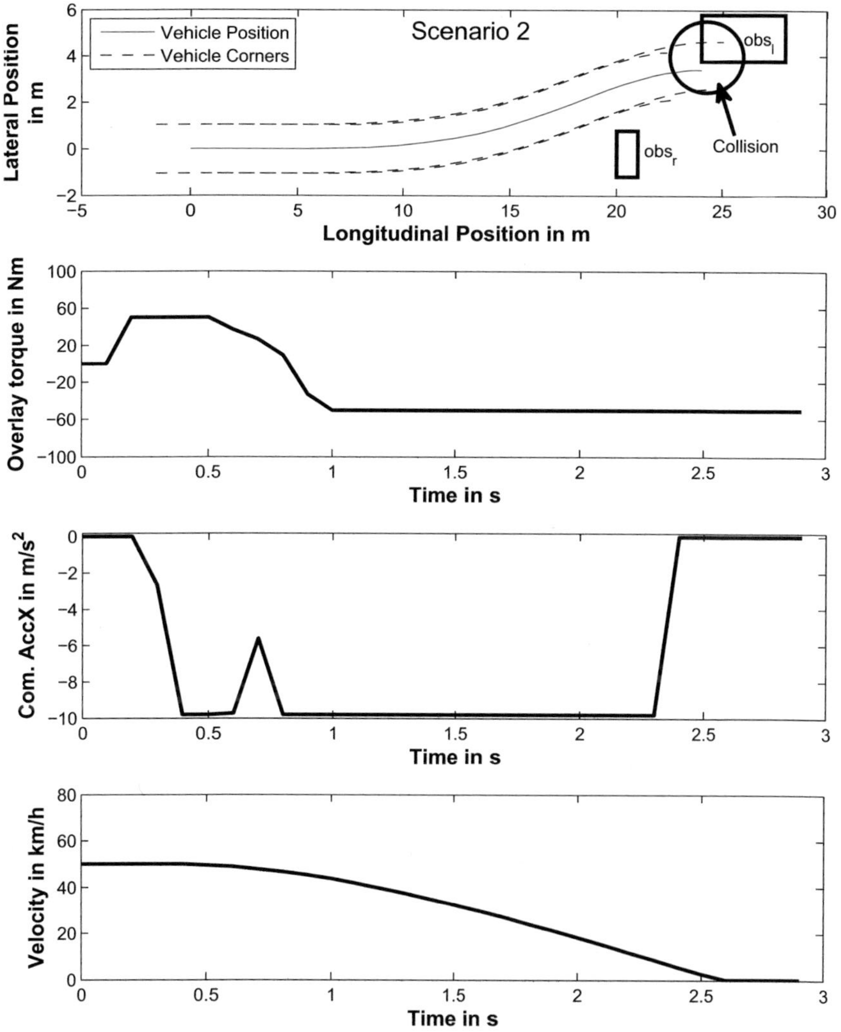

Figure 4.14: Simulation results of the nonlinear integrated trajectory planning and vehicle dynamics control method with parameter setup 3 in Scenario 2 without terminal collision avoidance constraint

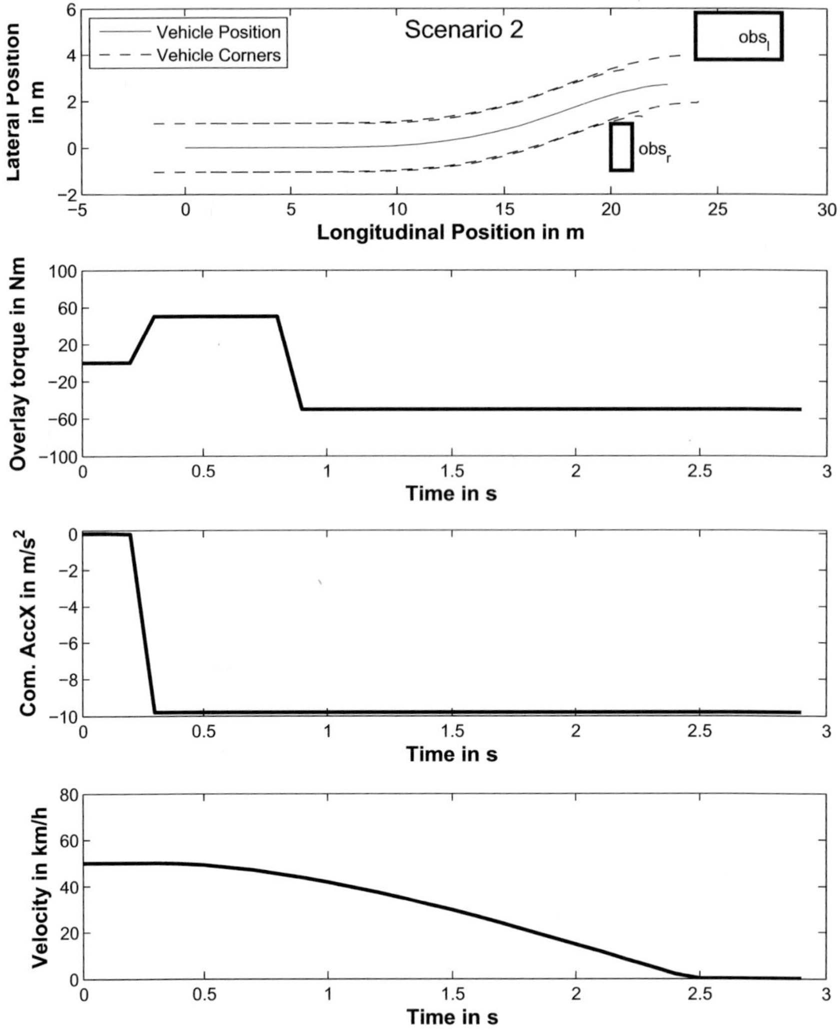

Figure 4.15: Simulation results of the nonlinear integrated trajectory planning and vehicle dynamics control method with parameter setup 3 in Scenario 2 with terminal collision avoidance constraint

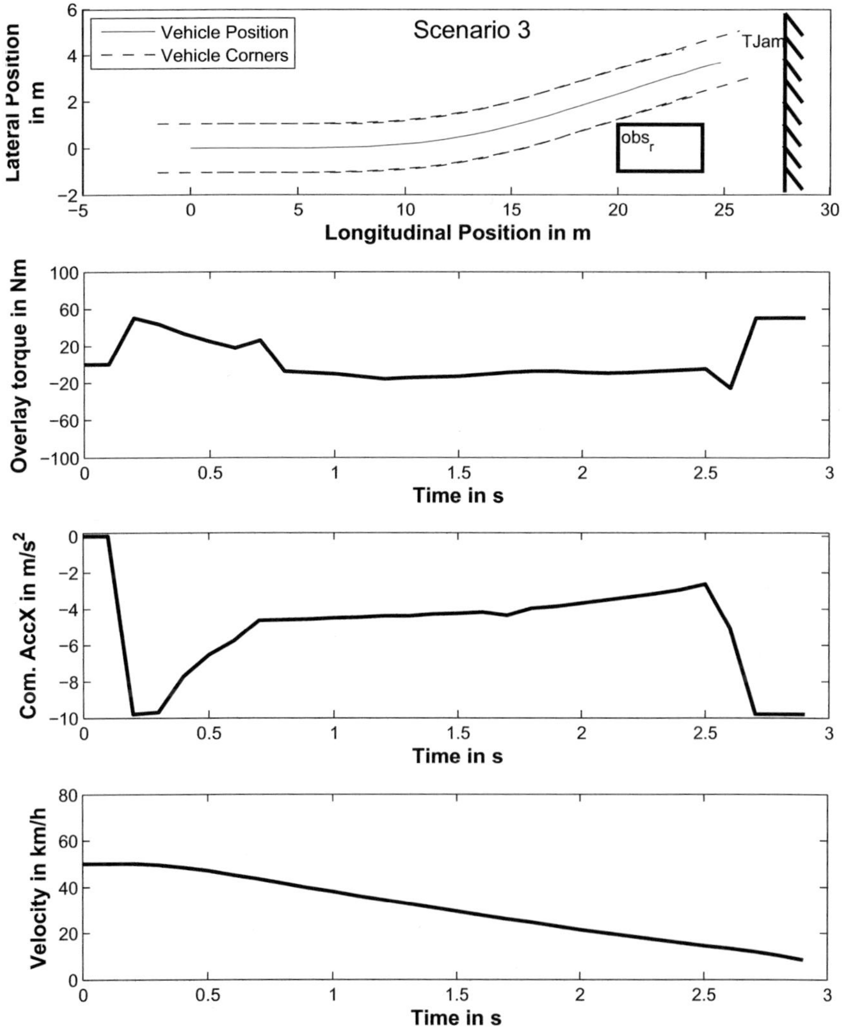

Figure 4.16: Simulation results of the nonlinear integrated trajectory planning and vehicle dynamics control method with parameter setup 3 in Scenario 3 with terminal collision avoidance constraint

4.8 Discussion

The integrated planning and control method considers a combined vehicle dynamics model and avoids collision by considering any obstacles as well as road information. The terminal collision avoidance constraint enables consideration of obstacles outside of the prediction horizon. The performance of the method on all scenarios has been discussed in Section 4.7. Table 4.4 shows the metric values introduced in Section 4.7.1, the relative velocity reduction at potential crash on the first obstacle $\mathrm{obs_r}$ and the final yaw angle, for the proposed method applied to Scenario 1 and 2 with parameter setup 1. The method successfully avoided the collision with the obstacles while reducing both the velocity and the yaw angle during the maneuver.

Table 4.4: Metric values of Nonlinear Integrated Planning and Control Method for typical scenarios in simulation

Scenarios	Velocity Reduction Δv_{red}	Final Yaw Angle φ_{fnl}	Computation Time t_{cmp}
Scenario 1	44.2 %	33 °	10 s - 15 s
Scenario 2	44.4 %	6 °	10 s - 15 s

The metric values of the proposed method and of the classical trajectory tracking method from Section 3.3 for Scenario 1 are shown in Table 4.5. The table demonstrates that the nonlinear integrated planning and control method achieves a higher performance level in terms of velocity reduction, but lower performance level for the final yaw angle in Scenario 1 compared to the classical concept. The performance of the integrated method is strongly influenced by the selection of the parameter setup as shown in section 4.7.3 and has been selected to deliever a maneuver performance with high velocity reduction in this thesis.

Table 4.5: Metric values of planning and control concepts for Scenario 1 in simulation

Metrics/ Requirements	Classical Tracking	Nonlinear Integrated
Velocity Reduction Δv_{red}	35 %	44.2 %
Final Yaw Angle φ_{fnl}	0 °	33 °
Computation Time t_{cmp}	0.006 s - 0.01 s	10 s - 15 s
Environmental Information	-	+
Tuning Effort	-	+

Furthermore, there are several key advantages of the proposed integrated planning and control concept compared to the classical control concept. First, the planning module is flexible and can adapt to scenarios with multiple obstacle and even newly detected obstacles to find a collision-free trajectory. Next, the terminal collision avoidance constraints enable planning of collision-free trajectories even with parameters not being tuned to the specific scenario and with obstacles outside the prediction horizon. Finally, the integrated concept reduces tuning and adaptation efforts of the trajectory planning module to the performance of the control module.

Challenges of the nonlinear integrated planning and control method are numerical weaknesses, such as convexity problems and problems at local minima, as well as the computation time. Due to the high computational burden required to solve the optimization problem, the method cannot be applied in real time on a test vehicle. In Chapter 5, these challenge will be addressed by suitable modifications of the proposed method.

5 Fast Integrated Planning and Control

The nonlinear integrated trajectory planning and vehicle dynamics control method presented in Chapter 4 achieves the desired goal of collision avoidance using nonlinear constraints to consider obstacles and a vehicle dynamics model for combined longitudinal and lateral dynamics in the simulative environment. As already mentioned before, a big weakness of the method is the high computational complexity which prevents its application in a real vehicle. This chapter addresses this weakness by exploiting linearization techniques and reformulating the nonlinear optimization problem into efficient linear optimization problems.

The structure of this enhanced method, to be called fast integrated planning and control method, is illustrated in Figure 5.1. The main difference to the structure in Chapter 4 is the linearization of all nonlinear constraints using the successive linearization technique. This technique and its application in this thesis will be explained in detail in Section 5.1. To compensate for linearization errors and uncertainties, a robust tube based approach is exploited to guarantee stability and constraint satisfaction in Section 5.2. The resulting MPC formulation is presented in Section 5.3. Finally, Section 5.4 shall demonstrate the performance of the fast integrated method for predefined scenarios in a simulation environment and in test drives with a real vehicle.

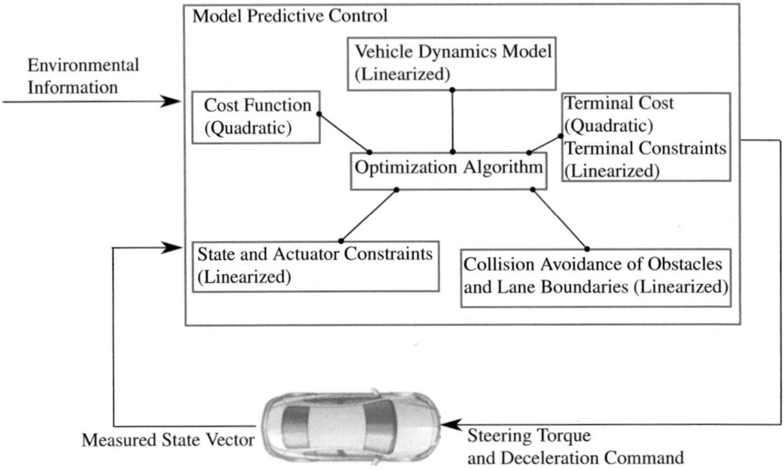

Figure 5.1: MPC structure of Fast Integrated Planning and Control Method

5.1 Successive linearization of nonlinear constraints

The successive linearization technique transforms nonlinear constraints into linear time varying constraints by linearizing nonlinearities around predicted operation points from the previous control step. It is a promising method for this task, allowing the use of efficient solvers [13][14][69][74].

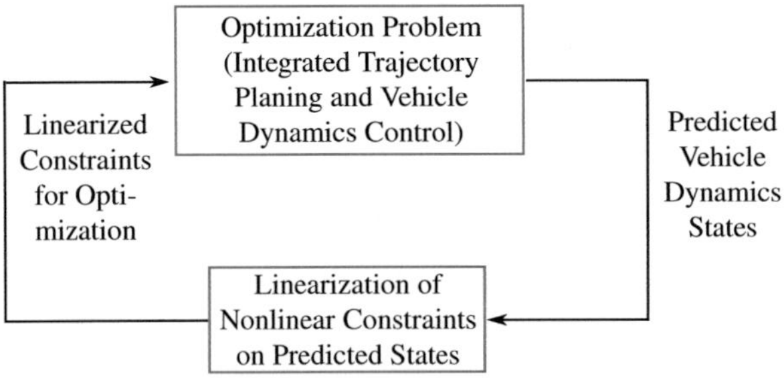

Figure 5.2: Principle of Successive Linerization

The principle of successive linearization in optimization problems is illustrated in Figure 5.2. The first module solves the optimization problem which may consist of the problem resulting from the integrated planning and control method. The outcomes are predicted vehicle dynamics states which can be used in a second step to linearize nonlinear constraints. With the linearized constraints, the optimization problem can be reformulated into an efficient convex optimization problem structure. Suitable solvers can then be applied to calculate the result. Successive linearization is based on valid prediction results in each consecutive control step and small control sampling time t_{ctrl}, so that consecutive predictions have little deviations. For the first control step, an appropriate trajectory must be chosen as base for the linearization of the nonlinear equations.
Several nonlinear constraints have been introduced for the nonlinear integrated planning and control method. The nonlinear vehicle dynamics model for combined lateral and longitudinal dynamics, with a single track model for the lateral dynamics and a PT1 characteristic for the longitudinal dynamics has been presented in Section 3.1.3. The nonlinear equation for the acceleration circle represents the maximum transferable forces of the tire according to the friction circle. Further, nonlinear collision avoidance constraints have been presented considering lateral position and yaw angle in dependence with the longitudinal position of the vehicle and of the obstacle.

5.1.1 Vehicle Dynamics Model

The vehicle dynamics model introduced in Section 3.1 is nonlinear due to the dependance of the lateral dynamics states with the longitudinal state, the velocity, (3.6). Assume the predicted state and input vector of the last control step $k-1$ for the prediction step i be given by $\mathbf{x}_{\text{i,k-1}}$ and $\mathbf{u}_{\text{i,k-1}}$, then the state space matrices of the linearized system for the prediction steps i (similarly to Section 3.3.2) are given by

$$\mathbf{A}_{c,i} = \frac{\partial \mathbf{f}(\mathbf{x}_{i,k-1}, \mathbf{u}_{i,k-1})}{\partial \mathbf{x}} \tag{5.1}$$

$$\mathbf{B}_{c,i} = \frac{\partial \mathbf{f}(\mathbf{x}_{i,k-1}, \mathbf{u}_{i,k-1})}{\partial \mathbf{u}} \tag{5.2}$$

$$\mathbf{E}_{c,i} = \mathbf{f}(\mathbf{x}_{i,k-1}, \mathbf{u}_{i,k-1}) - \mathbf{A}_{c,i}, \mathbf{x}_{i,k-1} - \mathbf{B}_{c,i}\mathbf{u}_{i,k-1} \tag{5.3}$$

and the state space equation is

$$\dot{\mathbf{x}} = \mathbf{A}_{c,i}\mathbf{x} + \mathbf{B}_{c,i}\mathbf{u} + \mathbf{E}_{c,i} \tag{5.4}$$

with $\mathbf{A}_{c,i} =$

$$\begin{bmatrix} A_{11} & A_{12} & 0 & 0 & 0 & \frac{c_{\alpha_f,i,k-1}}{m v_{i,k-1} i_L} & 0 & A_{18} & 0 \\ A_{21} & A_{22} & 0 & 0 & 0 & A_{26} & 0 & A_{28} & 0 \\ 0 & 1 & 0 & 0 & 0 & 0 & 0 & 0 & 0 \\ v & 0 & v & 0 & 0 & 0 & 0 & 0 & 0 \\ \frac{2D_f}{J_{st}} & \frac{2 l_f D_f}{v_{i,k-1} J_{st}} & 0 & 0 & -\frac{d_{st}}{J_{st}} & \frac{-2D_f}{i_L J_{st}} & 0 & A_{58} & 0 \\ 0 & 0 & 0 & 0 & 1 & 0 & 0 & 0 & 0 \\ 0 & 0 & 0 & 0 & 0 & 0 & 0 & 1 & 0 \\ 0 & 0 & 0 & 0 & 0 & 0 & 0 & 0 & -\frac{1}{t_{del}} \end{bmatrix}$$

$$A_{11} = -\frac{c_{\alpha_f,i,k-1} + c_{\alpha_r,i,k-1}}{m v_{i,k-1}}; A_{12} = \frac{c_{\alpha_r,i,k-1} l_r - c_{\alpha_f,i,k-1} l_f}{m v_{i,k-1}^2} - 1;$$

$$A_{18} = \frac{c_{\alpha_f,i,k-1} + c_{\alpha_r,i,k-1}}{m v_0^2} \beta_{i,k-1} - 2\frac{c_{\alpha_r,i,k-1} l_r - c_{\alpha_f,i,k-1} l_f}{m v_{i,k-1}^3} \dot{\varphi}_{i,k-1} - \frac{c_{\alpha_f,i,k-1}}{i_L m v_{i,k-1}^2} \delta_{i,k-1}$$

$$A_{21} = \frac{c_{\alpha_r,i,k-1} l_r - c_{\alpha_f,i,k-1} l_f}{J_z}; A_{22} = \frac{c_{\alpha_r,i,k-1} l_r^2 - c_{\alpha_f,i,k-1} l_f^2}{J_z v_{i,k-1}};$$

$$A_{26} = -\frac{c_{\alpha_\mathrm{r},\mathrm{i,k\text{-}1}} l_\mathrm{r}^2 + c_{\alpha_\mathrm{f},\mathrm{i,k\text{-}1}} l_\mathrm{f}^2}{J_z v_{\mathrm{i,k\text{-}1}}}$$

$$A_{28} = \frac{c_{\alpha_{f,i,k-1}} l_\mathrm{f}^2 + c_{\alpha_{r,i,k-1}} l_\mathrm{r}^2}{J_z v_{\mathrm{i,k\text{-}1}}^2} \dot{\varphi}_{\mathrm{i,k\text{-}1}}; A_{58} = \frac{-2 l_\mathrm{f} D_\mathrm{f}}{J_\mathrm{st} v_{\mathrm{i,k\text{-}1}}^2} \dot{\varphi}_{\mathrm{i,k\text{-}1}};$$

$$\mathbf{B}_{\mathrm{c,i}} = \begin{bmatrix} 0 & 0 \\ 0 & 0 \\ 0 & 0 \\ 0 & 0 \\ \frac{1}{J_\mathrm{st}} & 0 \\ 0 & 0 \\ 0 & 0 \\ 0 & 0 \\ 0 & \frac{1}{t_{del}} \end{bmatrix} \mathbf{E}_{\mathrm{c,i}} = \begin{bmatrix} \frac{F_{\mathrm{yf,0,i,k\text{-}1}} + F_{\mathrm{yr,0,i,k\text{-}1}}}{m v_{\mathrm{i,k\text{-}1}}} - A_{18} v_{\mathrm{i,k\text{-}1}} \\ \frac{F_{\mathrm{yf,0,i,k\text{-}1}} l_f + F_{\mathrm{yr,0,i,k\text{-}1}} l_r}{J_\mathrm{z}} - A_{28} v_{\mathrm{i,k\text{-}1}} \\ 0 \\ 0 \\ -A_{58} v_{\mathrm{i,k\text{-}1}} \\ 0 \\ 0 \\ 0 \\ 0 \end{bmatrix}$$

Note that $v_{\mathrm{i,k\text{-}1}}$, $\varphi_{i,k-1}$, $\dot{\varphi}_{\mathrm{i,k\text{-}1}}$ and $\beta_{\mathrm{i,k\text{-}1}}$ are the velocity, yaw angle, yaw rate and side slip angle predicted for the prediction step i at control time step $k-1$ and $c_{\alpha_\mathrm{f},\mathrm{i,k\text{-}1}}, c_{\alpha_\mathrm{r},\mathrm{i,k\text{-}1}}, F_{\mathrm{yf,0,i,k\text{-}1}}$ as well as $F_{\mathrm{yr,0,i,k\text{-}1}}$ correspond to predicted cornering stiffnesses and tire force biases for the predicted slip angles $\alpha_{\mathrm{r,i,k\text{-}1}}$ and $\alpha_{\mathrm{f,i,k\text{-}1}}$, the predicted wheel loads $F_{\mathrm{zf,i,k\text{-}1}}$ and $F_{\mathrm{zr,i,k\text{-}1}}$ as well as the predicted friction coefficient $\mu_{\mathrm{yf,i,k\text{-}1}}$ and $\mu_{\mathrm{yr,i,k\text{-}1}}$ with

$$c_{\alpha_\mathrm{f},\mathrm{i,k\text{-}1}} = F_{\mathrm{zf,i,k\text{-}1}} \frac{\partial \mu_\mathrm{y}}{\partial \alpha_\mathrm{f}} |_{\alpha_{\mathrm{f,i,k\text{-}1}}} \tag{5.5}$$

$$c_{\alpha_\mathrm{r},\mathrm{i,k\text{-}1}} = F_{\mathrm{zr,i,k\text{-}1}} \frac{\partial \mu_\mathrm{y}}{\partial \alpha_\mathrm{r}} |_{\alpha_{\mathrm{r,i,k\text{-}1}}}$$

$$F_{\mathrm{f,0,i,k\text{-}1}} = F_{\mathrm{zf,i,k\text{-}1}} \mu_{\mathrm{y,i,k\text{-}1}} - c_{\alpha_\mathrm{f},\mathrm{i,k\text{-}1}} \alpha_{\mathrm{f,i,k\text{-}1}} \tag{5.6}$$

$$F_{\mathrm{r,0,i,k\text{-}1}} = F_{\mathrm{zr,i,k\text{-}1}} \mu_{\mathrm{y,i,k\text{-}1}} - c_{\alpha_\mathrm{r},\mathrm{i,k\text{-}1}} \alpha_{\mathrm{r,i,k\text{-}1}}$$

To illustrate the effect of successive linearization for the vehicle dynamics model, Figure 5.3 shows the linearization of the tire equation for consecutive control steps. Assume the nonlinear tire model be given in Figure 5.3.a.
At maneuver start $k = 0$ in Figure 5.3.b, the tire equation for all predicted operation points are linearized at the same constant angle, the slip angle at the current time step α_0. The linearized tire equations for the predicted time steps are represented by g_1, g_2 and g_3. At the following step $k = 1$ in Figure 5.3.c, the predicted slip angle positions taken from the result of the optimization problem are considered for the linearization of the tire equation. Following this principle the predicted operation point will be adapted at each consecutive control step as shown exemplarily for $k = 2$ in Figure 5.3.d. The linearized tire parameters can then be used for the linearized vehicle dynamics model.
For the MPC method, the linearized continuous vehicle dynamics model equation 5.4 will be discretized, such that the discrete state space equation is

$$\mathbf{x}_{i+1} = \mathbf{A}_{\mathrm{i+1}} \mathbf{x}_i + \mathbf{B}_\mathrm{i} \mathbf{u}_i + \mathbf{E} \ \ \forall \mathrm{i} = 0, \cdots, \mathrm{N} \tag{5.7}$$

The affine time varying state space equation is continuous on the state $\mathbf{x}$ and input $\mathbf{u}$ at each prediction step, such that Assumption 1 is satisfied. This affine model will be transformed into a linear model in Section 5.2 to calculate the penalty matrix and to apply the robust control design in Section 3.2.2.

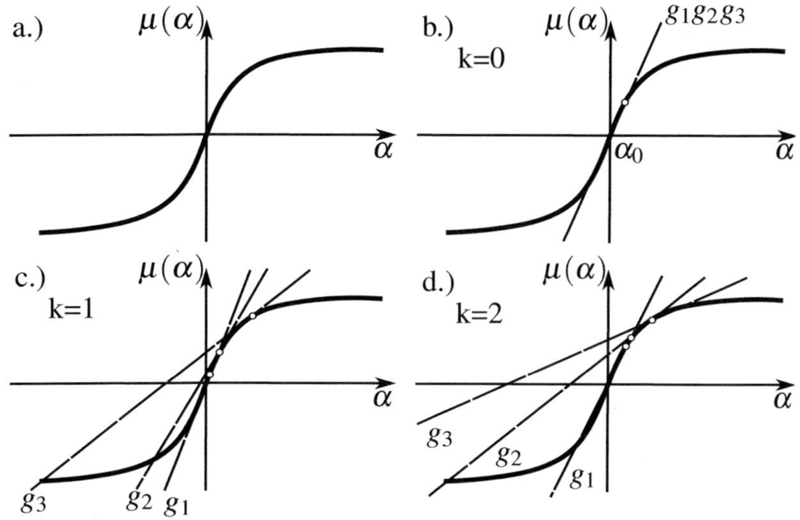

Figure 5.3: Successive linearization of tire equation - a.) Nonlinear Tire model b.) Linearization on current slip angle α_0 c.) Linearization on predicted operation points d.) Linearization on consecutive prediction points

5.1.2 Collision Avoidance Constraints

Collision avoidance has been achieved using nonlinear inequality constraints in Chapter 4. The nonlinear collision avoidance constraints depends on the longitudinal position, the lateral position and the yaw angle. In this section, the nonlinear constraints will be transformed to linear constraints for the lateral position and the yaw angle. According to the principles of successive linearization, the longitudinal position at each control step can be estimated using the longitudinal position prediction of the last control step. With the predicted positions, suitable constraints for the lateral position and the yaw angle can be determined.

In Chapter 4 the nonlinear inequality constraints in (4.5) required the calculation of y_{Br} and y_{Bl} given by

$$\begin{aligned} y_{\text{Br}} &= y - \frac{w}{2\cos\varphi(k)} + \Delta x \tan\varphi \\ y_{\text{Bl}} &= y + \frac{w}{2\cos\varphi(k)} + \Delta x \tan\varphi \end{aligned} \tag{5.8}$$

and the constraints

$$\begin{aligned} y_{\text{Br}} &> y_{\text{obsr}} + w_{\text{obsr}}/2 \\ y_{\text{Bl}} &< y_{\text{obsr}} - w_{\text{obsr}}/2 \;. \end{aligned} \tag{5.9}$$

In this section, the inequalities are first simplified using small angles approximation to

$$\begin{aligned} y_{\text{Br}} &= y - \frac{w}{2} + \Delta x \varphi \\ y_{\text{Bl}} &= y + \frac{w}{2} + \Delta x \varphi . \end{aligned} \tag{5.10}$$

Note that (4.5) and (4.7) involve nonlinearities such as if-else structures and min-max operations. The effects of these nonlinear operations need to be approximated by linear equations. First, the if-else structure, which depends on the vehicle states of the predicted step i, is replaced by an approximation using predicted vehicle states of time step $k-1$ and the corresponding prediction step i. The equations are given as

$$\begin{aligned} & y_{\text{Br}} \geq y_{\text{obsr}} + w_{\text{obsr}}/2 \\ & \forall (x_{\text{fr,i,k-1}} > x_{\text{obsr}}) \cap (x_{\text{rr,i,k-1}} < x_{\text{obsr}} + l_{\text{obsr}}) \end{aligned} \tag{5.11}$$

$$\begin{aligned} & y_{\text{Bl}} \leq y_{\text{obsl}} - w_{\text{obsl}}/2 \\ & \forall (x_{\text{fl,i,k-1}} > x_{obsl}) \cap (x_{\text{rl,i,k-1}} < x_{\text{obsl}} + l_{\text{obsl}}) \;. \end{aligned} \tag{5.12}$$

The parameters $x_{\text{fr,i,k-1}}$, $x_{\text{fl,i,k-1}}$, $x_{\text{rr,i,k-1}}$ and $x_{\text{rl,i,k-1}}$ are obtained from the preceding optimization step, such that the dependancy on the longitudinal position of the current control step k is eliminated and inequality constraints become linear. The collision avoidance set is transformed to

$$\begin{aligned}\mathbb{C} := \{\mathbf{x} \in \mathbb{R}^9 \mid\ & (y_{\text{Br}} \geq y_{\text{obsr}} + w_{\text{obsr}}/2, \forall x \in \mathbb{X}_{\text{or}}) \cap \\ & (y_{\text{Bl}} \leq y_{\text{obsl}} - w_{\text{obsl}}/2, \forall x \in \mathbb{X}_{\text{ol}})\}\end{aligned}$$

with

$$\begin{aligned}\mathbb{X}_{\text{or}} &:= \{x \in \mathbb{R} \mid (x_{\text{fr,i,k-1}} > x_{\text{obsr}}) \cap (x_{\text{rr,i,k-1}} < x_{\text{obsr}} + l_{\text{obsr}})\} \\ \mathbb{X}_{\text{ol}} &:= \{x \in \mathbb{R} \mid (x_{\text{fl,i,k-1}} > x_{\text{obsl}}) \cap (x_{\text{rl,i,k-1}} < x_{\text{obsl}} + l_{\text{obsl}})\} \, .\end{aligned}$$

The approach is illustrated in Figure 5.4. At maneuver start $k = 0$, when no longitudinal position predictions are available, the considered longitudinal position for the prediction steps are approximated by the constant velocity assumption (constant measured velocity $v_{0,\text{k}}$ for all prediction steps) illustrated with the equidistant position marks (vertical dashed lines) for the corresponding prediction time steps $i = 0, \cdots, N$ in Figure 5.4.a. After solving the optimization problem, a trajectory will result in Figure 5.4.b which satisfies the linearized constraints on the lateral position from the considered longitudinal positions in Figure 5.4.a. The predicted longitudinal positions extracted from the result of the optimization problem in Figure 5.4.b can then be used to update the linearized lateral position constraints. Note that the predicted longitudinal position between consecutive control steps change, in general, due to change in the predicted velocities between each control step. This procedure continues with each consecutive control step, shown exemplarily for $k = 0$ and $k = 1$ in Figure 5.4.b and 5.4.c.

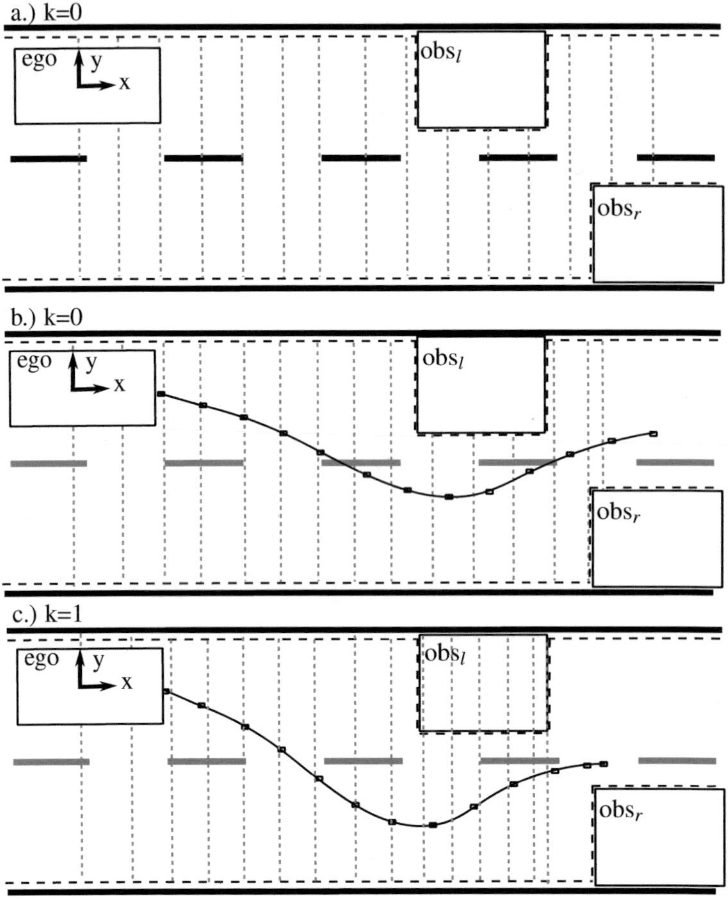

Figure 5.4: Successive linearization of collision avoidance constraints; a.) Constant velocity prediction at maneuver start leads to equidistant longitudinal position distances (vertical dashed lines) for predictive time steps; (b) Calculated trajectory satisfying constraints at lateral position; (c) Predicted longitudinal positions are used for further linearization of constraints in the next time step

Following this principle the constraints to keep the vehicle in the lane

$$\begin{aligned} \min(y_{\mathrm{fr}}, y_{\mathrm{rr}}) &> y_{\mathrm{R}}(x) \\ \max(y_{\mathrm{fl}}, y_{\mathrm{rl}}) &< y_{\mathrm{L}}(x) \\ \text{with } y_{\mathrm{fr}} &= y + l_{\mathrm{f}}\sin(\varphi) - \frac{w}{2}\cos(\varphi) \\ y_{\mathrm{rr}} &= y - l_{\mathrm{r}}\sin(\varphi) - \frac{w}{2}\cos(\varphi) \\ y_{\mathrm{fl}} &= y + l_{\mathrm{f}}\sin(\varphi) + \frac{w}{2}\cos(\varphi) \\ y_{\mathrm{rl}} &= y - l_{\mathrm{r}}\sin(\varphi) + \frac{w}{2}\cos(\varphi) \end{aligned} \tag{5.13}$$

can be transformed accordingly to

$$\begin{aligned} y + c_{\mathrm{r}}\varphi &> y_{\mathrm{R}}(x_{\mathrm{i,k\text{-}1}}) + \frac{w}{2} \\ y + c_{\mathrm{l}}\varphi &< y_{\mathrm{L}}(x_{\mathrm{i,k\text{-}1}}) - \frac{w}{2} \\ \text{with } c_{\mathrm{r}} &= \begin{cases} l_{\mathrm{f}} & y_{\mathrm{fr,i,k\text{-}1}} < y_{\mathrm{rr,i,k\text{-}1}} \\ -l_{\mathrm{r}} & y_{\mathrm{fr,i,k\text{-}1}} > y_{\mathrm{rr,i,k\text{-}1}} \end{cases} \\ c_l &= \begin{cases} -l_{\mathrm{r}} & y_{\mathrm{fl,i,k\text{-}1}} < y_{\mathrm{rl,i,k\text{-}1}} \\ l_{\mathrm{f}} & y_{\mathrm{fl,i,k\text{-}1}} > y_{\mathrm{rl,i,k\text{-}1}} \end{cases} \quad . \end{aligned} \tag{5.14}$$

Here $y_{\mathrm{fr,i,k\text{-}1}}$, $y_{\mathrm{fl,i,k\text{-}1}}$, $y_{\mathrm{rr,i,k\text{-}1}}$ and $y_{\mathrm{rl,i,k\text{-}1}}$ are predicted positions for the vehicle corners of the last control step $k-1$ for the prediction step i. The corresponding set is transformed to

$$\begin{aligned} \mathbb{L}_q = \{\mathbf{x} \in \mathbb{R}^9 \mid\ & y + c_{\mathrm{r}}\varphi \geq y_{\mathrm{R}}(x_{\mathrm{i,k\text{-}1}}) + \frac{w}{2}\,, \\ & y + c_{\mathrm{l}}\varphi \leq y_{\mathrm{L}}(x_{\mathrm{i,k\text{-}1}}) - \frac{w}{2}\} \\ \text{with } c_r = & \begin{cases} l_{\mathrm{f}} & y_{\mathrm{fr,i,k\text{-}1}} < y_{\mathrm{rr,i,k\text{-}1}} \\ -l_{\mathrm{r}} & y_{\mathrm{fr,i,k\text{-}1}} > y_{\mathrm{rr,i,k\text{-}1}} \end{cases} \end{aligned} \tag{5.15}$$

$$c_l = \begin{cases} -l_\mathrm{r} & y_{\mathrm{fl,i,k\text{-}1}} < y_{\mathrm{rl,i,k\text{-}1}} \\ l_\mathrm{f} & y_{\mathrm{fl,i,k\text{-}1}} > y_{\mathrm{rl,i,k\text{-}1}} \end{cases} .$$

The sets $\mathbb{C}_\mathrm{q}$ and $\mathbb{L}_\mathrm{q}$ are closed sets for the state vector, fulfilling Assumption 2. A fundamental difference between the nonlinear collision avoidance constraints in Section 4.3 and the collision avoidance constraints in this section is that the nonlinearity is linearized prior to the optimization in this section so that the quadratic solver calculates the solution based on linearized position constraints for each prediction step.

5.1.3 Terminal Collision Avoidance

Section 4.4 introduced nonlinear terminal collision avoidance constraints to consider relevant obstacles outside of the prediction horizon. Successive linearization will be exploited in this section to acquire linear terminal collision avoidance constraints.

The nonlinear constraint for Scenario 3 in Section 4.4 is

$$\begin{aligned} \Delta x &\geq x_\mathrm{B} \\ \equiv (x_\mathrm{Tjam} - x(N)) &\geq \frac{v(N)^2 \cos(\varphi(N))}{2a_\mathrm{x}(N)} . \end{aligned} \tag{5.16}$$

This inequality can be linearized to

$$\begin{aligned} &\equiv (x_\mathrm{Tjam} - x(N)) \geq \frac{v_0^2 \cos(\varphi_0(N))}{2a_{\mathrm{x}0}} + \frac{v_0 \cos(\varphi_0(N))}{a_{\mathrm{x}0}}(v(N) - v_0(N)) \\ &- \frac{v_0^2 \cos(\varphi_0(N))}{2a_{\mathrm{x},0}^2}(a_x(N) - a_{\mathrm{x},0}) - \frac{v_0^2 \sin(\varphi_0(N))}{2a_{\mathrm{x},0}}(\varphi(N) - \varphi_0(N)) \\ &\equiv \mathbf{L}_\mathrm{B}\mathbf{x}(N) \leq c_\mathrm{B} \end{aligned}$$

where $v_0(N), \varphi_0(N)$ and $a_{\mathrm{x},0}(N)$ correspond to predicted state values of the last control time for the N-th prediction step.

Similarly the nonlinear inequality for Scenario 2 is

$$\Delta x(N) \geq x_{\mathrm{E}}(N) \tag{5.17}$$

$$\equiv (x_{\mathrm{obsl}} - x(N)) \geq v(N)\sqrt{\frac{2y_{\mathrm{E}}(N)}{a_{\mathrm{y}}(N)}}$$

and the linear constraints are calculated using the Taylor linearization below. For convenience of calculation y_{E} has been set to the constant predicted value of previous control step $y_{\mathrm{E},0}$ and is not linearized as a variable.

$$\begin{aligned}
&\equiv (x_{\mathrm{obsl}} - x_0(N))^2(\dot{\varphi}(N) + \dot{\beta}(N)) + v_0(N)y_{\mathrm{E},0}\varphi(N) \\
&\quad - 2(\dot{\varphi}_0(N) + \dot{\beta}_0(N))(x_{\mathrm{obsl}} - x_0(N))x(N) \\
&\quad + \tan\varphi_0(N)y_{\mathrm{E},0}(N)v(N) \leq -v_0(N)y_{\mathrm{E},0}(N)\tan\varphi_0(N) \\
&\quad - 2(\dot{\varphi}_0(N) + \dot{\beta}_0(N))(x_{\mathrm{obsl}} - x_0(N))x_0(N) + 2\varphi_0(N)y_{\mathrm{E},0}(N)v(N) \\
&\equiv \mathbf{L}_{\mathrm{N}}\mathbf{x}_{\mathrm{N}} + \mathbf{L}_{\mathrm{N\text{-}1}}\mathbf{x}_{\mathrm{N\text{-}1}} \leq c_{\mathrm{E}}
\end{aligned}$$

The sets to consider for collision avoidance in the scenarios are

$$\mathbb{G}_{\mathrm{B}} := \{\mathbf{x} \in \mathbb{R}^9 | \mathbf{L}_{\mathrm{B}}\mathbf{x}(N) \leq c_{\mathrm{B}}\} \quad \text{and} \tag{5.18}$$

$$\mathbb{G}_{\mathrm{E}} := \{\mathbf{x} \in \mathbb{R}^9 | \mathbf{L}_{\mathrm{N}}\mathbf{x}(N) + \mathbf{L}_{\mathrm{N\text{-}1}}\mathbf{x}(N-1) \leq c_{\mathrm{E}}\}\,. \tag{5.19}$$

5.1.4 Acceleration Circle

The nonlinear acceleration circle, representing the effect of the friction circle on each wheel, introduced in Section 3.1.4 will be modified in this section. A similar approximation approach was first introduced in [19], where the nonlinear friction circle is replaced by several linear inequalities. In this thesis, the circle will be approximated by the linear inequalities illustrated in Figure 5.5.

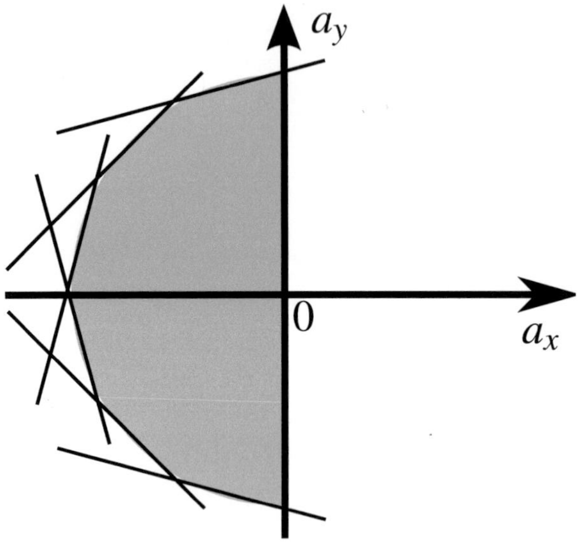

Figure 5.5: Approximation of the friction circle by linear inequalities

The considered model with the vehicle dynamics states does not directly represent the lateral acceleration but can be calculated by

$$a_{\mathrm{y}}(i,k) = \left(\frac{\beta(i+1,k) - \beta(i,k)}{t_{\mathrm{pred}}} + \dot{\varphi}(i,k) \right) \cdot v(i,k) \ . \tag{5.20}$$

This equation is not linear on the vehicle dynamic states in (3.21) and will again be modified by successive linearization. In this section, successive linearization is proposed to approximate the result. In the equation above $v(i,k)$ will be substituted by $v_{\mathrm{i,k\text{-}1}}$, which is the predicted speed value of the preceding control time $k-1$ for the prediction step i.

$$a_{\mathrm{y}}(i,k) = \left(\frac{\beta(i+1,k) - \beta(i,k)}{t_{\mathrm{pred}}} + \dot{\varphi}(i,k) \right) \cdot v_{\mathrm{i,k\text{-}1}} \tag{5.21}$$

The approximation of the acceleration circle with 6 linear inequalities is expressed by

$$\mathbf{L}_{\mathrm{y}}\left(\frac{\beta(i+1,k)-\beta(i,k)}{t_{\mathrm{pred}}}+\dot{\varphi}(i,k)\right)\cdot v_{\mathrm{i,k\text{-}1}}+\mathbf{L}_{\mathrm{x}}a_{\mathrm{x}}(i,k)<\mathbf{M} \tag{5.22}$$

$$\mathbf{L}_{\mathrm{y}}=\begin{bmatrix}\sin((4)\frac{2\pi}{8}-\frac{\pi}{8})\\ \sin((5)\frac{2\pi}{8}-\frac{\pi}{8})\\ \vdots\\ \sin((9)\frac{2\pi}{8}-\frac{\pi}{8})\end{bmatrix},\mathbf{L}_{\mathbf{x}}=\begin{bmatrix}\cos((4)\frac{2\pi}{n}-\frac{\pi}{8})\\ \cos((5)\frac{2\pi}{n}-\frac{\pi}{8})\\ \vdots\\ \cos((9)\frac{2\pi}{n}-\frac{\pi}{8})\end{bmatrix},\mathbf{M}=\begin{bmatrix}\lambda\mu a_{\max}\\ \lambda\mu a_{\max}\\ \vdots\\ \lambda\mu a_{\max}\end{bmatrix}$$

with the parameter $\lambda=\frac{5\pi}{12}$ as proposed by [19]. The set of possible vehicle dynamics states is transformed to

$$\mathbb{V}_{\mathrm{q}}=\{\mathbf{x}\in\mathbb{R}^{9}\mid v\geq 0,\ a_{\mathrm{x}}(k)\leq 0 \quad \text{and} \tag{5.23}$$

$$\mathbf{L}_{\mathrm{y}}\left(\frac{\beta(i+1,k)-\beta(i,k)}{t_{\mathrm{pred}}}+\dot{\varphi}(i,k)\right)\cdot v_{\mathrm{i,k\text{-}1}}+\mathbf{L}_{\mathrm{x}}a_{\mathrm{x}}(i,k)<\mathbf{M}. \tag{5.24}$$

The inequality constraints in the set are closed fulfilling Assumption 2.

5.2 Compensation of Linearization Errors

In the last section approximation techniques have been presented to linearize nonlinear constraints of the nonlinear integrated planning and control method. By exploiting these techniques the nonlinear optimization problem can be reformulated into the anticipated quadratic program to achieve an efficient, fast optimization problem. However, linearization of the nonlinear constraints leads to the side effects of linearization errors which can threaten the stability and the satisfaction of constraints. This section shall show how these errors can be considered to ensure stability and constraint satisfaction by exploiting the principles of robust tube based MPC presented in Section 3.2.2.

Following this method, a linear feedback vector needs to be calculated and constraint tightening shall be applied. The linear feedback vector **k**, similarly to Section 3.3.5, is calculated by solving the Ricatti equation for the matrix **P** with the linear state space model

$$\begin{bmatrix} \mathbf{x}_{k+1} \\ 1 \end{bmatrix} = \begin{bmatrix} \mathbf{A}(\theta) & \mathbf{E}(\theta) \\ 0 & 1 \end{bmatrix} \begin{bmatrix} \mathbf{x}_k \\ 1 \end{bmatrix} + \begin{bmatrix} \mathbf{B}(\theta) \\ 0 \end{bmatrix} \mathbf{u}_k \tag{5.25}$$

which have been transformed from (5.7). The Ricatti equation is then given by

$$\begin{bmatrix} \mathbf{A}(\theta) & \mathbf{E}(\theta) \\ 0 & 1 \end{bmatrix}^T \mathbf{P} + \mathbf{P} \begin{bmatrix} \mathbf{A}(\theta) & \mathbf{E}(\theta) \\ 0 & 1 \end{bmatrix} \leq \begin{bmatrix} -\mathbf{Q} & 0 \\ 0 & 0 \end{bmatrix} \tag{5.26}$$

for $\theta \in \Theta$ and **Q** as defined in (4.1). Here Θ is the set of 2^3 possible parameter combinations describing the extreme points of the uncertain parameter vector similar to the robust control design in [84].

$$\theta = [v, \alpha_{\mathrm{r}}, \alpha_{\mathrm{f}}] \tag{5.27}$$

with the set given by their limits

$$\Theta := \{\theta \in \mathbb{R}^3 | v \in [v_{\mathrm{min}}, v_{\mathrm{max}}], \alpha_{\mathrm{r}} \in [0, \alpha_{\mathrm{r,max}}], \alpha_{\mathrm{f}} \in [0, \alpha_{\mathrm{f,max}}]\} \tag{5.28}$$

This problem has been solved by linear matrix inequalities [44] in Section 3.3.5, so that the feedback vector can stabilize the system in any dynamics range of the maneuver defined by the limits $v_{\mathrm{min}}, v_{\mathrm{max}}, \alpha_{\mathrm{r,max}}$ and $\alpha_{\mathrm{f,max}}$. Here **Q** is positive semi-definite and **P** is symmetric positive semi-definite. The linear feedback vector can then be calculated by

$$\mathbf{k} = -\mathbf{R}^{-1} \begin{bmatrix} \mathbf{B}(\theta) \\ 0 \end{bmatrix}^T \mathbf{P}. \tag{5.29}$$

The terminal weight matrix $\mathbf{P}_\mathrm{f}$is calculated by

$$\mathbf{A}_\mathrm{k}(\theta)'\mathbf{P}_\mathrm{f}\mathbf{A}_\mathrm{k}(\theta) - \mathbf{P}_\mathrm{f} \leq -\mathbf{Q}, \tag{5.30}$$

$$\mathbf{A}_\mathrm{k}(\theta) = \begin{bmatrix} \mathbf{A}(\theta) & \mathbf{E}(\theta) \\ 0 & 1 \end{bmatrix}^T + \begin{bmatrix} \mathbf{B}(\theta) \\ 0 \end{bmatrix} \mathbf{k}$$

with $\mathbf{A}(\theta)$ and $\mathbf{B}(\theta)$ being the state space matrices with the parameter vector $\theta \in \Theta$. Equation (5.30) is derived from Assumption 6 with

$$V_\mathrm{f} = [\mathbf{x}; 1]^T \mathbf{P}_\mathrm{f}[\mathbf{x}; 1] \tag{5.31}$$

$$\mathbb{X}_\mathrm{f} := \{\mathbf{x} \in \mathbb{R}^9 | v_{\min} \leq v \leq v_{\max}, \tag{5.32}$$

$$-\alpha_{\mathrm{f,max}} \leq \alpha_\mathrm{f} \leq \alpha_{\mathrm{f,max}},$$

$$-\alpha_{\mathrm{r,max}} \leq \alpha_r \leq \alpha_{\mathrm{r,max}} \quad .$$

Let $\mathbb{X}$ and $\mathbb{U}$ be defined by

$$\mathbb{X} = \mathbb{S} \cap \mathbb{L} \cap \mathbb{C}_\mathrm{q} \cap \mathbb{V}_\mathrm{q}$$

$$\mathbb{X} := \{\mathbf{c}_\mathrm{i}\mathbf{x} \leq d_\mathrm{i}\}$$

$$\mathbb{U} := \{\mathbf{a}_\mathrm{i}\mathbf{u} \leq b_\mathrm{i}\}$$

with i being the index over all considered constraints, then the sets $\mathbb{Z}$ and $\mathbb{V}$ for the nominal system can be given by

$$\mathbb{Z} := \{\mathbf{c}_\mathrm{i}\mathbf{z} \leq d_\mathrm{i} - \max_\mathrm{j}\{\mathbf{c}_\mathrm{i}\mathbf{e}_\mathrm{ij} | \mathbf{w}_\mathrm{j} \in \mathbb{W}\}\} \tag{5.33}$$

$$\mathbb{V} := \{\mathbf{a}_\mathrm{i}\mathbf{v} \leq b_\mathrm{i} - \max_\mathrm{j}\{\mathbf{a}_\mathrm{i}\mathbf{k}\mathbf{e}_\mathrm{ij} | \mathbf{w}_\mathrm{i,j} \in \mathbb{W}\}\} \tag{5.34}$$

with the set $\mathbb{W}$ and the additive uncertainty vector $\mathbf{w}_\mathrm{j}$ calculated according to (3.2.2) with the parameter vector $\theta_\mathrm{j} \in \Theta$. In the equation above the additive uncertainty vector is chosen which has the maximum effect on the considered constraint with the linearization error $\mathbf{e}_\mathrm{ij}$. The dynamic of the linearization error is given by

$$\mathbf{e}_\mathrm{i,j} = \mathbf{A}_\mathrm{k}^i \mathbf{e}_0 + \sum_{l=0}^{i-1} \mathbf{A}_\mathrm{k}^l \mathbf{w}_\mathrm{j} \tag{5.35}$$

with

$$\begin{aligned} \mathbf{w}_\mathrm{j} &= (\mathbf{A}(\theta_\mathrm{j}) - \mathbf{A})\mathbf{x} + (\mathbf{B}(\theta_\mathrm{j}) - \mathbf{B})\mathbf{u} + \mathbf{E}(\theta_j) \\ \theta_j &= \theta_0 + \Delta\theta_j \in \Theta \\ (\mathbf{x}, \mathbf{u}) &\in \mathbb{X} \times \mathbb{U} \end{aligned}$$

Here θ_0 is the nominal parameter vector and $\Delta\theta_\mathrm{j}$ is the set of parameter deviations with

$$\begin{aligned} \Delta\theta_\mathrm{j} \in \{ & [\Delta v, \Delta\alpha_\mathrm{f}, \Delta\alpha_\mathrm{r}];\ [\Delta v, \Delta\alpha_\mathrm{f}, -\Delta\alpha_\mathrm{r}] \\ & [\Delta v, -\Delta\alpha_\mathrm{f}, \Delta\alpha_\mathrm{r}];\ [-\Delta v, \Delta\alpha_\mathrm{f}, \Delta\alpha_\mathrm{r}] \\ & [-\Delta v, -\Delta\alpha_\mathrm{f}, \Delta\alpha_\mathrm{r}];\ [\Delta v, -\Delta\alpha_\mathrm{f}, -\Delta\alpha_\mathrm{r}] \\ & [-\Delta v, \Delta\alpha_\mathrm{f}, -\Delta\alpha_\mathrm{r}];\ [-\Delta v, -\Delta\alpha_\mathrm{f}, -\Delta\alpha_\mathrm{r}] \} \end{aligned}$$

where Δv, $\Delta\alpha_\mathrm{f}$ and $\Delta\alpha_\mathrm{r}$ determine the range of parameter deviations (linearization errors). In this thesis the range is set to the maximum deviation between the predicted state at the current and the last control step.

$$\begin{aligned} \Delta\alpha_\mathrm{f} &= \max_i(\alpha_\mathrm{f,i,k} - \alpha_\mathrm{f,i,k\text{-}1}) \\ \Delta\alpha_\mathrm{r} &= \max_i(\alpha_\mathrm{r,i,k} - \alpha_\mathrm{r,i,k\text{-}1}) \\ \Delta v &= \max_i(v_\mathrm{i,k} - v_\mathrm{i,k\text{-}1}) \end{aligned}$$

5.3 MPC Formulation

The fast integrated planning and control problem is finally formulated in this section. The MPC has a similar structure as the formulation in (4.6). Instead of nonlinear constraints, linearized constraint equations are used, which have been derived by successive linearization in this chapter. The constraints are modified to consider the maximum effect of linearization errors in $\mathbb{V}$ and $\mathbb{Z}$.

$$\begin{aligned}
\min_{\mathbf{u}} \ & J = \sum_{i=1}^{N} \mathbf{z}(i,k)^T \mathbf{Q} \mathbf{z}(i,k) + \mathbf{v}(i,k)^T \mathbf{R} \mathbf{v}(i,k) \\
& \quad + [\mathbf{z}(N,k)^T \ 1] \mathbf{P}_{\mathrm{f}} [z(N,k)^T \ 1]^T + \mathbf{Q}_{\mathrm{s}} s_{\mathrm{obs}}(i,k)^2 \\
\text{s.t.} \ & \mathbf{z}(i+1,k) = \mathbf{A}_{\mathrm{i}} \mathbf{z}(i,k) + \mathbf{B}_{\mathrm{i}} \mathbf{v}(i,k) + \mathbf{E}_{\mathrm{i}} \\
& \mathbf{v}(i,k) \in \mathbb{V} \\
& \mathbf{z}(i,k) \in \mathbb{Z} \\
& \mathbf{z}(0,k) = \mathbf{z}_0 \\
& i = 1, \cdots, N \\
& \mathbf{z}(N,k) \in \mathbb{G}_{\mathrm{B/E}} \ \text{ if terminal collision avoidance is applied}
\end{aligned} \tag{5.36}$$

The MPC is calculated by the steps given as follows:

Robust Control Algorithm:
Initilization: At time $k = 0$ set $\mathbf{z}_0$=$\mathbf{x}_0$, where $\mathbf{x}(0)$ is the measured state.
Step 1 - Linearization: Linearize nonlinear constraints according to the sections 5.1.1, 5.1.2, 5.1.3 and 5.1.4.
Step 2 - Compute control: Compute control at time k and current states $(\mathbf{z}, \mathbf{v})$ according to the optimization problem in (5.36), solve the nominal optimal control problem to obtain nominal control vector $\mathbf{v}$ and apply control action $\mathbf{u} = \mathbf{v}(0,k) + \mathbf{k}([\mathbf{x}^T(0,k) \ 1]^T - [\mathbf{z}^T(0,k) \ 1])^T$. Save predicted state vectors $\mathbf{z}(i,k)$,$i = 1, \cdots, N$
Step 3 - Apply control: Apply control $\mathbf{u}$ to the system.

Step 4 - Update: Measure the successor state $\mathbf{x}^+$ of the system according to (3.35) and compute the successor state $\mathbf{z}^+ = f(\mathbf{z}(0,k), \mathbf{v}(0,k))$ of the nominal system.
Step 5: Set $(\mathbf{x}(0,k), \mathbf{z}(0,k)) = (\mathbf{x}^+, \mathbf{z}^+)$, set k=k+1, and go to Step 1.

5.4 Results

In previous sections, the fast model predictive approach for integrated trajectory planning and vehicle dynamics control has been presented. Furthermore, the robust tube MPC approach has been discussed to compensate linearization errors. In this section, the method will be demonstrated on Scenario 1 and 2 defined in Section 4.7.2 and compared with the nonlinear MPC approach from Chapter 4. Finally, experimental results will show the performance of the proposed method in the real vehicle.
In the simulation environment, the optimization is solved with an interior point algorithm implemented in MATLAB Simulink. The applied vehicle dynamics model is the nonlinear vehicle dynamcis model in Section 3.1.3 The computer is equipped with an Intel(R) Core(TM) i5-3380M with 2.9h GHz, 8 GB RAM and Windows 7 SP1 as the given operating system.
In the real vehicle, the algorithm runs on a dSpace MicroAutobox II from dSpace. It has a IBM PPC 750 GL 900Mhz processor and 16 MB main memory. The optimization is solved with the CVXGEN solver [46].

5.4.1 Simulation of Typical Scenarios

The performance of the fast integrated planning and control method with parameter setup 4 in Table 5.1 is demonstrated for Scenario 1 in Figure 5.6.
The parameter setup is the same as parameter setup 1 in Chapter 4 except for the slack variable. The difference in the slack variable penalty can be explained by the different characteristic of tube planning compared to the planning of the nominal trajectory in Chapter 4.
Figure 5.6 illustrates the position, the steering torque command, the deceleration command and the velocity of the simulated vehicle during the maneuver. Similarly to Chapter 4, the maneuver is marked at characteristic time marks. In the steering phase, starting at ①, the maximum torque limit and the maximum deceleration command is applied. The steering phase is interrupted shortly by an alternating switching behaviour of steering torque and deceleration command at ②. This effect can be explained by transient dynamics of successive linearization before the predicted dynamics converge. The counter steering phase starting at ③ is characterized by a negative steering torque and a maximum deceleration command. After passing the obstacle the velocity is reduced to a standstill at ④.
The performance in Scenario 2 with two obstacles is illustrated in Figure 5.7. The figure shows that the control method can successfully avoid collision with two obstacles. A difference lies in the longer counter steering phase in Figure 5.7 to avoid collision with the second obstacle obs_{l}.

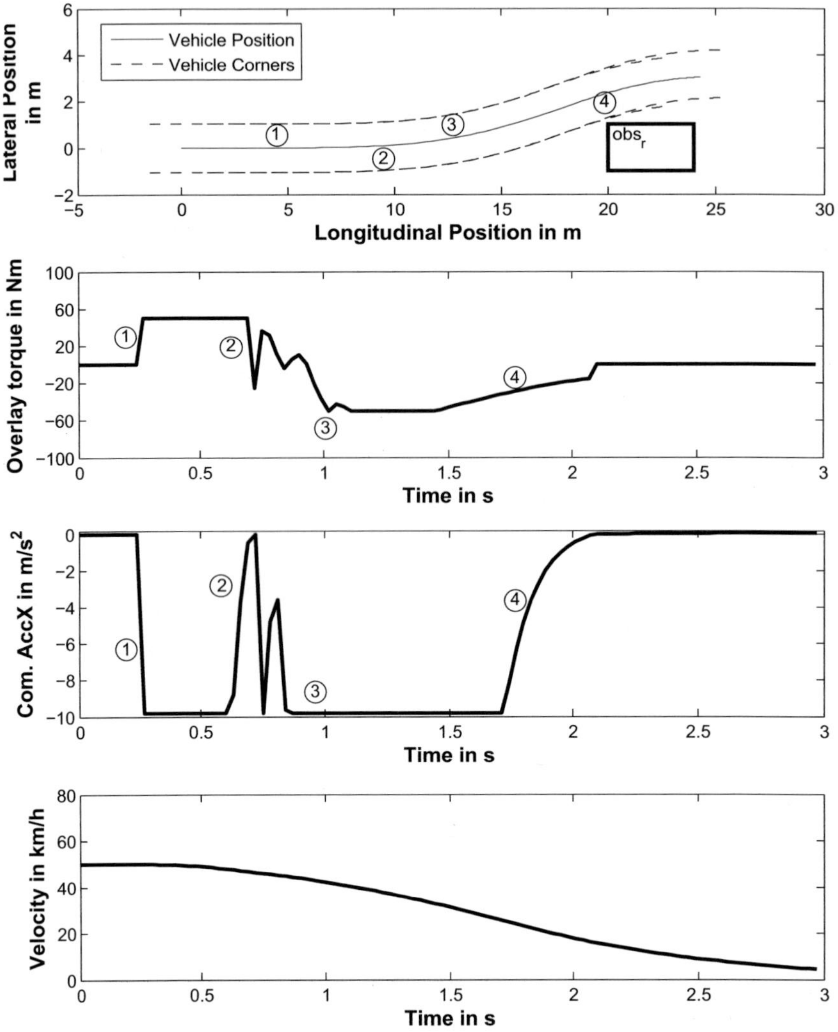

Figure 5.6: Simulation results of the fast integrated trajectory planning and vehicle dynamics control method with parameter setup 4 in Scenario 1

Table 5.1: Parameter setup 4 for the Fast Integrated Planning and Control Method

Parameter	Symbol	Value	Unit
Prediction horizon length	N	15	(none)
Prediction horizon step size	dt_{pred}	0.14	s
Controller time step size	dt_{ctrl}	0.1	s
Penalty of slack variable	q_{s}	10^6	$1/\mathrm{m}^2$
Penalty of yaw angle	q_{φ}	$3{\cdot}10^3$	$1/\mathrm{rad}^2$
Penalty of velocity	q_{v}	10	$\mathrm{s}^2/\mathrm{m}^2$
Penalty of steering torque command	r_{T}	0.1	$1/\mathrm{Nm}^2$
Penalty of deceleration command	$r_{\mathrm{a_x}}$	0.1	$\mathrm{s}^4/\mathrm{m}^2$
Limit of steering system	T_{max}	50	Nm
Limit of braking system	$a_{\mathrm{x,cmd,max}}$	-9.81	$\mathrm{m/s}^2$

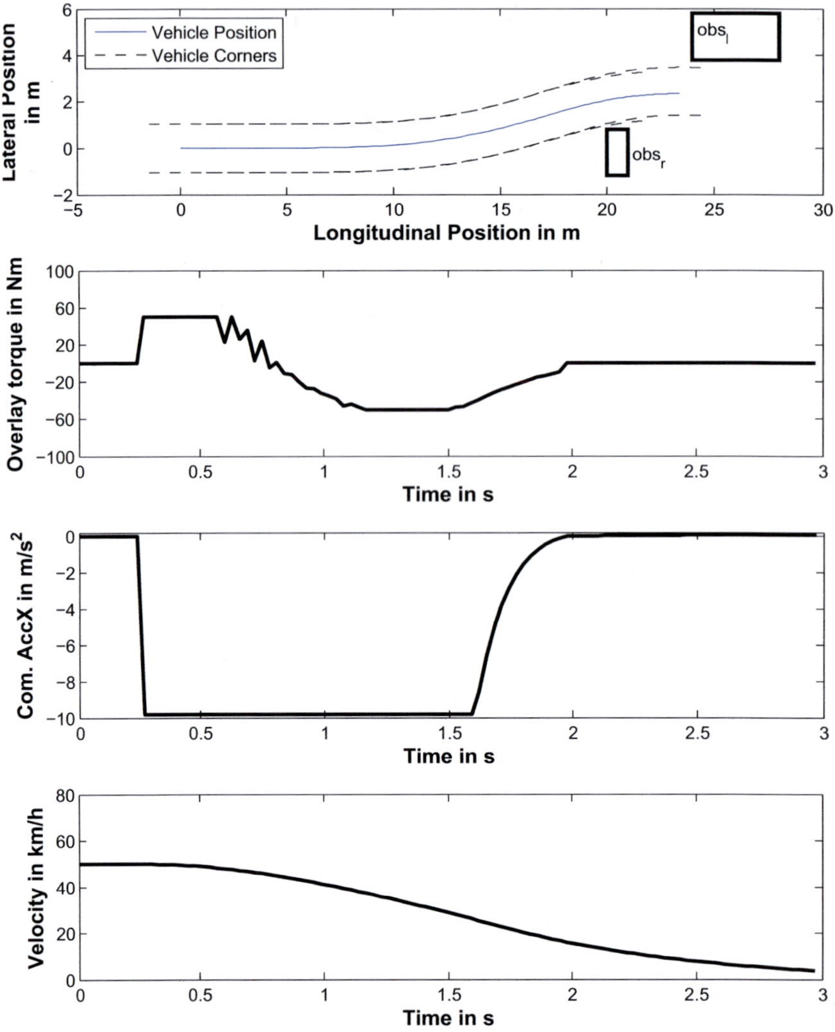

Figure 5.7: Simulation results of the fast integrated trajectory planning and vehicle dynamics control method with parameter setup 4 in Scenario 2

5.4.2 Simulation with Terminal Collision Avoidance

Similarly to Section 4.4, this section shall demonstrate the effect of the terminal collision avoidance constraint in the fast integrated planning and control method. In this section, parameter setup 5 in Table 5.2 is chosen which has the same parameter values as parameter setup 3 in Table 4.3. It has a shorter prediction horizon compared to the setup in Table 5.1 and higher penalty on the command signals. As has been demonstrated in Section 4.7.4, the method with this parameter setup was able avoid collision with the obstacle in Scenario 1 and does not reduce the velocity to standstill at the end of maneuver. Furthermore, the method with this parameter setup cannot avoid collision with the obstacle, as the obstacle has been detected too late.

Table 5.2: Parameter setup 5 for the Fast Integrated Planning and Control Method

Parameter	Symbol	Value	Unit
Prediction horizon length	N	15	(none)
Prediction horizon step size	dt_{pred}	0.07	s
Controller time step size	dt_{ctrl}	0.1	s
Penalty of slack variable	q_{s}	$3{\cdot}10^6$	$1/\mathrm{m}^2$
Penalty of yaw angle	q_{φ}	10^4	$1/\mathrm{rad}^2$
Penalty of velocity	q_{v}	0	$\mathrm{s}^2/\mathrm{m}^2$
Penalty of steering torque command	r_{T}	1	$1/\mathrm{Nm}^2$
Penalty of deceleration command	$r_{a_{\mathrm{x}}}$	10	$\mathrm{s}^4/\mathrm{m}^2$
Limit of steering system	T_{max}	50	Nm
Limit of braking system	$a_{\mathrm{x,cmd,max}}$	-9.81	$\mathrm{m/s}^2$

The performance of the terminal collision avoidance constraint for an evasive maneuver in (5.19) is shown in Figure 5.8. By considering the proposed linearized terminal collision avoidance constraint, the collision is avoided.

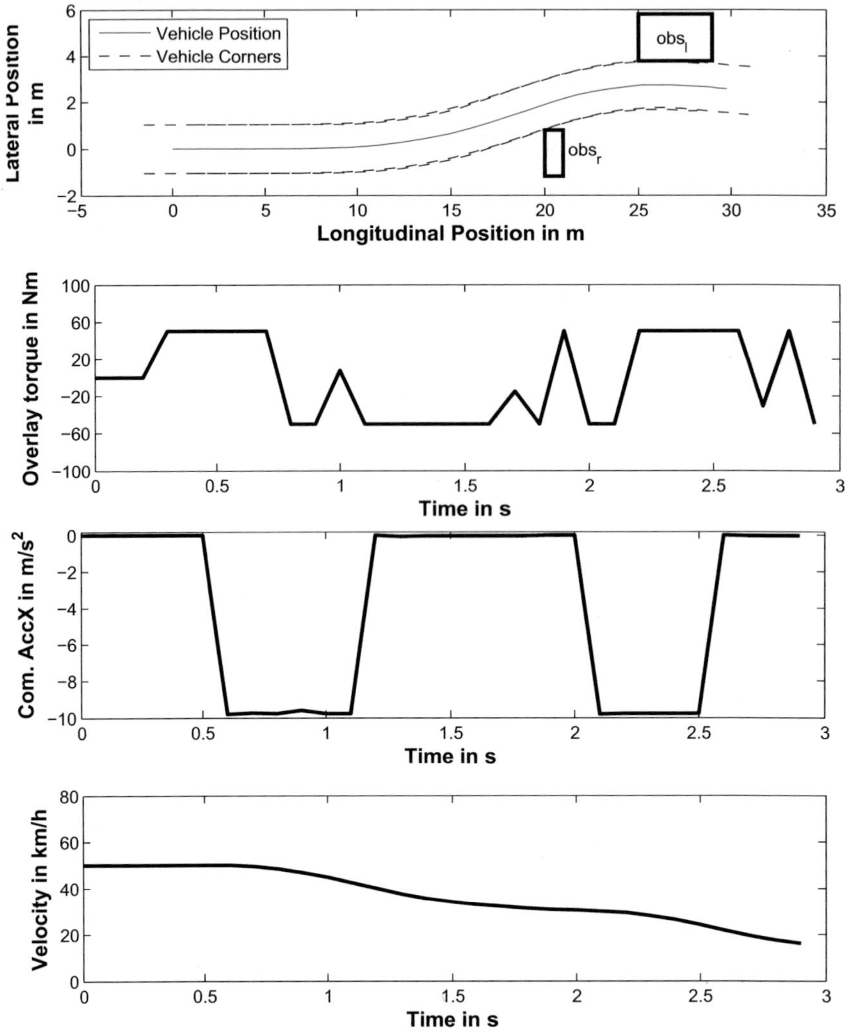

Figure 5.8: Simulation results of the fast integrated trajectory planning and vehicle dynamics control method with parameter setup 5 in Scenario 2 with terminal collision avoidance constraint

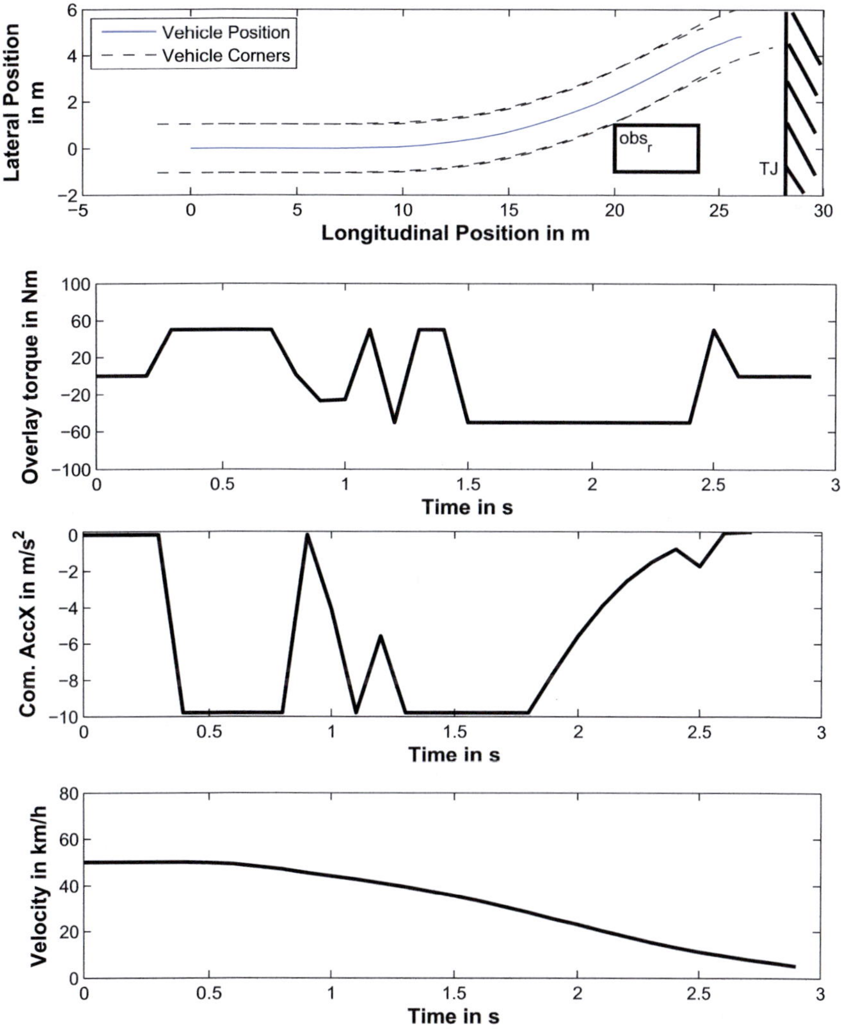

Figure 5.9: Simulation results of the fast integrated trajectory planning and vehicle dynamics control method with parameter setup 5 in Scenario 3 with terminal collision avoidance constraint

Finally, the terminal collision avoidance constraint for a braking maneuver in (5.18) will be used to ensure collision avoidance in Scenario 3, which requires braking in front of a traffic jam as shown in Figure 4.16. The figures illustrates that the implemented terminal collision avoidance constraint is able to reduce velocity to standstill before colliding with obstacles in the traffic jam and can thus be used to guarantee collision avoidance with the obstacle in the scenarios.

5.4.3 Real Vehicle Application

The proposed fast integrated planning and control method can be solved with fast and efficient solvers like [46] on a real time computer platform. It has been implemented in a dSpace MicroAutobox on a test vehicle and tested in the predefined scenarios. The obstacles have been simulated in a vehicle-in-the-loop (VIL) environment.

Figure 5.10 shows the result of the fast integrated planning and control method with parameter setup 6 in Table 5.3 for Scenario 1 in the test vehicle. The maneuver starts with the steering phase at ① where the maximum steering torque and deceleration command are commanded. At ② the counter-steering phase starts where the deceleration command is reduced for a short period. In the stabilization phase from ③ to ④ the method reorientates the vehicle to zero yaw angle and brakes the vehicle to a standstill. Comparing Figure 5.6 with Figure 5.10 indicates that the velocity reduction is stronger in the simulation results though the deceleration command characteristic is similar. The deviation can be explained by ESC and ABS interventions. As illustrated the ABS and the ESC are activated during a large time period of the maneuver, which weakens the velocity reduction by the intervention during the evasive maneuver.

Furthermore, the proposed method has been demonstrated in Scenario 2 shown in Figure 5.11. The vehicle successfully avoids the collision with the obstacles and brakes the velocity to zero at the end of the maneuver.

Table 5.3: Parameter setup 6 for the Fast Integrated Planning and Control Method in real vehicle application

Parameter	Symbol	Value	Unit
Prediction horizon length	N	10	(none)
Prediction horizon step size	dt_{pred}	0.2	s
Controller time step size	dt_{ctrl}	0.13	s
Penalty of slack variable	q_{s}	$3{\cdot}10^2$	$1/\text{m}^2$
Penalty of yaw angle	q_{φ}	$3{\cdot}10^3$	$1/\text{rad}^2$
Penalty of velocity	q_{v}	10	s^2/m^2
Penalty of steering torque command	r_{T}	0.1	$1/\text{Nm}^2$
Penalty of deceleration command	r_{a_x}	0.1	s^4/m^2
Limit of steering system	T_{max}	50	Nm
Limit of braking system	$a_{\text{x,cmd,max}}$	-9.81	m/s^2

5.5 Discussion

This chapter presented successive linearization as an efficient technique to approximate nonlinearities in MPC by linear equation and reformulate nonlinear optimization problems into quadratic programs. Linearization errors may threaten the stability of the system and must be considered in constraint formulations. The robust tube based MPC approach has been deployed to consider linearization errors as additive uncertainties and derive modified constraints to guarantee stability for the uncertain system. The performance of the planning and control method has been demonstrated in simulation and in experiments, showing the applicability of the algorithm in real time.

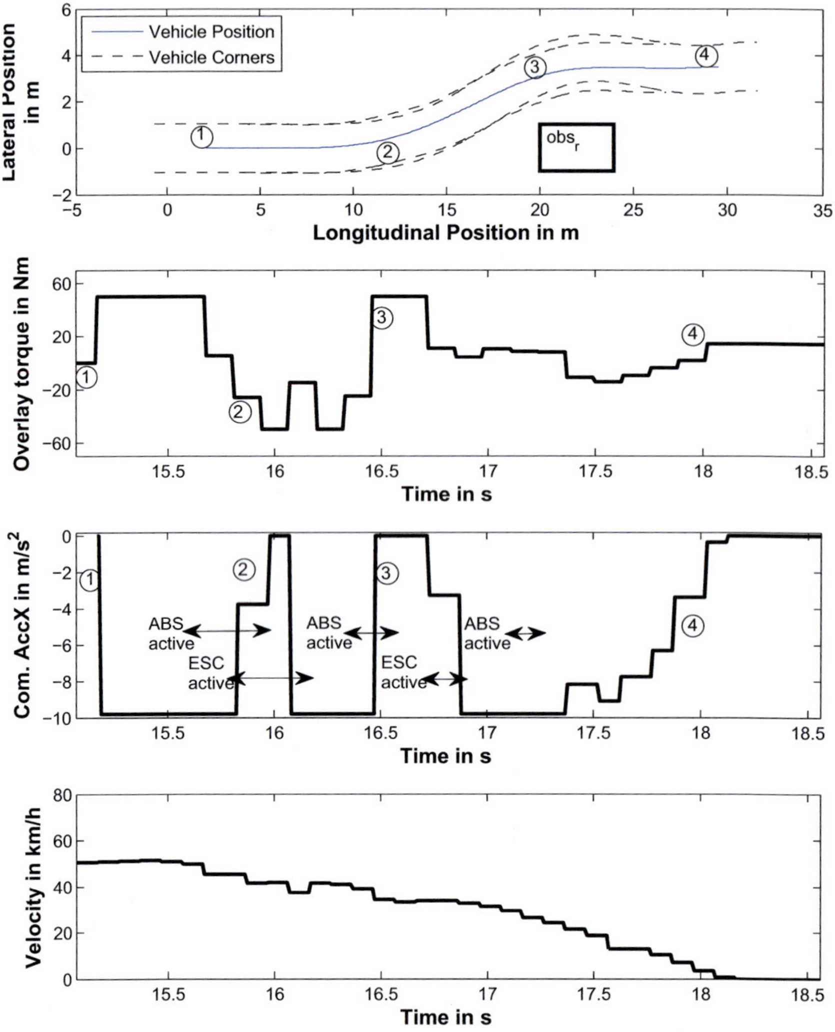

Figure 5.10: Results in real vehicle of the fast integrated trajectory planning and vehicle dynamics control method with parameter setup 4 in Scenario 1

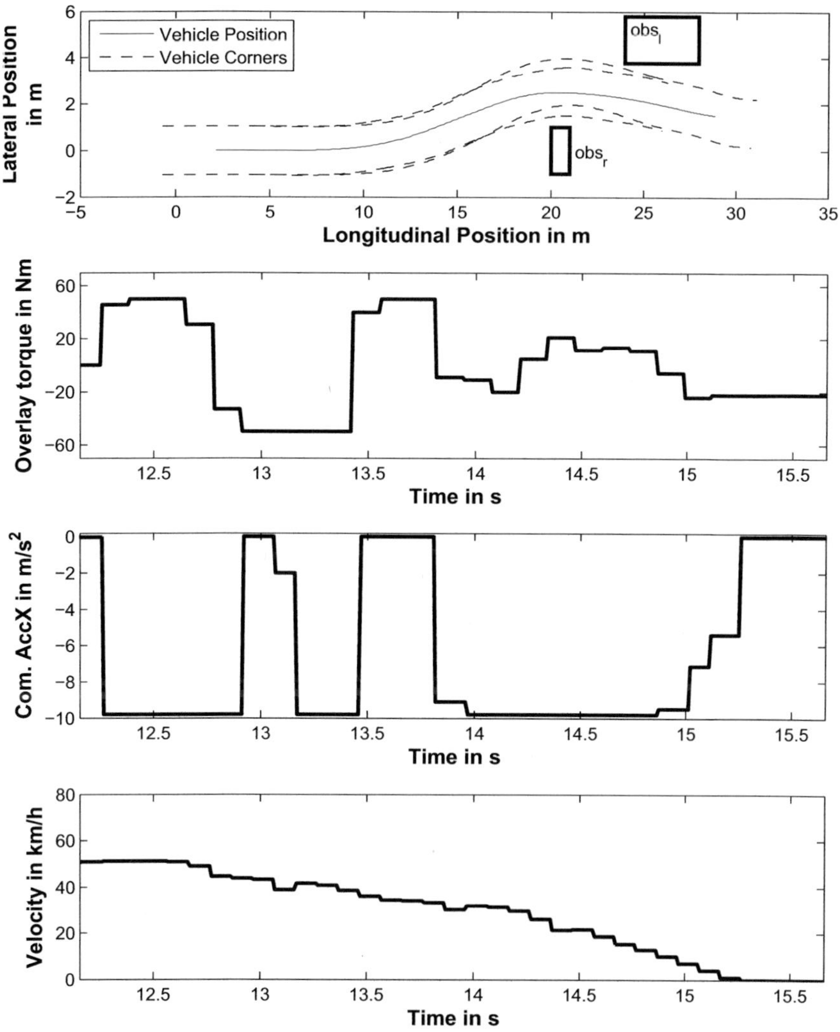

Figure 5.11: Results in real vehicle of the fast integrated trajectory planning and vehicle dynamics control method with parameter setup 4 in Scenario 2

Table 5.4: Metric values of planning and control concepts in simulation

Scenarios	Velocity Reduction Δv_{red}	Final Yaw Angle φ_{fnl}	Computation Time t_{cmp}
Scenario 1			
Nonlinear Integrated	44%	33°	$10s - 15s$
Fast Integrated	36%	14°	$0.15s$
Scenario 2			
Nonlinear Integrated	44.2%	6°	$10s - 15s$
Fast Integrated	45%	1°	$0.15s$

Table 5.4 shows the performance of the nonlinear and the fast integrated method in Scenario 1 and 2 in simulation. The nonlinear integrated planning and control method provides the flexibility needed to adapt to multiple obstacle scenarios and lead to reduced tuning effort of both planning and control module. The fast method inherits the flexibility required to plan in complex scenarios and can be calculated in approximately hundred of milliseconds. In Scenario 1, the fast method was able to drastically reduce the final yaw angle compared to the nonlinear method even though the same parameter setup have been chosen for both simulations. This has been explained by transient dynamics caused by successive linearization of the fast method, which disturb the deceleration command, extend the maneuver distance and thus enables the algorithm to re-orientate the vehicle. It is expected that the effect of the transient dynamics can be reduced with lower sampling time, which is possible when applied on computer platforms with higher processing power. In Scenario 2, the performance of both methods are very similar and the fast method shows has a significant of computation time.
Finally, Table 5.5 summarizes the performance metric values of the fast integrated planning and control method for the presented scenarios. The applied simulation environment showed to be sufficiently correlated with the test vehicle characteristic to develope and optimize the method in this thesis.

Table 5.5: Metric values of the Fast Integrated Planning and Control method for typical scenarios in simulation and in real vehicle

Scenarios	Velocity Reduction Δv_{red}	Final Yaw Angle φ_{fnl}	Computation Time t_{cmp}
Scenario 1			
Simulation	36.1%	14 °	0.15s
Real Vehicle	31.2%	1 °	0.13s
Scenario 2			
Simulation	42.5%	1 °	0.15s
Real Vehicle	28.2%	−9 °	0.13s

The Table shows little deviation in the final yaw angle and higher deviation in the velocity reduction metric between both environment. First, this can be explained with model deviations and sensor errors especially during phases with suspended braking intervention. Furthermore, the deviation may result from the intervention of the ABS and the ESC system which reduce the pressure in the brake line to improve the overall vehicle dynamics stability. Though the computation time in simulation and in the real vehicle are similar, it must be noted that the prediction horizon in simulation has 15 prediction steps and the horizon in the real vehicle has 10 prediction steps. All in all the proposed method is able to avoid collision in both environments with similar parameter setups.

6 Conclusion and Outlook

Recent developments in advanced driver assistance systems have increased the qualitative degree as well as the quantitative extent of system features in vehicles. Among these system features collision avoidance systems play a crucial role in protecting the vehicle and improve the overall safety of traffic participants. Automated emergency braking systems as an important representative have significantly reduced the rate of accidents by initiating a braking maneuver in critical situations.

To enhance the effectiveness of collision avoidance systems, the exploitation of the lateral maneuver space by evasive maneuvers shall be investigated.

Effective avoidance systems drive the need for holistic integrated approaches to cover planning and control in one method. The key research question as established in this thesis leads to an integrated method considering all known constraints. A key advantage comes along with the optimization based upon one physical model for both planning and control tasks. Another characteristic of this concept allows for reduction of the necessary tuning effort by reducing the number of design parameters in the overall system.

A nonlinear integrated planning and control method has been presented to design and realize optimal evasion maneuvers. Combined planning of longitudinal and lateral dynamics together with non-linear optimization supports the full exploitation of the available maneuver space at the limit of the given vehicle dynamics. The underlying model predictive control method provides the prerequisites to consider actuator limitations and stability limits through input and state constraints. Environmental information is explicitly considered by obstacle and maneuver space constraints. An extension of the method provides

a solution for the well-known issue associated limited planning horizons. The concept of terminal collision avoidance has been introduced in this work to take into account objects outside of the prediction horizon.
The interaction between the longitudinal and the lateral vehicle dynamics leads to nonlinear constraints and thus to a nonlinear optimization problem. The application of concepts for maneuver planning requires computationally efficient algorithms to cope with limited calculation power. The successive linearization technique transforms the nonlinear constraints into linear ones and enables the application of design principles in real vehicles. The robust tube based model predictive approach is exploited to explicitly consider linearization errors for the satisfaction of constraints and the guarantee of control stability.
The resulting fast integrated planning and control method shows a control performance comparable to the nonlinear method in a simulation environment for all characteristic scenarios. Furthermore, the applicability of the algorithm in real time has been verified in a test vehicle. The results underline a high degree of conformance in control characteristics for simulation and for real vehicle application.
The proposed integrated planning and control method may serve as the foundation for a number of potential future investigations.
The design of the collision avoidance constraints presented in this work represent a base for planning of trajectories in complex scenarios with arbitrary information about the road participants. In this thesis, simple scenarios have been chosen with static obstacles represented by rectangular shapes to demonstrate the applicability of the proposed planning and control method. In a next step, the proposed method can easily be modified to take into account moving obstacles as well as predicted obstacle trajectories based on statistical representations or Markov based models for the planning task.
Driver related factors have a huge impact in collision avoidance scenarios and require specific design of actuator components. To fully compensate for possible driver disturbances during the execution of collision avoidance maneuvers, suitable actuator systems such as steer-by-wire or steer-angle-overlay may be

investigated in future work to overcome weaknesses of existing actuator components. As characteristics and limitations of actuators have been reflected in this thesis, the proposed method can be easily modified to handle different and more actuators.

The design of the proposed methods in this thesis focused on the requirements and scenarios for evasive collision avoidance maneuvers in an advanced driver assistance system. In the future, the proposed framework has big potentials for general automated driving tasks where the driver is out of the control loop. Further, the design goals may be modified to integrate traffic rule information into the method such as the maximum allowed speed and the traffic light status.

High standards are set for trajectory planning and vehicle dynamics control modules in order to enable automated driving functionality. The proposed fast integrated planning and control method lays the foundation for the full exploitation of the available vehicle dynamics and the available space during the maneuver execution. The design concepts investigated in this thesis are promising to accelerate the realization of future design principles and thereby contribute to the overall development of automated driving technologies.

References

[1] J. Ackermann, J. Guldner, W. Sienel, R. Steinhauser, and V. I. Utkin. Linear and Nonlinear Controller Design for Robust Automatic Steering. In: *Control System Technology* 3 (1995), pp. 132–143.

[2] K. J. Astroem and T. Haegglund. *Advanced PID Control*. The Instrumentation, Systems, and Automation Society, North Carolina, 2006.

[3] E. Bauer and U. Konigorski. Ein modellprädiktiver Querplanungsansatz zur Kollisionsvermeidung. In: *Steuerung und Regelung von Fahrzeugen und Motoren*. Baden-Baden, 2013.

[4] E. Bauer, F. Lotz, M. Pfromm, M. Schreier, B. Abendroth, S. Cieler, A. Eckert, A. Hohm, S. Lüke, P. Rieth, et al. PRORETA 3: An Integrated Approach to Collision Avoidance and Vehicle Automation. In: *at-Automatisierungstechnik Methoden und Anwendungen der Steuerungs-, Regelungs-und Informationstechnik* 60 (12) (2012), pp. 755–765.

[5] C. E. Beal. Applications of model predictive control to vehicle dynamics for active safety and stability. PhD thesis. Stanford University, 2011.

[6] C. E. Beal and C. Gerdes. Model Predictive Control for Vehicle Stabilization at the Limits of Handling. In: *IEEE Transactions on Control Systems Technology* 21 (4) (2013), pp. 1258–1269.

[7] A. G. Beccuti, S. Mariéthoz, S. Cliquennois, S. Wang, and M. Morari. Explicit model predictive control of DC–DC switched-mode power supplies with extended Kalman filtering. In: *IEEE Transactions on Industrial Electronics* 56 (6) (2009), pp. 1864–1874.

[8] P. Bender, Ö. S. Tas, J. Ziegler, and C. Stiller. The combinatorial aspect of motion planning: Maneuver variants in structured environments. In: *IEEE Intelligent Vehicles Symposium, Seoul*. 2015.

[9] F. Borrelli, A. Bemporad, M. Fodor, and D. Hrovat. A hybrid approach to traction control. In: *International Workshop on Hybrid Systems: Computation and Control, Rome*. 2001.

[10] F. Borrelli, A. Bemporad, and M. Morari. Predictive Control for Linear and Hybrid Systems, 2015. In: *http://www.mpc.berkeley.edu/mpc-course-material* (2015). Accessed 09 April 2017.

[11] S. Boyd, L. E. Ghaoui, E. Feron, and V. Balakrishnan. *Linear Matrix Inequalities in System and Control Theory*. Society for Industrial and Applied Mathematics, Philadelphia, 1994.

[12] D. Burton, A. Delaney, S. Newstead, D. Logan, and B. Fildes. Effectiveness of ABS and vehicle stability control systems. In: *Royal Automobile Club of Victoria, Adelaide* (2004).

[13] M. Cannon. Efficient nonlinear model predictive control algorithms. In: *Annual Reviews in Control* 28 (2) (2004), pp. 229–237.

[14] M. Cannon, J. Buerger, B. Kouvaritakis, and S. Rakovic. Robust tubes in nonlinear model predictive control. In: *IEEE Transactions on Automatic Control* 56 (8) (2011), pp. 1942–1947.

[15] M. Choi and S. B. Choi. Model Predictive Control for Vehicle Yaw Stability With Practical Concerns. In: *IEEE Transactions on Vehicular Technology* 63 (8) (2014), pp. 3539–3548.

[16] C. R. Cutler and B. L. Ramaker. Dynamic matrix control - a computer control algorithm. In: *Joint automatic control conference, San Francisco*. 1980.

[17] T. A. Dingus, F. Guo, S. Lee, J. F. Antin, M. Perez, M. Buchanan-King, and J. Hankey. Driver crash risk factors and prevalence evaluation using naturalistic driving data. In: *Proceedings of the National Academy of Sciences, Washington*. 2016.

[18] S. Engel. Regelungstechnische Konzepte zum Umgang mit dem Fahrer in Kollisionsvermeidungssystemen. MA thesis. Technische Universität Darmstadt, 2016.

[19] S. M. Erlien. Shared Vehicle Control Using Safe Driving Envelopes for Obstacle Avoidance and Stability. PhD thesis. Stanford University, 2015.

[20] P. Falcone, M. Tufo, F. Borrelli, J. Asgari, and H. E. Tseng. A linear time varying model predictive control approach to the integrated vehicle dynamics control problem in autonomous systems. In: *IEEE Conference on Decision and Control, New Orleans*. 2007.

[21] J. Farinella, C. Lay, and S. Bhandari. UAV Collision Avoidance using a Predictive Rapidly-Exploring Random Tree. In: *AIAA SciTech Forum, San Diego*. 2016.

[22] C. E. Garcia. Quadratic dynamic matrix control of nonlinear processes. An application to a batch reaction process. In: *AIChE annual meeting, San Francisco*. 1984.

[23] T. M. Gasser. *Vehicle Automation: Definitions, legal aspects, research needs*. Tech. rep. German Federal Highway Research Institute, Bergisch Gladbach, 2014.

[24] T. M. Gasser et al. Legal consequences of an increase in vehicle automation: English translation of BASt-Report F83 (Part 1). In: *Bundesanstalt für Straßenwesen* (Bergisch Gladbach, 2013).

[25] T. M. Gasser, C. Arzt, M. Ayoubi, A. Bartels, L. Bürkle, J. Eier, F. Flemisch, D. Häcker, T. Hesse, W. Huber, et al. Legal consequences of an increase in vehicle automation. In: *Die Bundesanstalt für Straßenwesen (BASt) Report F* 83 (Bergisch Gladbach, 2009).

[26] T. M. Gasser and R. Auerswald. Vervollständigung der Landkarte Fahrzeugautomatisierung - Ein Diskussionsentwurf. In: *URBAN Konferenz, Munich*. 2016.

[27] S. Gottschling. Integrierte Fahrdynamikregelung und Trajektorienplanung zur Kollisionsvermeidung. MA thesis. Technische Universität Darmstadt, 2016.

[28] Y. Hattori, E. Ono, and S. Hosoe. Optimum vehicle trajectory control for obstacle avoidance problem. In: *IEEE/ASME Transactions on Mechatronics* 11 (5) (2006), pp. 507–512.

[29] R. Hayashi, J. Isogai, P. Raksincharoensak, and M. Nagai. Autonomous collision avoidance system by combined control of steering and braking using geometrically optimised vehicular trajectory. In: *Vehicle System Dynamics* 50 (1) (2012), pp. 151–168.

[30] W. Hulshof, I. Knight, A. Edwards, M. Avery, and C. Grover. Autonomous emergency braking test results. In: *International Technical Conference on the Enhanced Safety of Vehicles*. 2013.

[31] Insurance Institute for Highway Safety. ESC reduces multiple-vehicle crashes as well as single-vehicle ones. http://www.iihs.org/iihs/sr/status report/article/41/5/2. Accessed: 2016-03-19. 2006.

[32] Insurance Institute for Highway Safety. *Volvo's City Safety prevents low-speed crashes and cuts insurance costs*. Status Report 46 (6). 2011.

[33] S. Kammel, J. Ziegler, B. Pitzer, M. Werling, T. Gindele, D. Jagzent, J. Schröder, M. Thuy, M. Goebl, F. v. Hundelshausen, O. Pink, C. Frese, and C. Stiller. Team AnnieWAY's Autonomous System for the 2007 DARPA Urban Challenge. In: *J. Field Robot.* 25 (9) (Sept. 2008), pp. 615–639.

[34] S. Karush. *They're working. Special issue: Crash Avoidance*. Tech. rep. 47(3). Insurance Institute for Highway Safety, 2012.

[35] A. Katriniok. Optimal vehicle dynamics control and state estimation for a low-cost GNSS-based collision avoidance system. PhD thesis. Rheinisch-Westfälische Technische Hochschule Aachen, 2013.

[36] J. Keisler. *Elementary calculus: An infinitesimal approach*. University of Wisconsin, Wisconsin, 2012.

[37] M. Keller, C. Haß, A. Seewald, et al. A vehicle lateral control approach for collision avoidance by emergency steering maneuvers. In: *International Munich Chassis Symposium, Munich*. 2015.

[38] M. Keller, C. Hass, A. Seewald, and T. Bertram. Ein modellprädiktives Planungs-und Fahrzeugquerregelungsverfahren zur Kollisionsvermeidung durch Notausweichmanöver. In: *Automotive meets Electronics, Dortmund*. 2015.

[39] O. Khatib. Real-time obstacle avoidance for manipulators and mobile robots. In: *The International Journal of Robotics Research* 5 (1) (1986), pp. 90–98.

[40] T. Kranz, S. Hahn, and K. Zindler. Nonlinear lateral vehicle control in combined emergency steering and braking maneuvers. In: *IEEE Intelligent Vehicles Symposium, Gothenburg*. 2016.

[41] K. Kritayakirana and J. C. Gerdes. Autonomous vehicle control at the limits of handling. In: *International Journal of Vehicle Autonomous Systems* 10 (4) (2012), pp. 271–296.

[42] B. Krogh and C. Thorpe. Integrated path planning and dynamic steering control for autonomous vehicles. In: *IEEE International Conference on Robotics and Automation, San Francisco*. 1986.

[43] Y. Kuwata, G. A. Fiore, J. Teo, E. Frazzoli, and J. P. How. Motion planning for urban driving using RRT. In: *IEEE/RSJ International Conference on Intelligent Robots and Systems, Nice*. 2008.

[44] J. Löfberg. YALMIP : a toolbox for modeling and optimization in MATLAB. In: *IEEE International Conference on Robotics and Automation, Barcelona*. 2004.

[45] T. N. Matsko. Internal model control for chemical recovery. In: *Chemical Engineering Progress* 81 (12) (1985), pp. 46–51.

[46] J. Mattingley and S. Boyd. CVXGEN: A code generator for embedded convex optimization. In: *Optimization and Engineering* 13 (1) (2012), pp. 1–27.

[47] R. K. Mehra, R. Rouhani, J. Eterno, J. Richalet, and A. Rault. Model algorithmic control: review and recent developments. In: *Engineering Foundation Conference on Chemical Process Control, New York*. 1982.

[48] H. S. Meinert. *Untersuchung von robusten modellprädiktiven Regelungen zur Anwendung in der Flugführung*. Verlag Dr. Hut, Munich, 2011.

[49] M. Mitschke and H. Wallentowitz. *Dynamik der Kraftfahrzeuge*. Springer Berlin, 2004.

[50] R. Möbus, M. Baotic, and M. Morari. Multi-object adaptive cruise control. In: *International Workshop on Hybrid Systems: Computation and Control, Prague*. 2003.

[51] M. Morari. Predicting the future of model predictive control. In: *AIChE Annual Meeting, Systems and Process Control Centennial Session, Philadelphia*. 2008.

[52] N. Moshchuk. Collision avoidance with steering: Towards autonomous driving: ADAS and role of steering systems. In: *Steering Systems, Detroit* (2014).

[53] N. Moshchuk, S. K. Chen, C. Zagorski, and A. Chatterjee. Optimal braking and steering control for active safety. In: *International IEEE Conference on Intelligent Transportation Systems, Crete*. 2012.

[54] O. Nelles. *Nonlinear System Identification: From Classical Approaches to Neural Networks and Fuzzy Models*. Springer, Berlin, 2001.

[55] G. C. Nunes. Design and analysis of multivariable predictive control applied to an oil-water-gas separator: A polynomial approach. PhD thesis. University of Florida, 2001.

[56] H. B. Pacejka. *Tyre and Vehicle Dynamics*. Automotive engineering. Butterworth-Heinemann, Waltham, 2006.

[57] F. G. Papi. Control Strategies for Collision Avoidance of Ground Vehicles. Bachelor's thesis. Universidade Federal de Santa Catarina, 2015.

[58] C. Pfeiffer. Fahrdynamikanalyse und Entwicklung eines Entscheideralgorithmus für ein Kollisionsvermeidungssystem. Bachelor's thesis. Hochschule Darmstadt, 2014.

[59] D. Pfeiffer and U. Franke. Efficient representation of traffic scenes by means of dynamic stixels. In: *IEEE Intelligent Vehicles Symposium, San Diego*. 2010.

[60] D. Pfeiffer and U. Franke. Towards a Global Optimal Multi-Layer Stixel Representation of Dense 3D Data. In: *British Machine Vision Conference, Dundee*. 2011.

[61] J. B. Rawlings and D. Q. Mayne. *Model predictive control: Theory and design*. Nob Hill Pub., Madison, 2009.

[62] J. Richalet, A. Rault, J. L. Testud, and J. Papon. Model Predictive heuristic control: application to an Industrial Process. In: *Automatica* 14 (5) (1978), pp. 413–428.

[63] P. Riekert and T. E. Schunck. *Zur Fahrmechanik des gummibereiften Kraftfahrzeugs*. Deutsche Kraftfahrtforschung. Technische Hochschule Stuttgart, 1940.

[64] On-Road Automated Vehicle Standards Committee. SAE J3016: Taxonomy and Definitions for Terms Related to On-Road Motor Vehicle Automated Driving Systems. In: *SAE International, Warrendale* (2014).

[65] N. R. Ruchika. Model predictive control: History and development. In: *International Journal of Engineering Trends and Technology (IJETT)* 4 (6) (2013), pp. 2600–2602.

[66] F. Schindler. Parameter Study and Robustness Analysis of Control Concepts for Collision Avoidance. Bachelor's thesis. Hochschule RheinMain, 2016.

[67] C. Schmidt. Fahrstrategien zur Unfallvermeidung im Straßenverkehr für Einzel- und Mehrobjektszenarien. PhD thesis. Karlsruhe Institute of Technology, 2013.

[68] M. Schorn. Quer- und Längsregelung eines Personenkraftwagens für ein Fahrerassistenzsystem zur Unfallvermeidung. PhD thesis. Technische Universität Darmstadt, 2007.

[69] H. Seki, S. Ooyama, and M. Ogawa. Nonlinear model predictive control using successive linearization-Application to chemical reactors. In: *Transactions-society of instrument and control engineers* 38 (1) (2002), pp. 61–66.

[70] S. Skogestad. Probably the best simple PID tuning rules in the world. In: *AIChE Annual Meeting, Reno*. 2001.

[71] U. Stählin. *Eingriffsentscheidung für ein Fahrerassistenzsystem zur Unfallvermeidung*. Berichte aus dem Institut für Automatisierungstechnik der TU Darmstadt. VDI-Verlag, Düsseldorf, 2008.

[72] H. Stübing. Entwicklung von Kriterien zur Bewertung von Kollisionsvermeidungssystemen. Bachelor's thesis. Hochschule Darmstadt, 2015.

[73] S. Thrun, M. Montemerlo, H. Dahlkamp, D. Stavens, A. Aron, J. Diebel, P. Fong, J. Gale, M. Halpenny, G. Hoffmann, et al. Stanley: The robot that won the DARPA Grand Challenge. In: *Journal of field Robotics* 23 (9) (2006), pp. 661–692.

[74] J. P. Timings and D. J. Cole. Minimum maneuver time calculation using convex optimization. In: *Journal of Dynamic Systems, Measurement, and Control* 135 (3) (2013).

[75] A. T. Van Zanten. *Bosch ESP systems: 5 years of experience*. Tech. rep. SAE Technical Paper, 2000.

[76] R. Vilanova and A. Visioli. *PID control in the third millennium*. Springer, London, 2012.

[77] M. Werling. Ein neues Konzept für die Trajektoriengenerierung und -stabilisierung in zeitkritischen Verkehrsszenarien. PhD thesis. Karlsruhe Institute of Technology, 2011.

[78] M. Werling, B. Gutjahr, S. Galler, and L. Gröll. Riccati-Trajektorienoptimierung für den aktiven Fußgängerschutz. In: *at-Automatisierungstechnik* 63 (3) (2015), pp. 202–210.

[79] M. Werling, J. Ziegler, S. Kammel, and S. Thrun. Optimal trajectory generation for dynamic street scenarios in a frenet frame. In: *IEEE International Conference on Robotics and Automation, Anchorage*. 2010.

[80] H. Wu. Fahrdynamikregelung von kombinierten Brems- und Ausweichmanövern. MA thesis. Technische Universität Darmstadt, 2014.

[81] B. Yi. Kombiniertes Ausweichen und Bremsen zum letztmöglichen Moment. In: *URBAN Konferenz*. Munich, 2016.

[82] B. Yi, P. Bender, F. Bonarens, and C. Stiller. Model Predictive Trajectory Planning for Automated Driving. In: *IEEE Transactions on Intelligent Vehicles* ((submitted on 27.06.2017)).

[83] B. Yi, J. Ferdinand, N. Simm, and F. Bonarens. Application of Local Linear Steering Models with Model Predictive Control for Collision Avoidance Maneuvers. In: *IFAC Symposium on Intelligent Autonomous Vehicles, Leipzig*. 2016.

[84] B. Yi, J. Ferdinand, and H. Wu. Robuste Fahrdynamikregelung zur Kollisionsvermeidung. In: *Automotive meets Electronics, Dortmund* (2015).

[85] B. Yi, S. Gottschling, J. Ferdinand, N. Simm, F. Bonarens, and C. Stiller. Real Time Integrated Vehicle Dynamics Control and Trajectory Planning with MPC for Critical Maneuvers. In: *IEEE Intelligent Vehicles Symposium*. Gothenburg, 2016.

[86] J. Ziegler, P. Bender, T. Dang, and C. Stiller. Trajectory planning for Bertha 2014; A local, continuous method. In: *IEEE Conference on Intelligent Vehicles Symposium, Ypsilanti*. 2014.

[87] J. Ziegler, P. Bender, M. Schreiber, H. Lategahn, T. Strauss, C. Stiller, T. Dang, U. Franke, N. Appenrodt, C. G. Keller, et al. Making bertha drive - an autonomous journey on a historic route. In: *IEEE Intelligent Transportation Systems Magazine* 6 (2) (2014), pp. 8–20.

Publications by the author

[81] B. Yi. Kombiniertes Ausweichen und Bremsen zum letztmöglichen Moment. In: *URBAN Konferenz*. Munich, 2016.

[82] B. Yi, P. Bender, F. Bonarens, and C. Stiller. Model Predictive Trajectory Planning for Automated Driving. In: *IEEE Transactions on Intelligent Vehicles* ((submitted on 27.06.2017)).

[83] B. Yi, J. Ferdinand, N. Simm, and F. Bonarens. Application of Local Linear Steering Models with Model Predictive Control for Collision Avoidance Maneuvers. In: *IFAC Symposium on Intelligent Autonomous Vehicles, Leipzig*. 2016.

[84] B. Yi, J. Ferdinand, and H. Wu. Robuste Fahrdynamikregelung zur Kollisionsvermeidung. In: *Automotive meets Electronics, Dortmund* (2015).

[85] B. Yi, S. Gottschling, J. Ferdinand, N. Simm, F. Bonarens, and C. Stiller. Real Time Integrated Vehicle Dynamics Control and Trajectory Planning with MPC for Critical Maneuvers. In: *IEEE Intelligent Vehicles Symposium*. Gothenburg, 2016.

Supervised theses

[18] S. Engel. Regelungstechnische Konzepte zum Umgang mit dem Fahrer in Kollisionsvermeidungssystemen. MA thesis. Technische Universität Darmstadt, 2016.

[27] S. Gottschling. Integrierte Fahrdynamikregelung und Trajektorienplanung zur Kollisionsvermeidung. MA thesis. Technische Universität Darmstadt, 2016.

[57] F. G. Papi. Control Strategies for Collision Avoidance of Ground Vehicles. Bachelor's thesis. Universidade Federal de Santa Catarina, 2015.

[58] C. Pfeiffer. Fahrdynamikanalyse und Entwicklung eines Entscheideralgorithmus für ein Kollisionsvermeidungssystem. Bachelor's thesis. Hochschule Darmstadt, 2014.

[66] F. Schindler. Parameter Study and Robustness Analysis of Control Concepts for Collision Avoidance. Bachelor's thesis. Hochschule Rhein-Main, 2016.

[72] H. Stübing. Entwicklung von Kriterien zur Bewertung von Kollisionsvermeidungssystemen. Bachelor's thesis. Hochschule Darmstadt, 2015.

[80] H. Wu. Fahrdynamikregelung von kombinierten Brems- und Ausweichmanövern. MA thesis. Technische Universität Darmstadt, 2014.

Schriftenreihe
Institut für Mess- und Regelungstechnik
Karlsruher Institut für Technologie
(1613-4214)

Band 001 Hans, Annegret
Entwicklung eines Inline-Viskosimeters auf Basis eines magnetisch-induktiven Durchflussmessers. 2004
ISBN 3-937300-02-3

Band 002 Heizmann, Michael
Auswertung von forensischen Riefenspuren mittels automatischer Sichtprüfung. 2004
ISBN 3-937300-05-8

Band 003 Herbst, Jürgen
Zerstörungsfreie Prüfung von Abwasserkanälen mit Klopfschall. 2004
ISBN 3-937300-23-6

Band 004 Kammel, Sören
Deflektometrische Untersuchung spiegelnd reflektierender Freiformflächen. 2005
ISBN 3-937300-28-7

Band 005 Geistler, Alexander
Bordautonome Ortung von Schienenfahrzeugen mit Wirbelstrom-Sensoren. 2007
ISBN 978-3-86644-123-1

Band 006 Horn, Jan
Zweidimensionale Geschwindigkeitsmessung texturierter Oberflächen mit flächenhaften bildgebenden Sensoren. 2007
ISBN 978-3-86644-076-0

Die Bände sind unter www.ksp.kit.edu als PDF frei verfügbar oder als Druckausgabe bestellbar.

Band 007 Hoffmann, Christian
Fahrzeugdetektion durch Fusion monoskopischer Videomerkmale. 2007
ISBN 978-3-86644-139-2

Band 008 Dang, Thao
Kontinuierliche Selbstkalibrierung von Stereokameras. 2007
ISBN 978-3-86644-164-4

Band 009 Kapp, Andreas
Ein Beitrag zur Verbesserung und Erweiterung der Lidar-Signalverarbeitung für Fahrzeuge. 2007
ISBN 978-3-86644-174-3

Band 010 Horbach, Jan
Verfahren zur optischen 3D-Vermessung spiegelnder Oberflächen. 2008
ISBN 978-3-86644-202-3

Band 011 Böhringer, Frank
Gleisselektive Ortung von Schienenfahrzeugen mit bordautonomer Sensorik. 2008
ISBN 978-3-86644-196-5

Band 012 Xin, Binjian
Auswertung und Charakterisierung dreidimensionaler Messdaten technischer Oberflächen mit Riefentexturen. 2009
ISBN 978-3-86644-326-6

Band 013 Cech, Markus
Fahrspurschätzung aus monokularen Bildfolgen für innerstädtische Fahrerassistenzanwendungen. 2009
ISBN 978-3-86644-351-8

Band 014 Speck, Christoph
Automatisierte Auswertung forensischer Spuren auf Patronenhülsen. 2009
ISBN 978-3-86644-365-5

Band 015 Bachmann, Alexander
Dichte Objektsegmentierung in Stereobildfolgen. 2010
ISBN 978-3-86644-541-3

Band 016 Duchow, Christian
Videobasierte Wahrnehmung markierter Kreuzungen mit lokalem Markierungstest und Bayes'scher Modellierung. 2011
ISBN 978-3-86644-630-4

Band 017 Pink, Oliver
Bildbasierte Selbstlokalisierung von Straßenfahrzeugen. 2011
ISBN 978-3-86644-708-0

Band 018 Hensel, Stefan
Wirbelstromsensorbasierte Lokalisierung von Schienenfahrzeugen in topologischen Karten. 2011
ISBN 978-3-86644-749-3

Band 019 Carsten Hasberg
Simultane Lokalisierung und Kartierung spurgeführter Systeme. 2012
ISBN 978-3-86644-831-5

Band 020 Pitzer, Benjamin
Automatic Reconstruction of Textured 3D Models. 2012
ISBN 978-3-86644-805-6

Band 021 Roser, Martin
Modellbasierte und positionsgenaue Erkennung von Regentropfen in Bildfolgen zur Verbesserung von videobasierten Fahrerassistenzfunktionen. 2012
ISBN 978-3-86644-926-8

Band 022 Loose, Heidi
Dreidimensionale Straßenmodelle für Fahrerassistenzsysteme auf Landstraßen. 2013
ISBN 978-3-86644-942-8

Band 023 Rapp, Holger
Reconstruction of Specular Reflective Surfaces using Auto-Calibrating Deflectometry. 2013
ISBN 978-3-86644-966-4

Band 024 Moosmann, Frank
Interlacing Self-Localization, Moving Object Tracking and Mapping for 3D Range Sensors. 2013
ISBN 978-3-86644-977-0

Band 025 Geiger, Andreas
Probabilistic Models for 3D Urban Scene Understanding from Movable Platforms. 2013
ISBN 978-3-7315-0081-0

Band 026 Hörter, Marko
Entwicklung und vergleichende Bewertung einer bildbasierten Markierungslichtsteuerung für Kraftfahrzeuge. 2013
ISBN 978-3-7315-0091-9

Band 027 Kitt, Bernd
Effiziente Schätzung dichter Bewegungsvektorfelder unter Berücksichtigung der Epipolargeometrie zwischen unterschiedlichen Ansichten einer Szene. 2013
ISBN 978-3-7315-0105-3

Band 028 Lategahn, Henning
Mapping and Localization in Urban Environments Using Cameras. 2013
ISBN 978-3-7315-0135-0

Band 029 Tischler, Karin
Informationsfusion für die kooperative Umfeldwahrnehmung vernetzter Fahrzeuge. 2014
ISBN 978-3-7315-0166-4

Band 030 Schmidt, Christian
Fahrstrategien zur Unfallvermeidung im Straßenverkehr für Einzel- und Mehrobjektszenarien. 2014
ISBN 978-3-7315-0198-5

Band 031 Firl, Jonas
Probabilistic Maneuver Recognition in Traffic Scenarios. 2014
ISBN 978-3-7315-0287-6

Band 032 Schönbein, Miriam
Omnidirectional Stereo Vision for Autonomous Vehicles. 2015
ISBN 978-3-7315-0357-6

Band 033 Nicht erschienen

Band 034 Liebner, Martin
Fahrerabsichtserkennung und Risikobewertung für warnende Fahrerassistenzsysteme. 2016
ISBN 978-3-7315-0508-2

Band 035 Ziegler, Julius
Optimale Trajektorienplanung für Automobile. 2017
ISBN 978-3-7315-0553-2

Band 036 Harms, Hannes
Genauigkeitsuntersuchung von binokularen Normalenvektoren für die Umfeldwahrnehmung. 2017
ISBN 978-3-7315-0628-7

Band 037 Ruhhammer, Christian
Inferenz von Kreuzungsinformationen aus Flottendaten. 2017
ISBN 978-3-7315-0721-5

Band 038 Stein, Denis
Mobile laser scanning based determination of railway network topology and branching direction on turnouts. 2018
ISBN 978-3-7315-0743-7

Band 039 Yi, Boliang
Integrated Planning and Control for Collision Avoidance Systems. 2018
ISBN 978-3-7315-0785-7